Geoscience: Concepts and Applied Principles

Geoscience: Concepts and Applied Principles

Edited by Jacques Howard

SYRAWOOD
PUBLISHING HOUSE

New York

Published by Syrawood Publishing House,
750 Third Avenue, 9th Floor,
New York, NY 10017, USA
www.syrawoodpublishinghouse.com

Geoscience: Concepts and Applied Principles
Edited by Jacques Howard

International Standard Book Number: 978-1-64740-115-3 (Hardback)

Cataloging-in-Publication Data

Geoscience : concepts and applied principles / edited by Jacques Howard.
     p. cm.
Includes bibliographical references and index.
ISBN 978-1-64740-115-3
1. Geology. 2. Earth sciences. I. Howard, Jacques.
QE26.3 .G46 2022
550--dc23

TABLE OF CONTENTS

Preface..VII

Chapter 1   **Dynamic response based empirical liquefaction model**.................................1
             Snehal Rajeev Pathak and Asita Nilesh Dalvi

Chapter 2   **Un-differenced precise point positioning model using triple GNSS constellations**...22
             Akram Afifi and Ahmed El-Rabbany

Chapter 3   **Petrology and crustal inheritance of the Cloudy Bay Volcanics as derived from a fluvial conglomerate, Papuan Peninsula (Papua New Guinea): An example of geological inquiry in the absence of *in situ* outcrop**..................... 35
             Robert J. Holm and Benny Poke

Chapter 4   **Advantages of unmanned aerial vehicle (UAV) photogrammetry for landscape analysis compared with satellite data**.. 61
             Kotaro Iizuka, Masayuki Itoh, Satomi Shiodera, Takashi Matsubara, Mark Dohar and Kazuo Watanabe

Chapter 5   **Sedimentary facies and evolution of the lower Urho Formation in the 8th area of Karamay oilfield**.. 75
             Hongwei Kuang, Guangchun Jin and Zhenzhong Gao

Chapter 6   **Numerical modeling of virus transport through unsaturated porous media**............... 91
             Kandala Rajsekhar, Pramod Kumar Sharma and Sanjay Kumar Shukla

Chapter 7   **Effect of shear rate on the residual shear strength of pre-sheared clays**................ 103
             Farzad Habibbeygi and Hamid Nikraz

Chapter 8   **La Pintada landslide – A complex double-staged extreme event**..................... 112
             María Teresa Ramírez-Herrera and Krzysztof Gaidzik

Chapter 9   **A feasible way to increase carbon sequestration by adding dolomite and K-feldspar to soil**.. 132
             Leilei Xiao, Qibiao Sun, Huatao Yuan, Xiaoxiao Li, Yue Chu, Yulong Ruan, Changmei Lu and Bin Lian

Chapter 10   **Regression models for intrinsic constants of reconstituted clays**........................143
             Farzad Habibbeygi, Hamid Nikraz and Bill K Koul

Chapter 11   **Strike-slip deformation in the Inkisi Formation**...161
             Timothée Miyouna, Hardy Medry Dieu-Veill Nkodia, Olivier Florent Essouli, Moussa Dabo, Florent Boudzoumou and Damien Delvaux

Chapter 12   **Characteristics of soil exchangeable potassium according to soil color and landscape in Ferralsols environment**..**192**
Brahima Koné, Traoré Lassane, Sehi Zokagon Sylvain and
Kouassi Kouassi Jacques

Permissions

List of Contributors

Index

PREFACE

This book was inspired by the evolution of our times; to answer the curiosity of inquisitive minds. Many developments have occurred across the globe in the recent past which has transformed the progress in the field.

Geoscience is a natural science which deals with the study of the Earth, its physical constitution and its atmosphere. The understanding of the Earth's interior provides insights into plate tectonics, and the formation of volcanoes, mountain ranges and earthquakes. The mechanisms responsible for the sustenance of life on Earth, and water and atmosphere retention can be understood from a study of the Earth's atmosphere and magnetic field. Geoscience encompasses various domains such as atmospheric chemistry, hydrometeorology, hydrology, biogeochemistry, paleontology, oceanography, geology, geophysics, geomorphology, etc. These studies rely on a combination of observational, theoretical and experimental approaches. This book is compiled in such a manner, that it will provide in-depth knowledge about the concepts and applied principles of geosciences. The aim of this book is to present researches that have transformed this discipline and aided its advancement. Students, researchers, experts and all associated with geoscience will benefit alike from this book.

This book was developed from a mere concept to drafts to chapters and finally compiled together as a complete text to benefit the readers across all nations. To ensure the quality of the content we instilled two significant steps in our procedure. The first was to appoint an editorial team that would verify the data and statistics provided in the book and also select the most appropriate and valuable contributions from the plentiful contributions we received from authors worldwide. The next step was to appoint an expert of the topic as the Editor-in-Chief, who would head the project and finally make the necessary amendments and modifications to make the text reader-friendly. I was then commissioned to examine all the material to present the topics in the most comprehensible and productive format.

I would like to take this opportunity to thank all the contributing authors who were supportive enough to contribute their time and knowledge to this project. I also wish to convey my regards to my family who have been extremely supportive during the entire project.

Editor

Dynamic response based empirical liquefaction model

Snehal Rajeev Pathak[1] and Asita Nilesh Dalvi[2]*

*Corresponding author: Dalvi Asita Nilesh, Department of Civil Engineering, SITS, Pune, Maharashtra, India
E-mail: asitadalvi@gmail.com
Reviewing editor: Craig O'Neill, Macquarie University, Australia

Abstract: Dynamic response-based methodology, wherein integrated effect of dynamic soil properties and ground motion parameters proposed by authors, has been found to detect liquefaction susceptibility. The present work necessarily deals with the formulation of a comprehensive empirical liquefaction model (ELM) using this methodology. The absolute form of the ELM is dimensionally homogeneous and yields a correlation between proposed "liquefaction potential term" and "normalized standard penetration blow count corrected for fines content, $(N_1)_{60cs}$." The developed ELM demonstrates unbiased performance when verified over a wide range of significant parameters. One of the prominent features of the present ELM is accurate prediction of possibility of liquefaction. The proposed ELM has proven to work well on varied data-sets of more than 1000 case records within the given range of model parameters. Moreover, the dynamic response-based ELM proves its ability when compared with other liquefaction evaluation procedures. Thus, a generalized and optimistic ELM simulating realistic field conditions is formulated. It is anticipated that for accurate prediction of liquefaction occurrence, it would be more appropriate to employ the proposed ELM which will minimize the enormous losses caused due to liquefaction.

Subjects: Civil, Environmental and Geotechnical Engineering; Georisk & Hazards; Soil Mechanics

Keywords: seismic soil liquefaction; liquefaction susceptibility; empirical approach

ABOUT THE AUTHOR

Pathak is working in this field for more than a decade; the various areas dealt so far are 1-g shake table testing, Static and Cyclic Triaxial testing, and analytical modeling to assess susceptibility of liquefaction by varying soil and seismic parameters. The publications include 20 plus international and national journal papers in the field of seismic soil liquefaction; to name a few: ASCE, Natural Hazards, Geomechanics and Geoengineering, and equivalent papers presented and published in international and national conferences on the research theme.

PUBLIC INTEREST STATEMENT

Liquefaction is a phenomenon wherein a loose sandy soil loses its strength and flows like a fluid and is a topic of interest around the world. It was more thoroughly brought to the attention of engineers after 1964 Niigata and 1964 Alaska earthquakes which have witnessed major damage due to seismic soil liquefaction. Accordingly, prediction of liquefaction potential is the key step which will minimize the enormous losses caused due to liquefaction. The focus of the present work is thus formulation of a comprehensive and cogent empirical liquefaction model (ELM) and validating the same to ensure realistic dynamic soil response to earthquake-induced liquefaction. The final form of the ELM thus developed, engenders realistic yet optimistic prediction of liquefaction at a particular site within a given range of model parameters. ELM shows its versatility by performing on over 1000 case records and accurately predicts liquefaction occurrence.

1. Introduction

The use of empirical relationships based on correlation of observed field behavior with various *in situ* "index" tests is the dominant approach in common engineering practice to assess liquefaction susceptibility. Based on extensive literature review of empirical liquefaction procedures, Pathak and Dalvi (2011) inferred that inclusion of dynamic soil properties in conjunction with ground motion parameters would enhance the predictability of the model. Furthermore, it is observed that although there exist several studies to assess the liquefaction potential (LP), it is well understood that a large number of seismic as well as soil parameters affect the occurrence of liquefaction during an earthquake, making it necessary to identify the significant parameters based on their contribution to liquefaction phenomenon to expedite the assessment of liquefaction. Accordingly, Dalvi, Pathak, and Rajhans (2014) extracted significant parameters responsible for the phenomenon of liquefaction by employing multi-criteria decision-making tools (MCDM), namely: AHP and Entropy analysis. Such an attempt of applying MCDM tools in the field of earthquake-induced liquefaction has been firstly introduced by the authors.

Subsequently, Pathak and Dalvi (2012) presented a preliminary approach referred as, "dynamic response based methodology" to evaluate LP by incorporating thus extracted significant parameters to represent realistic dynamic response of soils. Based on this dynamic response-based methodology, Pathak and Dalvi (2013) established an elementary empirical liquefaction model (ELM) to identify susceptibility of liquefaction which is given by the following Equation. 1 as

$$(N_1)_{60} = 1.24(LPterm)^{0.31} \tag{1}$$

Where $(N_1)_{60}$ = normalized SPT blow count and,

$$LP \text{ term} = \left[\frac{v_{max} * G_{max} * dur}{\sigma'_v}\right]$$

where, "v_{max}" is the peak ground velocity (m/s),

"G_{max}" is the small strain shear modulus (kPa),

"dur" is duration of strong ground motion (sec), and

"σ'_v" is effective overburden pressure (kPa).

The performance of thus developed elementary model ascertained the potential of using the functional form LP term in assessing liquefaction susceptibility (Pathak & Dalvi, 2013). However, in order to predict liquefaction in practical situations, a dimensionless relationship might be required. Moreover, it is also required to take into account all types of soils ranging from sandy soils to silty sands and also site conditions to simulate actual field conditions. Thus, in the present work, an endeavor has been made to establish an exhaustive generalized ELM. The main motive behind the present work is to detect the occurrence of liquefaction at a particular site based on dynamic response. A sequential procedure as elaborated in preceding sections is thus adopted to arrive at the final form of ELM.

2. Development of ELM

In order to develop an exhaustive ELM to replicate pragmatic dynamic response which will take into account these issues to be applicable in realistic situations, the fundamental relationship as in Equation. 1 is employed. Hence, initially, the elementary model is transformed into a dimensionally homogeneous form using Fourier's principle of dimensional homogeneity, expressing the dimensions of the LP term in terms of primary quantities M–L–T, i.e. mass, length and time as

$$LPterm \text{ dim } ensions = \frac{\frac{m}{s} * \frac{kN}{m^2} * s}{\frac{kN}{m^2}} = m,$$

Thus, the term remains with the units of length; accordingly, the epicentral distance is chosen as the dimensional homogeneity factor to convert it into a homogeneous form. Now, in order to formulate the model, a total of 314 case records (Appendix I) have been utilized which are extracted from Boulanger, Wilson, and Idriss (2012), Cetin et al. (2004), Davis and Berrill (1982), Hamada and Wakamatsu (1996), Hanna, Ural, and Saygili (2007), and Zhang (1998). Out of the total parameters involved in formulating the present model, the values of earthquake magnitude (M_w), peak ground acceleration (a_{max}), epicentral distance (r), normalized SPT blow count $(N_1)_{60}$, and effective overburden pressure (σ_v) are used directly as available in the database, whereas the parameters, namely, peak ground velocity (v_{max}), small strain shear modulus (G_{max}), and duration of strong ground motion (dur) have been computed as detailed in Pathak and Dalvi (2013). Sampling bias and class balance have been maintained to ensure realistic predictability as illustrated by Oommen, Baise, and Vogel (2011) and accordingly, 220 cases have been employed for formulating the model and 94 cases have been used to validate the ELM. Thus, regression analysis has been performed in MATLAB using the training data-set of 220 cases out of the total 314 cases using the functional form as in Equation 1 and regression coefficients are obtained after applying dimensional homogeneity factor which resulted in Equation 2 as:

$$(N_1)_{60} = 7.47(LP_m term)^{0.28} \tag{2}$$

where LP_m term = modified LP term which is dimensionally homogeneous

   The boundary curve representing Equation 2 thus obtained is found to accurately discriminate the cases of liquefaction and non-liquefaction when validated on remaining 94 cases. All the 47 liquefaction cases are correctly predicted by Equation 2, hence signifies the theme underlying the present work. However, this proposed model is observed to predict "no" liquefaction conservatively, as some of the non-liquefied cases have been identified as liquefied by the model. Few such misclassified non-liquefaction case records indicate the normalized SPT values < 5 which actually illustrate high risk of liquefaction. Further, it is observed that for many of the remaining wrongly predicted "NO" cases, the value of normalized SPT blow count is within the range of 10–20, indicating intermediate risk of liquefaction. Thus, though the cases are of "no" liquefaction, the model rightly predicts them as "yes" cases. The model thereby achieves an overall success rate of 78% with accurate prediction for liquefied cases. Such a dimensionally homogeneous proposed model ensures it with physical significance in liquefaction assessment studies.

   Further to simulate different soil conditions on site, range of soils is then incorporated through the compositional factor "fines content" which is known to affect both cyclic shear strength and penetration resistance of soils. To do so, initially, the effect of fines content on the model performance is verified by categorizing the whole data-set of 314 case records as: Class (A) Low FC, clean sand (FC ≤ 5%), Class (B) Intermediate FC, silty sand (6 ≤ FC ≤ 35%), and Class (C) High FC, silty sand to sandy silt (FC > 35%), to distinguish the soil types. Consequently, three separate equations each representing a particular type of soil have been obtained. The regression coefficients as evaluated for Classes A, B, and C are as tabulated below.

   These three categories are also associated with corresponding relative risk of liquefaction (Dalvi et al., 2014) as stated in Table 1. It can also be observed from Table 1 that the regression coefficients corresponding to intermediate fines content are the same as those proposed through Equation 2. Interestingly, the other two ranges of fines content define the upper and lower bounds of boundary curves as illustrated in Figure 1.

   It is worth mentioning that despite considering a different form to represent LP, the trend of boundary curves as proposed in present work (Figure 1) indicates similarity with those established by previous researchers such as (Andrus & Stokoe II, 2000; Bolton Seed, Tokimatsu, Harder, & Chung, 1985; Rezania, Javadi, & Giustolisi, 2010). The trend of boundary curves as depicted in Figure 1 clearly indicates that fines content is an important parameter which affects the performance of the

Table 1. Summary of model parameters for range of fines content

Class	Fines content	Soil type	Relative risk of liquefaction	α_1	β_1
A	Low (FC ≤ 5%)	Clean sand	High	8.53	0.26
B	Intermediate (6 ≤ FC ≤ 35%)	Silty sand	Intermediate	7.47	0.28
C	High (FC > 35%)	Silty sand to sandy silt	Low	5.39	0.34

Note:*The algebraic form of the ELM is* $(N_1)_{60} = \alpha_1 (LP_m term)^{\beta_1}$.

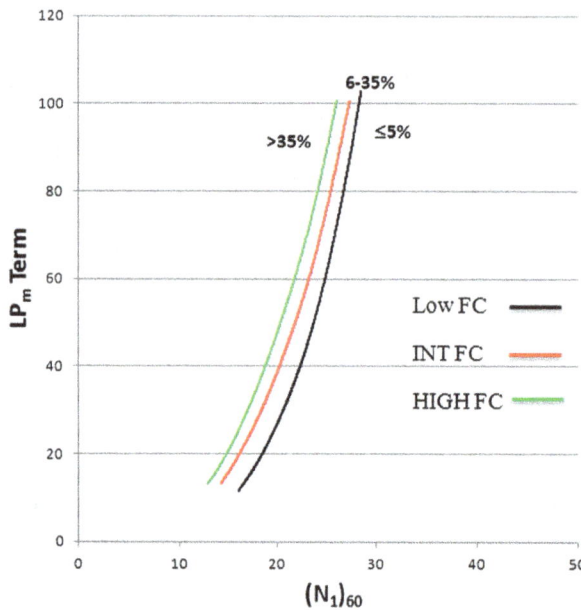

Figure 1. Fines content-based curves.

proposed model (Equation 2). Subsequently, the normalized SPT blow count is corrected for fines content presented as $(N_1)_{60cs}$ through Equation 3 to include its effect directly into the model,

$$(N_1)_{60cs} = (N_1)_{60} * C_{FINES} \tag{3}$$

where C_{FINES} is as given by Seed et al. (2003); thus, the final form of the model culminates into Equation 4:

$$(N_1)_{60cs} = 5.22(LP_m term)^{0.37} \tag{4}$$

This liquefaction triggering correlation obtained is in terms of dimensionally homogeneous "liquefaction potential term" (LP_m term) and the normalized SPT blow count corrected for fines content $(N_1)_{60cs}$. After verifying the predictive performance of the developed Equation. 4, on remaining 94 cases, it is found that by inclusion of fines content as model parameter, the predictive performance improves by over 4%, giving an overall success rate of 82%. Most importantly, all the 47 liquefaction cases are correctly predicted and 30 cases out of the 47 non-liquefaction cases have been identified correctly as shown in Figure 2.

From Figure 2, it can be seen that the scatter of non-liquefaction cases lying above the boundary curve is lesser when correction for fines content is applied. It is also noticed that the wrongly predicted "no" liquefaction cases having $(N_1)_{60cs}$ between 15 and 20, lie in the vicinity of the proposed curve. Thus, the transformed form of the empirical model stated via Equation. 4 certainly signifies remarkable potential in correctly identifying liquefaction-prone sites. *Thus, Equation. 4 has been proposed as the dynamic response-based ELM.*

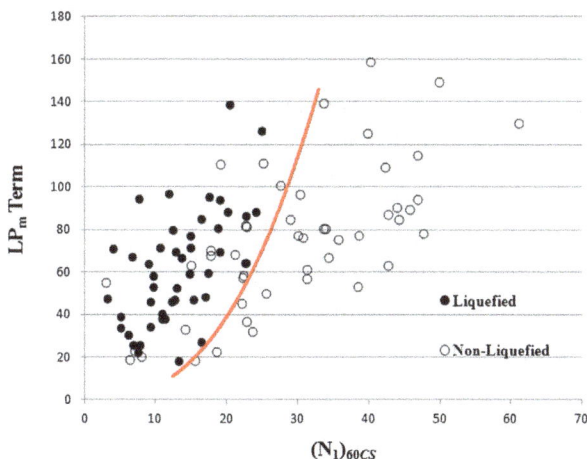

Figure 2. Validation of the proposed model with inclusion of fines content.

3. Performance of ELM

Now in order to justify the comprehensiveness of the proposed ELM, it is projected to assess its performance over diversified applications such as:

(3.1) Performance of ELM over the range of significant parameters,

(3.2) Predictability of ELM over an entirely new data-set, and

(3.3) Comparison of ELM with similar empirical procedures.

3.1. Performance of ELM over the range of significant parameters

Based on MCDM analysis, the effect of range of significant parameters, namely: a_{max}, M_w, and effective overburden pressure, is studied in present work. It has been demonstrated that these parameters are not only statistically significant but also possess physical relevance with the actual phenomenon (Dalvi et al., 2014). As a wide range of these parameters is included in the proposed ELM, the ranges are classified such that a particular range of respective parameter reflects the actual field behavior, indicating possible relative risk of liquefaction. Accordingly, the entire data-set of 314 cases is classified into three categories, namely: Class (A) High risk of liquefaction, Class (B) Intermediate risk of liquefaction, and Class (C) Low risk of liquefaction. These ranges of the respective parameters and their corresponding potential risk of liquefaction are summarized in Table 2.

3.1.1. Effect of peak ground acceleration (a_{max})

a_{max} has been commonly used to describe the ground motion because of its inherent relationship with inertial forces; indeed, the largest dynamic forces are assumed to be closely related to the a_{max} (Kramer, 1996). The scatter of data points relative to each class corresponding to ranges of acceleration as stated in Table 2 is represented graphically through Figure 3(a–c). Figure 3(a) is indicating low range of a_{max} ($a_{max} < 0.2$ g) categorized as Class C. Figure 3(b) demonstrates variation of ELM over

Table 2. Range and degree of risk of liquefaction			
Parameter	Peak ground acceleration (a_{max}) (g)	Magnitude (M_w)	Effective overburden pressure (σ_v') (kPa)
Relative risk of liquefaction			
High risk of liquefaction, class A	> 0.5	< 6.5	< 61
Intermediate risk of liquefaction, class B	0.2–0.5	6.5–7.5	61–81
Low risk of liquefaction, class C	< 0.2	> 7.5	> 81

case records falling under Class B, i.e. intermediate acceleration level (a_{max} = 0.2 to 0.5 g). Figure 3(c) depicts the variation of ELM for Class A, i.e. high risk of liquefaction (a_{max} > 0.5 g).

From these Figures (3 (a–c), it can be noted that the overall distribution of liquefaction and non-liquefaction points relative to the proposed ELM across these ranges as specified above appears to be fairly balanced, indicating efficiency of the developed model. Moreover, for intermediate and high a_{max} ranges, (Figure 3(b) and (c), respectively), more than 60% of correct prediction of non-liquefied sites is observed. It is worth mentioning that the positive predictability of the proposed ELM through Equation (4) remains unaltered, although a particular range of a_{max} is considered.

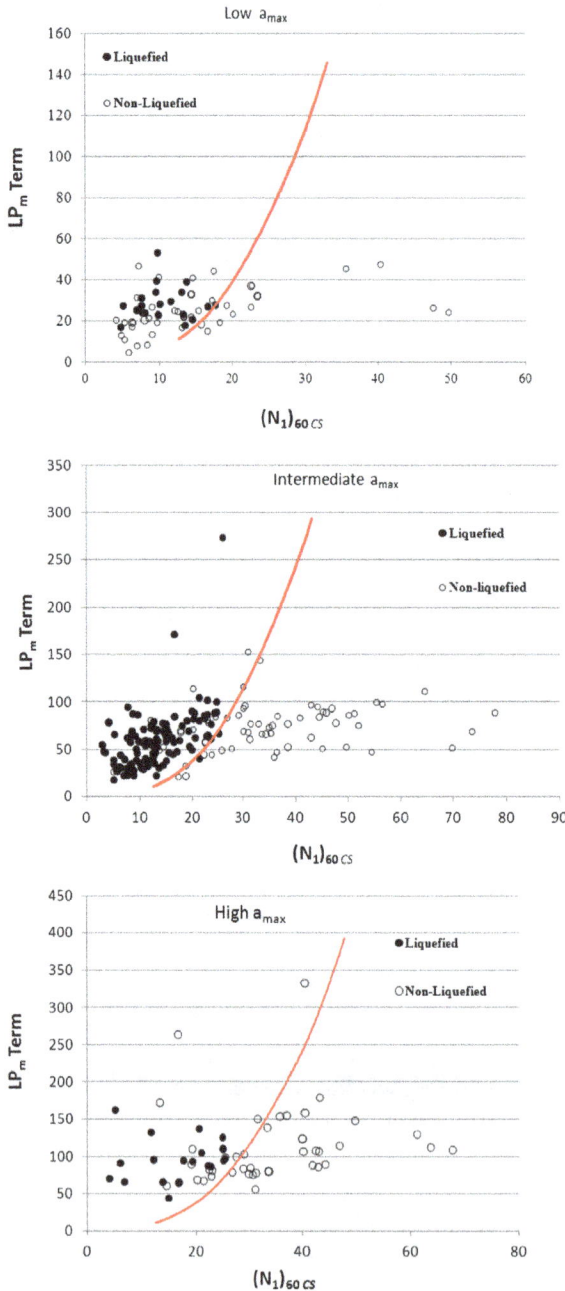

Figure 3. Variation of ELM with peak ground acceleration (a_{max}).

3.1.2. Effect of earthquake magnitude (M_w)

It is a known fact that for liquefaction to occur, there must be ground shaking and the potential for liquefaction increases with the increase in earthquake shaking represented by M_w. A similar observation as above is found when the model is verified relative to ranges of M_w as: Class C ($M_w < 6.5$), Class B ($M_w = 6.5$ to 7.5), and Class A ($M_w > 7.5$) as depicted in Figure 4(a–c), respectively.

The scatter of data points shown in Figure 4 (a) indicates that most of the misclassified non-liquefaction cases bear the $(N_1)_{60cs}$ value less than 10 which is indicative of high risk of liquefaction. Figure 4(b) represents Class B data points having intermediate risk of liquefaction and few of the liquefied cases could be seen in the vicinity of demarcation of curve. Figure 4 (c) clearly indicates that all the liquefaction cases of Class A are correctly predicted by the ELM. It can be stated that the model performance remains impartial at defined ranges of magnitude as well as acceleration of an earthquake as the overall success rate of accurate prediction remains unchanged. The functional form of LP term as employed in the present study includes these two parameters through v_{max}. This ensures the realistic seismic soil response as frequency content gets included.

3.1.3. Effect of vertical effective overburden pressure

Further, representing geological setting of soil strata, overburden stress effects for liquefaction analysis procedures have been investigated by Boulanger (2003), Boulanger and Idriss (2012) and Seed et al. (2003). Moreover, it is known that soil behaves as a deformable body with increasing depth; hence, depth is a site condition parameter and has a huge impact on soil response against its capacity to resist liquefaction. In addition to this, for liquefaction to occur, groundwater table should be at sufficient depth to create saturated soil conditions. Thus, the depth to groundwater table is an another important consideration in identifying soils that are susceptible to liquefaction. The vertical effective overburden pressure obviously takes into account the effect of depth as well as groundwater table level. For this purpose, data have been categorized w.r.t σ_v' in three different classes to represent the pertinent degree of risk of liquefaction (Table 2). The variation of ELM w.r.t these classes is illustrated through Figure 5 (a) to (c).

Figure 5 (a) to (c) implies that the proposed ELM presented through Equation (4) is unbiased, relative to the variation in vertical effective overburden pressure. Moreover, it rightly indicates that the susceptibility of liquefaction decreases (LP_m term) with increase in vertical effective overburden pressure. Thus, it is inferred that the observed behavior on field is replicated through the proposed model. In summary, the deterministic ELM not only yields accurate prediction of liquefied sites within a given range of model parameters but also replicates actual field behavior, thus an optimistic yet realistic ELM is developed.

3.2. Performance of proposed ELM using varied data-sets

Based on the investigation of predictive performance of recently developed empirical models, it is inferred that the success rate of accurate prediction reduces if the developed model is tested upon a totally different data-set and thus in general, the models are data specific (Pathak & Dalvi, 2011).

In this work, the predictability of the present ELM is ascertained on a varied data-set of around reported 740 cases. Among these, 386 case records are obtained from Hanna et al. (2007) and Boulanger et al. (2012). Based on verification over these data points, it is prominently observed that the developed ELM succeeds to accurately detect the occurrence of liquefaction for all the 227 "yes" cases out of the total 386 cases under consideration.

As the remaining 354 case records extracted from Kayen et al. (2013) include database of corrected shear wave velocity (V_{s1}), the $(N_1)_{60}$ value has been computed using the established correlation of Andrus and Stokoe II (2000) as given by:

$$V_{s1} = B_1[(N_1)_{60}]^{B_2} \qquad (5)$$

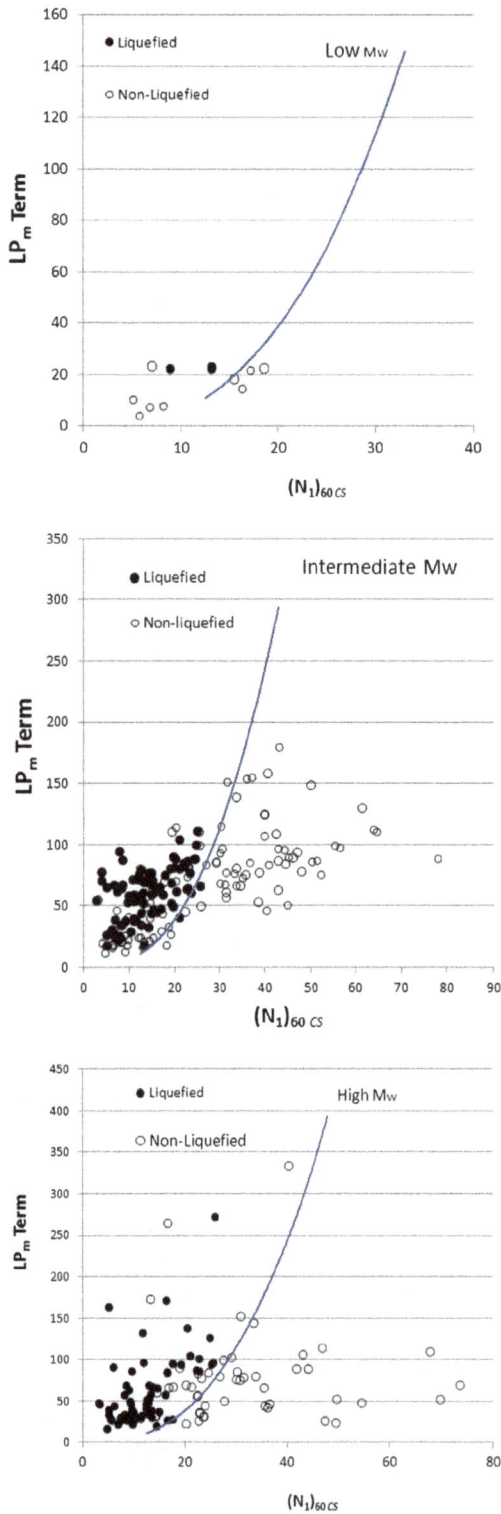

Figure 4. Variation of ELM with earthquake magnitude (M_w).

where V_{s1} = overburden stress-corrected shear wave velocity, $B_1 = 93.2 \pm 6.5$ and $B_2 = 0.231 \pm 0.022$. For this data-set, the predictability of ELM is verified against Equation (2) which gives the relation between LP_m term and $(N_1)_{60}$. The scatter of liquefied cases is as depicted in Figure 6 below.

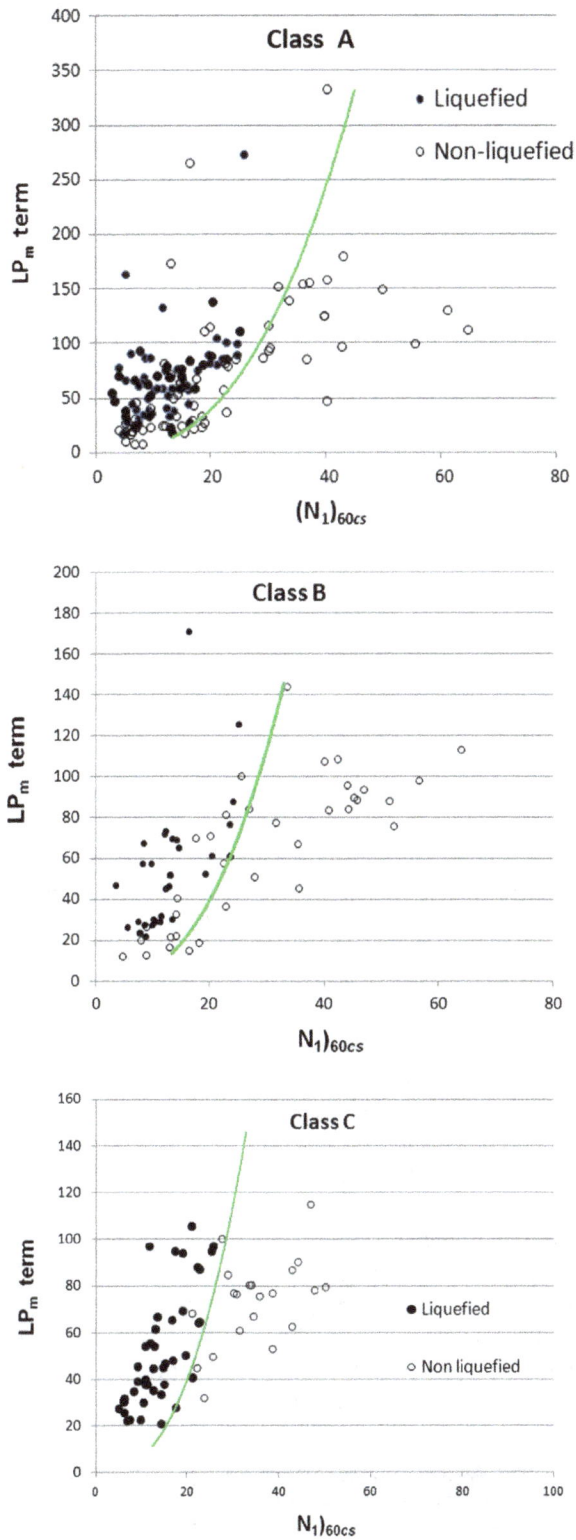

Figure 5. Variation of ELM with vertical effective pressure.

It is interesting to note that most of the 238 "liquefied" cases are accurately predicted as could be seen lying within the zone of liquefaction as shown in Figure 6. Although few of the data points could

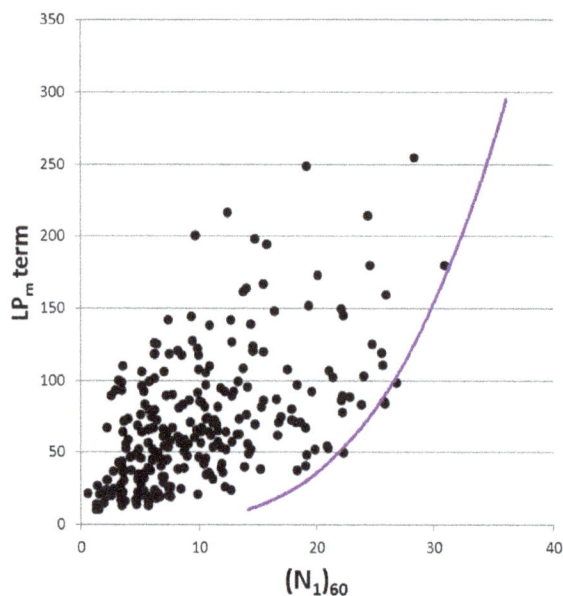

Figure 6. Scatter of liquefied cases in LP_m term-$(N_1)_{60}$ space.

be seen lying just on the boundary curve, but these are the records with $(N_1)_{60}$ value more than 23 and thus indicative of equal risk of liquefaction.

   Thus, overall, the ELM performance is verified on 740 case records apart from the 314 cases employed in formulating and validating the model. It is prominently observed that the developed ELM succeeds to accurately predict the occurrence of liquefaction for all the 465 "yes" cases out of the total 740 cases under consideration. Although conservative results are observed for non-liquefaction cases, it is to be noted that such wrongly predicted "no" liquefaction cases bear the properties which are prone to liquefaction. Finally, the ELM has performed well on over 1,054 case history records within the given range of model parameters, covering worldwide 46 number of earthquakes occurred during 1906–2011. Thus, the proposed final form ELM (Equation 4) certainly bears remarkable potential in detecting liquefaction by correlating the field conditions with the dynamic soil properties and ground motion parameters. The performance of the developed ELM to predict liquefaction is then compared with existing empirical models to demonstrate its applicability in the field as discussed in the next section. The comparison is purely meant to focus the inclusion of dynamic response in liquefaction evaluation procedures as proposed by the authors.

3.3. Comparison of proposed ELM with other approaches
The proposed ELM being fundamental in its nature has to be verified for its performance with SPT- and V_s-based simplified procedures. The SPT-based approach as originally proposed by Bolton Seed et al. (1985) is based on empirical evaluations of field observations. Although this method has been modified and improved periodically since that time, it presents the fundamental concept of inclusion of fines content into liquefaction evaluation procedures. Thus, the model performance when

Table 3. Comparison of predictive performance of ELM with other empirical approaches		
Approach	OA (%)	SRL (%)
I	82.6	81.6
II	83.1	72.4
III	85.2	89.9
IV	66.8	71.8
Present study	85.6	100

compared with these SPT-based approaches indicates that the developed ELM is efficient in correctly identifying the liquefied sites which are misclassified by these approaches. Further, it is known that Andrus and Stokoe (1997, 2000) pioneered in the use of V_s, a dynamic soil property as a field index of liquefaction resistance. Similar to the simplified procedure, they proposed CRR-V_{s1} curves corresponding to the range of fines content using compiled case histories. As stated in Andrus and Stokoe II (2000), only two evidences from 1989, Loma Prieta event (M_w = 6.93), lie incorrectly in "No" liquefaction zone; however, these two misclassifications lie very near to the demarcation curve; interestingly, the proposed ELM rightly predicts liquefaction occurrence for these two evidences which were actually reported as liquefied. This also indicates that success rate of correct prediction is more when dynamic soil properties are taken into account.

The performance of ELM is verified using the approaches by Youd et al. (2001): (Approach I) and Cetin et al. (2004): (Approach II) and also two recently developed approaches, namely: SPT-based approach by Oommen, Baise, and Vogel (2010): (Approach III) and V_s-based approach by Zhang (2010): (Approach IV). The research carried out by various researchers is undeniably exemplary. It is to be noted that the basis for this comparison is to verify the performance of various ELMs over a common data-set which in present work is as extracted from Cetin et al. (2004). The values of overall accuracy (OA) and success rate for liquefaction (SRL) by approaches I, II, and III have been mentioned as reported in Oommen et al. (2010), whereas the values of the same for approach IV and ELM are computed. The predictive performance in terms of OA and SRL is then summarized in Table 3.

From Table 3 it is observed that the OA in correct prediction can be seen to be at par with other four approaches. Moreover, the proposed ELM succeeds to detect liquefaction occurrence more accurately. It is to be noted that the comparison of ELM with other models is cited purely to symbolize the effectiveness of using the proposed methodology, although the basic framework of liquefaction assessment for each of these models differs significantly. Overall, the above discussion justifies the extensiveness of the developed ELM and thus for accurate prediction of liquefaction occurrence, it would be more appropriate to employ the proposed ELM which will minimize the enormous losses caused due to liquefaction.

4. Conclusions

Formulation, validation, and verification of a fundamental yet comprehensive ELM for detection of liquefaction occurrence is presented. Although it follows the general format of simplified procedure based on SPT, the liquefaction potential is characterized by integrating the effect of dynamic soil properties and ground motion parameters. The final form of the model is not only dimensionally homogeneous but also takes into account the fines content correction. Realistic predictability is achieved as sampling bias and class balance is maintained while formulating the ELM. Based on the performance of the ELM relative to soil and seismic parameters, it bears the potential to replicate the actual field behavior. A final correlation thus established is found to perform well for the diversified 1,000 case history records. Accurate prediction of liquefied cases for the specified range of model parameters is the prominent feature of the developed ELM.

Funding
The authors received no direct funding for this research.

Author details
Snehal Rajeev Pathak[1]
E-mail: srp.civil@coep.ac.in
Asita Nilesh Dalvi[2]
E-mail: asitadalvi@gmail.com
ORCID ID: http://orcid.org/0000-0003-2812-0242
[1] Department of Civil Engineering, College of Engineering Pune, Pune, Maharashtra, India.
[2] Department of Civil Engineering, SITS, Pune, Maharashtra, India.

References
Andrus, R., & Stokoe, K. H. (1997). Liquefaction resistance based on shear wave velocity. *NCEER workshop on evaluation of liquefaction resistance of soils, national center for Earthquake Engineering Research* (pp. 89–128). Buffalo, NY: State University of New York.
Andrus, R. D., & Stokoe II, K. H. (2000). Liquefaction resistance of soils from shear-wave velocity. *Journal of Geotechnical and Geoenvironmental Engineering, 126*, 1015–1025. http://dx.doi.org/10.1061/(ASCE)1090-0241(2000)126:11(1015)
Bolton Seed, H., Tokimatsu, K., Harder, L. F., & Chung, R. M. (1985). Influence of SPT procedures in soil liquefaction resistance evaluations. *Journal of Geotechnical*

Engineering, 111, 1425–1445.
http://dx.doi.org/10.1061/
(ASCE)0733-9410(1985)111:12(1425)

Boulanger, R. W. (2003). High overburden stress effects in liquefaction analyses. *Journal of Geotechnical and Geoenvironmental Engineering, 129*, 1071–1082. http://dx.doi.org/10.1061/
(ASCE)1090-0241(2003)129:12(1071)

Boulanger, R. W., & Idriss, I. M. (2012). Probabilistic standard penetration test-based liquefaction–triggering procedure. *Journal of Geotechnical and Geoenvironmental Engineering, 138*, 1185–1195. http://dx.doi.org/10.1061/(ASCE)GT.1943-5606.0000700

Boulanger, R. W., Wilson, D. W., & Idriss, I. M. (2012). Examination and reevalaution of spt-based liquefaction triggering case histories. *Journal of Geotechnical and Geoenvironmental Engineering, 138*, 898–909. http://dx.doi.org/10.1061/(ASCE)GT.1943-5606.0000668

Cetin, K. O., Seed, R. B., Der Kiureghian, A., Tokimatsu, K., Harder, L. F., Jr, Kayen, R. E., ... Moss, R. E. (2004). Standard penetration test-based probabilistic and deterministic assessment of seismic soil liquefaction potential. *Journal of Geotechnical and Geoenvironmental Engineering, 130*, 1314–1340. http://dx.doi.org/10.1061/
(ASCE)1090-0241(2004)130:12(1314)

Dalvi, A. N., Pathak, S. R., & Rajhans, N. R. (2014). Entropy analysis for identifying significant parameters for seismic soil liquefaction. *Geomechanics and Geoengineering, 9*(1), 1–8. doi:10.1080/17486025.2013.805255

Davis, R. O., & Berrill, J. B. (1982). Energy dissipation and seismic liquefaction in sands. *Earthquake Engineering Structural Dynamics, 10*, 51–68.

Hamada, M., & Wakamatsu, K. (1996). Liquefaction, ground deformation and their caused damage to structures. *The 1995 Hyogoken-Nanbu earthquake. Investigation into damage to Civil Engineering structures, committee of earthquake Engineering.* Japan Society of Civil Engineering, pp. 45–99.

Hanna, A. M., Ural, D., & Saygili, G. (2007). Neural network model for liquefaction potential in soil deposits using Turkey and Taiwan earthquake data. *Soil Dynamics and Earthquake Engineering, 27*, 521–540. http://dx.doi.org/10.1016/j.soildyn.2006.11.001

Kayen, R., Moss, R. E. S., Thompson, E. M., Seed, R. B., Cetin, K. O., Kiureghian, D., ... Tokimatsu, K. (2013). Shear-wave velocity-based probabilistic and deterministic assessment of seismic soil liquefaction potential. *Journal of Geotechnical

and Geoenvironmental Engineering, 139*, 407–419. http://dx.doi.org/10.1061/(ASCE)GT.1943-5606.0000743

Kramer, S. L. (1996). *Geotechnical earthquake Engineering.* Upper Saddle River, NJ: Prentice Hall .

Oommen, T., Baise, L. G., & Vogel, R. (2010). Validation and application of empirical liquefaction models. *Journal of Geotechnical and Geoenvironmental Engineering, 136*, 1618–1633.

Oommen, T., Baise, L. G., & Vogel, R. M. (2011). Sampling bias and class imbalance in maximum-likelihood logistic regression. *Mathematical Geosciences, 43*, 99–120. http://dx.doi.org/10.1007/s11004-010-9311-8

Pathak, S. R. & Dalvi, A. N. (2011). Performance of empirical models for assessment of seismic soil liquefaction. *International Journal of Earth Sciences and Engineering, 4*, 83–86.

Pathak, S. R., & Dalvi, A. N. (2012). Liquefaction potential assessment: An elementary approach. *International Journal of Innovative Research in Science, Engineering and Technology, 1*, 253–255.

Pathak, S. R., & Dalvi, A. N. (2013). Elementary empirical liquefaction model. *Natural HAZARDS, Journal of the International Society for the Prevention and Mitigation of Natural Hazards, 69*, 425–440. doi:10.1007/s11069-013-0723-x.

Rezania, M., Javadi, A. A., & Giustolisi, O. (2010). Evaluation of liquefaction potential based on CPT results using evolutionary polynomial regression. *Computers and Geotechnics, 37*, 82–92. http://dx.doi.org/10.1016/j.compgeo.2009.07.006

Seed, R. B, Cetin, K. O, Moss R, E, Kammerer A, M, Wu J, Pestana J. M, ... Faris, A. (2003). *Recent advances in soil liquefaction engineering: A unified and consistent framework.* 26th Annual Los Angeles Geotechnical Spring Seminar. Keynote Presentation, California.

Youd, T. L., Idriss, I. M., Andrus, R. D., Arango, I., Castro, G., Christian, J. T., ... Stokoe, II, K. H. (2001). Liquefaction resistance of soils: Summary report from the 1996 NCEER/NSF workshops on evaluation of liquefaction resistance of soils. *Journal of Geotechnical and Geoenvironmental Engineering, 127*, 817–833.

Zhang, L. (1998). Assessment of liquefaction potential using optimum seeking method. *Journal of Geotechnical and Geoenvironmental Engineering, 124*, 739–748. http://dx.doi.org/10.1061/
(ASCE)1090-0241(1998)124:8(739)

Zhang, L. (2010). A simple method for evaluating liquefaction potential from shear wave velocity. *Frontiers of Architecture and Civil Engineering in China, 4*, 178–195.

Appendix I

314 case records used for ELM development

No	Site	Mag.	σ'_v (kPa)	FC(%)	$(N_1)_{60}$	$(N_1)_{60cs}$	G_{max} (MPa)	v_{max} (m/s)	dur (s)	LP_m term	Obs. Liq?
		Extracted parameters				Computed parameters					
1	Ienaga	8.10	68.00	10.00	8.20	9.03	58.98	0.67	19.88	58.12	Y
2	Komei	8.10	61.00	30.00	3.40	5.31	42.87	0.67	19.88	47.09	Y
3	Shonenji temple	7.00	48.00	0.00	11.80	11.80	54.99	1.14	8.97	58.74	Y
4	Takaya 45	7.00	104.00	4.00	21.10	21.10	94.73	1.00	8.97	40.87	Y
5	Arayamotomachi	7.60	41.00	5.00	4.70	4.70	38.81	0.28	12.54	16.70	Y
6	Cc17-1	7.60	72.00	2.00	9.90	9.90	64.08	0.50	12.54	27.92	Y
7	Rail road-2	7.60	100.00	2.00	17.50	17.50	88.43	0.50	12.54	27.74	Y/N
8	Cc17-2	7.60	43.00	8.00	12.70	13.51	53.13	0.50	12.54	38.76	Y
9	Old town-1	7.60	81.00	2.00	22.70	22.70	85.20	0.56	12.54	37.12	N
10	Old town-2	7.60	109.00	2.00	23.50	23.50	99.72	0.56	12.54	32.29	N
11	Road site	7.60	79.00	0.00	14.10	14.10	74.12	0.56	12.54	33.12	N
12	Showa Br 4	7.60	67.00	0.00	35.50	35.50	86.54	0.56	12.54	45.59	N
13	Aomori station	8.30	38.00	3.00	16.50	16.50	53.66	0.74	16.24	84.80	Y
14	Nanaehama1-2-3	8.30	45.00	20.00	7.60	9.21	46.93	0.69	13.97	50.61	Y
15	Hachinohe- 2	8.30	76.00	5.00	35.30	35.30	92.04	0.80	13.97	67.59	N
16	Hachinohe- 4	8.30	45.00	5.00	23.00	23.00	63.72	0.80	13.97	79.02	N
17	Juvenilehall	6.61	96.00	55.00	3.90	6.20	56.10	1.21	8.56	30.33	Y
18	Panjin Ch. F. P.	7.00	89.00	67.00	7.60	10.41	66.00	0.57	14.06	29.79	Y
19	Amatitlan B-1	7.50	86.00	3.00	5.00	5.00	57.27	0.42	19.76	27.35	Y
20	Luan Nan L2	7.60	32.00	3.00	8.50	8.50	40.89	0.69	15.89	69.84	Y
21	Coastal region	7.60	54.00	12.00	11.70	12.86	58.19	0.41	15.58	34.13	Y
22	Le Ting L8-14	7.60	53.00	12.00	11.50	12.65	57.37	0.63	15.58	52.74	Y
23	Qing Jia Ying	7.60	59.00	20.00	20.10	22.71	70.46	1.09	15.58	101.82	Y
24	Amatitlan B-3&4	7.50	71.00	3.00	14.30	14.30	70.54	0.42	19.76	40.81	N
25	Luan Nan L1	7.60	38.00	5.00	24.40	24.40	59.44	0.69	15.89	85.50	N
26	Tangshan City	7.60	75.00	10.00	31.60	33.36	89.04	1.56	15.58	144.61	N
27	San Juan B-6	7.50	56.00	50.00	5.80	8.36	48.32	0.62	12.99	34.52	Y
28	Amatitlan B-2	7.50	55.00	3.00	9.70	9.70	55.68	0.42	19.76	41.58	Y/N
29	San Juan B-4	7.50	39.00	4.00	14.30	14.30	52.28	0.62	12.99	53.62	N
30	San Juan B-5	7.50	44.00	3.00	13.60	13.60	54.77	0.62	12.99	49.79	N
31	Nakamura 4	6.50	30.00	5.00	6.90	6.90	37.24	0.32	12.82	25.32	Y
32	Nakamura 5	7.70	42.00	4.00	9.60	9.60	48.51	1.02	14.83	87.00	Y
33	Ishinomaki-2	7.70	45.00	10.00	5.50	6.22	42.63	0.63	14.83	44.60	Y
34	Arahama	6.50	67.00	0.00	12.80	12.80	66.46	0.27	12.82	16.86	N
35	Kitawabuchi-2	6.50	59.00	5.00	12.30	12.30	61.68	0.37	12.82	24.88	N
36	Nakajima-18	6.50	79.00	3.00	14.10	14.10	74.12	0.37	12.82	22.33	N
37	Nakamura 5	6.50	42.00	4.00	9.60	9.60	48.51	0.32	12.82	23.56	N
38	Oiiri-1	6.50	85.00	5.00	9.40	9.40	68.60	0.37	12.82	19.21	N
39	Yuriage Br-3	6.50	42.00	2.00	11.80	11.80	51.44	0.32	12.82	24.98	N
40	Yuriagekami-2	6.50	47.00	0.00	15.10	15.10	58.26	0.32	12.82	25.28	N
41	Ishinomaki-2	6.50	45.00	10.00	5.50	6.22	42.63	0.32	12.82	19.32	N

Appendix (*Continued*)

No	Site	Mag.	σ'_v (kPa)	FC(%)	$(N_1)_{60}$	$(N_1)_{60cs}$	G_{max} (MPa)	v_{max} (m/s)	*dur* (s)	LP_m term	Obs. Liq?
		Extracted parameters				**Computed parameters**					
42	Shiomi-6	6.50	60.00	10.00	7.50	8.30	53.98	0.37	12.82	21.41	N
43	Yuriage Br-1	6.50	56.00	10.00	5.40	6.12	47.30	0.32	12.82	17.23	N
44	Yuriage Br-2	6.50	34.00	7.00	16.20	17.00	50.50	0.32	12.82	30.30	N
45	Hiyori-1 B	6.50	71.00	20.00	11.10	12.99	65.74	0.37	12.82	22.03	N
46	Yuriagekami-1	6.50	63.00	60.00	2.50	4.60	39.59	0.32	12.82	12.82	N
47	Radio tower b1	6.53	50.00	64.00	2.90	5.06	36.94	0.53	9.24	18.18	Y
48	Riverpark a	6.53	20.00	80.00	4.60	6.99	26.93	0.64	9.24	39.76	Y
49	Radio tower b2	6.53	38.00	30.00	15.20	18.52	52.48	0.53	9.24	33.98	N
50	Wildlifeb	6.53	54.00	30.00	10.30	13.04	56.13	0.45	9.24	21.74	N
51	Kombloom b	6.53	62.00	92.00	6.20	8.82	51.87	0.35	9.24	13.38	N
52	Owi 1	6.00	57.00	13.00	7.10	8.12	51.77	0.23	7.94	8.43	N
53	Owi 2	6.00	123.00	27.00	3.90	5.67	63.50	0.23	7.94	4.79	N
54	Wildlifeb	5.90	54.00	30.00	10.30	13.04	56.13	0.63	6.84	22.40	Y
55	Kombloom b	5.90	62.00	92.00	6.20	8.82	51.87	0.78	6.84	22.19	Y
56	Mckim ranch a	5.90	32.00	31.00	4.60	6.72	34.06	0.22	6.84	7.94	N
57	Riverparkc	5.90	45.00	18.00	15.20	17.19	57.11	0.51	6.84	22.09	N
58	Gaiko wharf B2	7.70	53.00	1.00	12.40	12.40	58.59	0.72	10.38	41.32	Y
59	Noshiro section N7	7.70	38.00	1.00	16.20	16.20	53.39	0.79	10.38	57.84	Y
60	Takeda elememtary school	7.70	42.00	0.00	13.30	13.30	53.18	0.90	10.38	59.01	Y
61	Giako 1	7.70	79.00	3.00	8.70	8.70	64.67	0.65	10.38	27.63	Y
62	Gaiko 2	7.70	107.00	4.00	6.90	6.90	70.34	0.65	10.38	22.19	Y
63	Nakajima no 2(1)	7.70	81.00	3.00	13.50	13.50	74.16	0.65	10.38	30.91	Y
64	Nakajima no 3(3)	7.70	71.00	2.00	10.20	10.20	64.18	0.65	10.38	30.51	Y
65	Nakajima no 3(4)	7.70	68.00	2.00	11.50	11.50	64.98	0.65	10.38	32.26	Y
66	Ohama no 2-1	7.70	56.00	2.00	5.40	5.40	47.30	0.65	10.38	28.51	Y
67	Ohama no 3-1	7.70	63.00	2.00	7.40	7.40	55.09	0.65	10.38	29.52	Y
68	Ohama no 3-3	7.70	64.00	2.00	5.60	5.60	51.12	0.65	10.38	26.96	Y
69	Ohama no 3-4	7.70	49.00	2.00	8.10	8.10	49.89	0.65	10.38	34.37	Y
70	Nakajima no 1(5)	7.70	74.00	8.00	9.90	10.62	64.96	0.65	10.38	29.63	Y
71	Nakajima no 2(2)	7.70	48.00	7.00	9.30	9.91	51.39	0.65	10.38	36.14	Y
72	Arayamotomachi	6.80	37.00	5.00	5.10	5.10	37.79	0.42	9.10	19.33	N
73	Arayamotomachi coarse sand	6.80	77.00	0.00	18.10	18.10	78.29	0.42	9.10	19.25	N
74	Heber road A1	6.54	42.00	12.00	37.80	40.21	69.53	0.42	13.77	47.42	N
75	Heber road A2	6.54	50.00	18.00	2.90	4.01	36.94	0.40	13.77	20.35	N
76	Heber road A3	6.54	56.00	25.00	16.20	19.07	64.81	0.35	13.77	27.63	N
77	mckim ranch a	6.54	32.00	31.00	4.60	6.72	34.06	0.43	13.77	31.27	N
78	River park C	6.54	45.00	18.00	15.20	17.19	57.11	0.51	13.77	44.27	N
79	Radio tower B1	6.54	50.00	64.00	2.90	5.06	36.94	0.53	13.77	27.13	N
80	River park A	6.54	20.00	80.00	4.60	6.99	26.93	0.51	13.77	46.98	N
81	Kornbloom B	6.54	62.00	92.00	6.20	8.82	51.87	0.46	13.77	26.73	N
82	Wildlife B	6.20	61.85	75.00	12.80	16.34	63.86	0.23	12.85	15.14	N
83	Radio tower B1	6.22	50.00	64.00	2.90	5.06	36.94	0.23	13.09	11.06	N

Appendix (*Continued*)

No	Site	Mag.	σ'_v (kPa)	FC(%)	$(N_1)_{60}$	$(N_1)_{60cs}$	G_{max} (MPa)	v_{max} (m/s)	dur (s)	LP_m term	Obs. Liq?
		Extracted parameters				Computed parameters					
84	Wildlife B	6.22	54.00	30.00	10.30	13.04	56.13	0.34	13.09	22.99	Y
85	Marine laboratory UC B1	6.93	65.00	3.00	13.10	13.10	65.88	0.79	13.02	52.24	Y
86	Sandholdt UC B10	6.93	43.00	2.00	15.30	15.30	55.92	0.79	13.02	67.02	Y
87	State Beach UC B1	6.93	46.00	1.00	10.30	10.30	51.80	0.79	13.02	58.04	Y
88	Marine laboratory UC B2	6.93	44.00	3.00	15.90	15.90	57.16	0.74	13.02	62.17	Y
89	Farris farm	6.93	92.00	8.00	10.20	10.93	73.06	1.05	13.02	54.08	Y
90	SFOBB-1,2	6.93	86.00	8.00	8.60	9.28	67.25	0.76	13.02	38.86	Y
91	Miller Farm CMF3	6.93	101.00	32.00	9.90	12.77	75.90	1.10	13.02	53.94	Y
92	Miller Farm CMF8	6.93	95.00	25.00	9.80	12.03	73.39	1.10	13.02	55.46	Y
93	Treasure Island	6.93	67.00	20.00	6.40	7.91	54.43	0.45	13.02	23.92	Y
94	Miller farm	6.93	66.00	22.00	10.00	11.98	61.53	1.19	13.02	72.07	Y
95	Woodmarine UC B4	6.93	25.00	35.00	9.10	12.12	36.86	0.79	13.02	75.98	Y
96	POR 2,3 4	6.93	73.00	50.00	5.10	7.56	53.08	0.51	13.02	24.09	Y
97	MBARI NO 3 EB5	6.93	47.00	1.00	14.90	14.90	58.04	0.79	13.02	63.65	N
98	Alameda BF dike	6.93	91.00	7.00	43.30	44.86	105.61	0.68	13.02	51.26	N
99	Perez B1v B11	7.70	90.00	19.00	13.00	14.94	77.36	0.79	11.01	37.54	Y
100	Cereenan St.B12	7.70	68.00	19.00	24.90	27.74	79.93	0.79	11.01	51.33	N
101	Kushiro Port site A	7.60	68.00	2.00	16.40	16.40	71.66	1.25	26.04	171.60	Y
102	Kushiro Port Seismo st	7.60	47.00	5.00	25.90	25.90	67.11	1.47	26.04	273.21	Y
103	Kushiro Port site D	7.60	118.00	0.00	30.90	30.90	111.08	1.25	26.04	153.29	N
104	Wynne Ave Unit C1	6.69	105.00	33.00	11.60	14.78	80.94	1.39	8.37	44.89	Y
105	Balboa B1vUnit C	6.69	143.00	50.00	13.10	16.68	97.72	2.29	8.37	65.54	Y
106	Ashiyama CDE (marine sand)	6.90	115.00	2.00	12.50	12.50	86.50	1.13	10.52	44.56	Y
107	Kobe alluvial site number 5	6.90	116.00	1.00	6.10	6.10	70.60	0.99	10.52	31.55	Y
108	Kobe alluvial site number 7	6.90	60.00	0.00	10.90	10.90	60.12	1.13	10.52	59.36	Y
109	Kobe alluvial site number 9	6.90	64.00	2.00	12.20	12.20	64.09	1.41	10.52	74.16	Y
110	Kobe alluvial site number 11	6.90	62.00	5.00	8.50	8.50	56.91	1.41	10.52	67.97	Y
111	Kobe alluvial site number 16	6.90	60.00	5.00	25.00	25.00	75.15	1.69	10.52	111.30	N/Y
112	Kobe alluvial site number 17	6.90	43.00	5.00	21.10	21.10	60.92	1.41	10.52	104.90	Y
113	Kobe alluvial site number 24	6.90	51.00	0.00	24.60	24.60	69.01	1.41	10.52	100.20	Y
114	Kobe alluvial site number 29	6.90	49.00	0.00	17.90	17.90	62.27	1.13	10.52	75.29	Y
115	Kobe alluvial site number 37	6.90	79.00	0.00	19.30	19.30	80.66	0.99	10.52	52.93	Y
116	Kobe alluvial site number 44	6.90	43.00	5.00	8.30	8.30	47.07	1.13	10.52	64.85	Y
117	Kobe alluvial site number 13	6.90	74.00	15.00	12.70	14.21	69.69	1.41	10.52	69.74	Y
118	Kobe alluvial site number 28	6.90	44.00	8.00	21.10	22.18	61.62	1.13	10.52	82.96	Y

Appendix (*Continued*)

No	Site	Mag.	σ'_v (kPa)	FC(%)	$(N_1)_{60}$	$(N_1)_{60cs}$	G_{max} (MPa)	v_{max} (m/s)	dur (s)	LP_m term	Obs. Liq?
			Extracted parameters				Computed parameters				
119	Kobe alluvial site number 34	6.90	73.00	9.00	24.20	25.52	82.22	1.13	10.52	66.72	Y
120	Kobe alluvial site number 35	6.90	55.00	6.00	18.90	19.65	66.93	1.41	10.52	90.12	Y
121	Kobe alluvial site number 42	6.90	46.00	10.00	12.10	13.08	54.21	1.13	10.52	69.82	Y
122	ashiyama CDE (mountain sand 2)	6.90	80.00	18.00	21.10	23.52	83.09	1.13	10.52	61.53	Y
123	Port island borehole array st	6.90	96.00	20.00	6.80	8.34	66.34	0.96	10.52	34.79	Y
124	Port island site I	6.90	123.00	20.00	10.80	12.66	85.86	0.96	10.52	35.15	Y
125	Rokko Island building D	6.90	107.00	25.00	16.80	19.73	90.47	1.13	10.52	50.09	Y
126	Rokko island site G	6.90	146.00	20.00	12.30	14.28	97.03	0.96	10.52	33.46	Y
127	Torishma dike	6.90	46.00	20.00	14.00	16.12	56.45	0.70	10.52	45.44	Y
128	Kobe alluvial site number 6	6.90	72.00	21.00	17.80	20.35	75.37	1.13	10.52	62.02	Y
129	Kobe alluvial site number 1	6.90	80.00	3.00	52.00	52.00	103.14	1.13	10.52	76.38	N
130	Kobe alluvial site number 40	6.90	59.00	0.00	39.70	39.70	83.36	1.69	10.52	125.55	N
131	Kobe alluvial site number 36	6.90	36.00	3.00	31.60	31.60	61.69	1.69	10.52	152.27	N
132	Kobe alluvial site number 4	6.90	54.00	1.00	36.60	36.60	78.25	1.13	10.52	85.85	N
133	Kobe alluvial site number 25	6.90	50.00	3.00	35.80	35.80	74.91	1.97	10.52	155.31	N
134	Kobe alluvial site number 26	6.90	37.00	0.00	37.00	37.00	64.94	1.69	10.52	155.96	N
135	Kobe alluvial site number 20	6.90	75.00	0.00	63.70	63.70	104.30	1.55	10.52	113.28	N
136	Kobe alluvial site number 22	6.90	79.00	6.00	38.60	39.83	95.83	1.69	10.52	107.80	N
137	Kobe alluvial site number 12	6.90	72.00	14.00	24.70	26.78	82.08	1.41	10.52	84.41	N
138	Kobe alluvial site number 19	6.90	124.00	10.00	21.30	22.65	103.70	1.69	10.52	74.31	N
139	Kobe alluvial site number 23	6.90	72.00	10.00	24.00	25.46	81.48	1.69	10.52	100.56	N
140	Kobe alluvial site number 27	6.90	29.00	10.00	40.80	42.93	58.81	1.69	10.52	180.21	N
141	Kobe alluvial site number 32	6.90	41.00	6.00	29.10	30.10	64.52	1.41	10.52	116.54	N
142	Ashiyama A (mountain sand 1)	6.90	80.00	18.00	21.10	23.52	83.09	1.13	10.52	61.53	N
143	Adapzari	7.40	57.40	4.00	22.00	22.00	71.15	1.21	11.38	85.59	Y
144	Adapzari	7.40	84.10	3.00	13.00	13.00	74.78	1.21	11.38	61.41	Y
145	Adapzari	7.40	25.90	2.00	4.00	4.00	29.37	1.21	11.38	78.30	Y
146	Adapzari	7.40	61.30	14.00	12.00	13.37	62.44	1.21	11.38	70.34	Y
147	Yuanlin	7.60	160.00	13.00	13.00	14.33	103.15	0.56	11.36	20.60	Y
148	Yuanlin	7.60	68.10	15.00	10.00	11.35	62.50	0.56	11.36	29.33	Y
149	Nantou	7.60	73.00	6.00	14.00	14.64	71.11	1.19	11.36	65.72	Y

Appendix (*Continued*)

No	Site	Mag.	σ'_v (kPa)	FC(%)	$(N_1)_{60}$	$(N_1)_{60cs}$	G_{max} (MPa)	v_{max} (m/s)	dur (s)	LP_m term	Obs. Liq?
			Extracted parameters				Computed parameters				
150	Wufeng	7.60	104.10	15.00	19.00	20.89	92.21	2.09	11.36	105.37	Y
151	Wufeng	7.60	145.30	8.00	24.00	25.17	115.75	2.09	11.36	94.76	Y
152	Wufeng	7.60	157.80	9.00	21.00	22.21	116.55	2.09	11.36	87.85	Y
153	Yuanlin	7.60	33.60	18.00	8.00	9.48	41.16	0.56	11.36	39.15	Y
154	Yuanlin	7.60	45.70	19.00	6.00	7.41	44.10	0.56	11.36	30.84	Y
155	Wufeng	7.60	157.50	21.00	20.00	22.73	114.97	2.09	11.36	86.83	Y
156	Wufeng	7.60	43.10	31.00	9.00	11.67	48.24	2.09	11.36	133.14	Y
157	Wufeng	7.60	56.90	31.00	4.00	6.05	43.53	2.09	11.36	90.99	Y
158	Wufeng	7.60	132.50	26.00	22.00	25.59	108.09	2.09	11.36	97.04	Y
159	Adapzari	7.40	65.60	70.00	19.00	23.41	73.20	1.21	11.38	77.06	Y
160	Adapzari	7.40	97.30	75.00	23.00	27.97	93.69	1.21	11.38	66.50	Y
161	Adapzari	7.40	21.50	87.00	1.00	2.89	17.26	1.21	11.38	55.45	Y
162	Adapzari	7.40	50.30	39.00	20.00	24.55	64.97	1.21	11.38	89.20	Y
163	Adapzari	7.40	56.40	54.00	17.00	21.13	65.89	1.21	11.38	80.68	Y
164	Adapzari	7.40	16.80	51.00	2.00	4.03	19.05	1.21	11.38	78.32	Y
165	Adapzari	7.40	26.50	97.00	6.00	8.59	33.58	1.21	11.38	87.51	Y
166	Adapzari	7.40	53.10	99.00	16.00	19.99	62.90	1.21	11.38	81.81	Y
167	Adapzari	7.40	30.30	57.00	3.00	5.17	29.06	1.21	11.38	66.24	Y
168	Adapzari	7.40	48.00	91.00	11.00	14.29	53.91	1.21	11.38	77.56	Y
169	Adapzari	7.40	47.00	65.00	5.00	7.45	42.34	1.21	11.38	62.21	Y
170	Yuanlin	7.60	51.80	90.00	5.00	7.45	44.45	0.56	11.36	27.42	Y
171	Yuanlin	7.60	92.60	94.00	7.00	9.73	65.71	0.56	11.36	22.68	Y
172	Wufeng	7.60	14.80	65.00	3.00	5.17	20.31	2.09	11.36	163.24	Y
173	Adapzari	7.40	85.40	5.00	33.00	33.00	96.01	1.21	11.38	77.64	N
174	Adapzari	7.40	106.20	5.00	31.00	31.00	105.46	1.21	11.38	68.58	N
175	Adapzari	7.40	96.60	4.00	35.00	35.00	103.56	1.21	11.38	74.03	N
176	Adapzari	7.40	79.10	1.00	20.00	20.00	81.47	1.21	11.38	71.13	N
177	Adapzari	7.40	63.40	4.00	44.00	44.00	88.47	1.21	11.38	96.37	N
178	Adapzari	7.40	55.40	4.00	30.00	30.00	75.57	1.21	11.38	94.19	N
179	Adapzari	7.40	82.50	4.00	50.00	50.00	103.84	1.21	11.38	86.92	N
180	Adapzari	7.40	79.00	8.00	39.00	40.65	96.06	1.21	11.38	83.97	N
181	Adapzari	7.40	79.20	9.00	49.00	51.21	101.29	1.21	11.38	88.32	N
182	Adapzari	7.40	92.50	8.00	75.00	77.80	119.81	1.21	11.38	89.45	N
183	Yuanlin	7.60	139.80	8.00	19.00	20.01	106.86	0.56	10.98	23.62	N
184	Adapzari	7.40	97.00	19.00	27.00	30.00	97.42	1.21	11.38	69.36	N
185	Yuanlin	7.60	113.10	20.00	20.00	22.60	97.42	0.56	10.98	26.62	N
186	Yuanlin	7.60	195.30	29.00	43.00	49.44	154.47	0.56	10.98	24.44	N
187	Nantou	7.60	158.90	35.00	42.00	49.63	138.59	1.19	10.36	53.65	N
188	Nantou	7.60	160.10	20.00	21.00	23.68	117.40	1.19	10.36	45.11	N
189	Nantou	7.60	177.20	19.00	33.00	36.46	138.30	1.19	10.36	48.01	N
190	Nantou	7.60	195.30	26.00	62.00	69.75	167.36	1.19	10.36	52.72	N
191	Nantou	7.60	210.50	22.00	49.00	54.41	165.14	1.19	10.36	48.26	N
192	Nantou	7.60	217.00	29.00	31.00	36.05	150.75	1.19	10.36	42.74	N

Appendix (*Continued*)

No	Site	Mag.	σ'_v (kPa)	FC(%)	$(N_1)_{60}$	$(N_1)_{60cs}$	G_{max} (MPa)	v_{max} (m/s)	*dur* (s)	LP_m term	Obs. Liq?
			Extracted parameters			Computed parameters					
193	Wufeng	7.60	129.70	35.00	36.00	42.79	120.80	2.09	11.02	107.53	N
194	Wufeng	7.60	13.10	23.00	14.00	16.44	30.13	2.09	11.02	265.51	N
195	Wufeng	7.60	13.10	33.00	34.00	40.14	37.87	2.09	11.02	333.80	N
196	Adapzari	7.40	75.20	58.00	26.00	31.39	84.97	1.21	11.38	78.03	N
197	Adapzari	7.40	62.80	53.00	48.00	56.47	89.79	1.21	11.38	98.73	N
198	Adapzari	7.40	52.60	87.00	10.00	13.15	54.93	1.21	11.38	72.12	N
199	Adapzari	7.40	60.40	71.00	47.00	55.33	87.64	1.21	11.38	100.20	N
200	Adapzari	7.40	57.70	97.00	24.00	29.11	72.94	1.21	11.38	87.30	N
201	Adapzari	7.40	51.70	99.00	55.00	64.45	83.93	1.21	11.38	112.11	N
202	Adapzari	7.40	26.90	99.00	16.00	19.99	44.77	1.21	11.38	114.94	N
203	Adapzari	7.40	56.60	88.00	36.00	42.79	79.80	1.21	11.38	97.37	N
204	Adapzari	7.40	104.80	97.00	28.00	33.67	102.18	1.21	11.38	67.33	N
205	Adapzari	7.40	67.60	41.00	38.00	45.07	88.33	1.21	11.38	90.23	N
206	Adapzari	7.40	38.40	99.00	9.00	12.01	45.53	1.21	11.38	81.89	N
207	Yuanlin	7.60	162.40	47.00	40.00	47.35	138.54	0.56	10.98	26.36	N
208	Nantou	7.60	111.70	75.00	63.00	73.57	126.99	1.19	10.36	69.94	N
209	Wufeng	7.60	153.40	38.00	58.00	67.87	146.24	2.09	11.02	110.06	N
210	Wufeng	7.60	25.40	58.00	10.00	13.15	38.17	2.09	11.02	173.50	N
211	Wufeng	7.60	169.20	67.00	25.00	30.25	126.20	2.09	11.02	86.12	N
212	Wufeng	7.60	180.60	52.00	22.00	26.83	126.20	2.09	11.02	80.68	N
213	Wufeng	7.60	116.90	72.00	15.00	18.85	91.71	2.09	11.02	90.58	N
214	Wufeng	7.60	183.10	49.00	35.00	41.65	142.58	2.09	11.02	89.90	N
215	Wufeng	7.60	113.40	51.00	24.00	29.11	102.26	2.09	11.02	104.11	N
216	Wufeng	7.60	197.70	76.00	13.00	16.57	114.66	2.09	11.02	66.96	N
217	Wufeng	7.60	213.30	82.00	11.00	14.29	113.65	2.09	11.02	61.52	N
218	Wufeng	7.60	150.60	74.00	18.00	22.27	109.33	2.09	11.02	83.82	N
219	Wufeng	7.60	202.00	67.00	16.00	19.99	122.69	2.09	11.02	70.12	N
220	Wufeng	7.60	202.00	47.00	26.00	31.39	139.26	2.09	11.02	79.60	N
221	Meiko	8.10	39.00	27.00	1.70	3.23	27.57	0.67	19.88	47.37	Y
222	River Site	7.60	47.00	0.00	9.40	9.40	51.01	0.50	12.54	34.05	Y
223	Old Town -1	7.60	81.00	2.00	22.70	22.70	85.20	0.56	12.54	37.12	N
224	Old Town -2	7.60	109.00	2.00	23.50	23.50	99.72	0.56	12.54	32.29	N
225	Road Site	7.60	79.00	0.00	14.10	14.10	74.12	0.56	12.54	33.12	N
226	Aomori station	8.30	38.00	3.00	16.50	16.50	53.66	0.74	16.24	84.80	Y
227	Hachinohe-6	8.30	42.00	5.00	9.10	9.10	47.77	0.80	13.97	63.48	Y
228	Juvenile Hall	6.61	96.00	55.00	3.90	6.20	56.10	1.21	8.56	30.33	Y
229	Van Norman	6.61	96.00	50.00	8.10	10.98	69.83	1.21	8.56	37.75	Y
230	Ying Kou P. P.	7.00	92.00	5.00	11.00	11.00	74.64	0.86	11.50	39.99	Y
231	Amatitlan B-2	7.50	34.00	3.00	9.70	9.70	43.78	0.42	19.76	52.89	Y/N
232	Yao Yuan village	7.60	67.00	20.00	10.50	12.34	62.86	0.63	15.58	45.71	Y
233	San Juan B-3	7.50	156.00	5.00	7.60	7.60	87.37	0.62	12.99	22.40	Y
234	San Juan B-1	7.50	106.00	20.00	6.30	7.80	68.14	0.62	12.99	25.71	Y
235	Nakamura 4	6.50	30.00	5.00	6.90	6.90	37.24	0.32	12.82	25.32	Y
236	Arahama	7.70	67.00	0.00	12.80	12.80	66.46	0.63	14.83	46.69	Y

Appendix (*Continued*)

No	Site	Mag.	σ'_v (kPa)	FC(%)	$(N_1)_{60}$	$(N_1)_{60cs}$	G_{max} (MPa)	v_{max} (m/s)	dur (s)	LP_m term	Obs. Liq?
			Extracted parameters				Computed parameters				
237	Oiiri-1	7.70	85.00	5.00	9.40	9.40	68.60	0.76	14.83	45.59	Y
238	Hiyori-18	7.70	71.00	20.00	11.10	12.99	65.74	0.76	14.83	52.30	Y
239	yuriagebr-3	7.70	42.00	12.00	11.80	12.97	51.44	0.76	14.83	69.18	Y
240	Kitawabuchi-3	7.70	73.00	0.00	17.60	17.60	75.67	0.89	14.83	68.31	N
241	Ishinomaki-2	6.50	45.00	10.00	5.50	6.22	42.63	0.32	12.82	19.32	N
242	Ishinomaki-4	7.70	57.00	10.00	20.90	22.24	69.96	0.63	14.83	57.78	N
243	yuriagebr-5	7.70	78.00	17.00	20.10	22.32	81.01	0.76	14.83	58.67	N
244	heber road a2	6.53	50.00	18.00	2.90	4.01	36.94	2.08	9.24	70.91	Y
245	mckim ranch a	6.53	32.00	31.00	4.60	6.72	34.06	1.36	9.24	66.80	Y
246	heber road a1	6.53	42.00	12.00	37.80	40.21	69.53	2.08	9.24	158.89	N
247	heber road a3	6.53	56.00	25.00	16.20	19.07	64.81	2.08	9.24	111.08	N
248	Radiotowerb1	5.90	5.00	64.00	2.90	5.06	11.68	0.48	6.84	38.73	Y
249	Radiotowerb2	5.90	38.00	30.00	15.20	18.52	52.48	0.48	6.84	22.89	N
250	Riverpark a	5.90	20.00	80.00	4.60	6.99	26.93	0.51	6.84	23.43	N
251	Takeda elememtary school	6.80	42.00	0.00	13.30	13.30	53.18	0.31	9.10	17.74	Y
252	Aomori station	7.70	38.00	3.00	16.50	16.50	53.66	0.37	10.38	26.97	Y
253	Arayamotomachi	7.70	37.00	5.00	5.10	5.10	37.79	0.63	10.38	33.64	Y
254	Radio tower B1	6.20	39.79	75.00	12.00	15.43	50.30	0.23	12.85	18.53	N
255	Marine laboratory UC B2	6.93	55.00	3.00	14.90	14.90	62.79	0.79	13.02	58.83	Y
256	POO7-1	6.93	89.00	3.00	15.40	15.40	80.60	0.79	13.02	46.67	Y
257	POO7-3	6.93	89.00	3.00	17.00	17.00	82.78	0.79	13.02	47.93	Y/N
258	Miller Farm CMF5	6.93	108.00	13.00	20.90	22.64	96.30	1.10	13.02	64.01	Y
259	Miller Farm CMF10	6.93	105.00	20.00	20.20	22.82	94.11	1.10	13.02	64.34	Y
260	MBARI NO 3 EB1	6.93	35.00	1.00	22.60	22.60	55.94	0.79	13.02	82.37	N
261	MBARI NO 3 EB5	6.93	47.00	1.00	14.90	14.90	58.04	0.79	13.02	63.65	N
262	Hall Avenue	6.93	64.00	30.00	5.70	7.88	51.39	0.40	13.02	20.69	N
263	Potrero Canyon C1	6.69	88.00	64.00	8.50	11.44	67.80	1.17	8.37	37.82	Y
264	Malden street Unit D	6.69	101.00	25.00	27.20	31.17	99.59	1.39	8.37	57.42	N
265	Kobe alluvial site number 8	6.90	65.00	0.00	24.10	24.10	77.50	1.41	10.52	88.29	Y
266	Kobe alluvial site number 38	6.90	94.00	5.00	19.10	19.10	87.75	1.41	10.52	69.13	Y
267	Kobe alluvial site number 41	6.90	50.00	0.00	15.00	15.00	59.98	1.13	10.52	71.06	Y
268	Kobe alluvial site number 43	6.90	55.00	20.00	15.20	17.42	63.13	0.99	10.52	59.50	Y
269	Ashiyama A (marine sand)	6.90	97.99	2.00	31.30	31.30	101.54	1.13	10.52	61.39	N
270	Kobe alluvial site number 18	6.90	171.00	0.00	42.60	42.60	144.23	1.97	10.52	87.44	N
271	Kobe alluvial site number 31	6.90	46.00	0.00	49.70	49.70	77.44	1.69	10.52	149.59	N
272	Kobe alluvial site number 39	6.90	66.00	0.00	61.00	61.00	96.95	1.69	10.52	130.54	N

Appendix (*Continued*)

No	Site	Mag.	σ'_v (kPa)	FC(%)	$(N_1)_{60}$	$(N_1)_{60cs}$	G_{max} (MPa)	v_{max} (m/s)	*dur* (s)	LP_m term	Obs. Liq?
			Extracted parameters				Computed parameters				
273	Kobe alluvial site number 40	6.90	59.00	0.00	39.70	39.70	83.36	1.69	10.52	125.55	N
274	Kobe alluvial site number 21	6.90	44.00	0.00	33.50	33.50	69.17	1.69	10.52	139.69	N
275	Kobe alluvial site number 3	6.90	77.00	3.00	49.80	49.80	100.23	1.13	10.52	111.30	N
276	Port island improved site (Ikegaya)	6.90	125.00	20.00	22.70	25.52	105.84	1.13	10.52	50.16	N
277	Port island improved site (Tanahashi)	6.90	140.00	20.00	19.50	22.06	107.67	1.13	10.52	45.56	N
278	Port island improved site (Watanabe)	6.90	135.00	20.00	34.60	38.37	122.09	1.13	10.52	53.58	N
279	Kobe alluvial site number 2	6.90	103.00	15.00	39.50	42.62	110.01	1.13	10.52	63.27	N
280	Kobe alluvial site number 10	6.90	107.00	9.00	27.40	28.84	102.69	1.69	10.52	85.29	N
281	Kobe alluvial site number 14	6.90	69.00	19.00	20.30	22.79	76.39	1.41	10.52	81.98	N
282	Kobe alluvial site number 30	6.90	78.00	10.00	40.10	42.20	96.07	1.69	10.52	109.45	N
283	Kobe alluvial site number 33	6.90	83.00	50.00	27.90	33.56	90.85	1.41	10.52	81.06	N
284	Adapzari city	7.40	43.20	28.00	10.00	12.52	49.78	1.21	11.38	79.58	Y
285	Adapzari city	7.40	48.50	18.00	18.00	20.20	62.05	1.21	11.38	88.35	Y
286	Adapzari city	7.40	55.00	13.00	21.00	22.74	68.81	1.21	11.38	86.39	Y
287	Adapzari city	7.40	47.40	33.00	8.00	10.71	48.89	1.21	11.38	71.23	Y
288	Adapzari city	7.40	55.90	9.00	14.00	14.95	62.23	1.21	11.38	76.88	Y
289	Adapzari city	7.40	25.10	9.00	7.00	7.70	34.21	1.21	11.38	94.12	Y
290	Adapzari city	7.40	76.40	9.00	9.00	9.77	64.23	1.21	11.38	58.05	Y
291	Wufeng	7.60	80.10	14.00	23.00	24.99	85.01	2.09	11.36	126.24	Y
292	Wufeng	7.60	91.50	10.00	11.00	11.94	74.44	2.09	11.36	96.77	Y
293	Wufeng	7.60	112.80	23.00	15.00	17.53	90.09	2.09	11.36	95.00	Y
294	Wufeng	7.60	123.90	18.00	17.00	19.12	97.67	2.09	11.36	93.76	Y
295	Wufeng	7.60	202.40	17.00	12.00	13.67	113.45	2.09	11.36	66.68	Y
296	Wufeng	7.60	58.70	20.00	18.00	20.44	68.26	2.09	11.36	138.32	Y
297	Adapzari city	7.40	53.00	35.00	15.00	18.85	61.75	1.21	11.38	80.46	Y
298	Adapzari	7.40	66.20	34.00	14.00	17.60	67.72	1.21	11.38	70.64	N
299	Adapzari	7.40	65.40	17.00	43.00	46.77	89.39	1.21	11.38	94.39	N
300	Adapzari	7.40	71.70	17.00	42.00	45.71	93.09	1.21	11.38	89.66	N
301	Adapzari	7.40	88.00	20.00	32.00	35.56	96.74	1.21	11.38	75.92	N
302	Adapzari	7.40	96.30	11.00	45.00	47.53	109.59	1.21	11.38	78.59	N
303	Adapzari	7.40	77.80	20.00	40.00	44.20	95.89	1.21	11.38	85.12	N
304	Adapzari	7.40	86.90	24.00	34.00	38.46	97.55	1.21	11.38	77.52	N
305	Adapzari	7.40	106.80	32.00	29.00	34.31	104.05	1.21	11.38	67.28	N
306	Wufeng	7.60	118.00	32.00	23.00	27.54	103.18	2.09	11.02	100.96	N
307	Wufeng	7.60	216.60	10.00	32.00	33.78	151.78	2.09	11.02	80.90	N

Appendix (*Continued*)

No	Site	Mag.	σ'_v (kPa)	FC(%)	$(N_1)_{60}$	$(N_1)_{60cs}$	G_{max} (MPa)	v_{max} (m/s)	dur (s)	LP_m term	Obs. Liq?
			Extracted parameters				**Computed parameters**				
308	Wufeng	7.60	120.00	27.00	41.00	46.78	119.77	2.09	11.02	115.23	N
309	Wufeng	7.60	213.50	26.00	26.00	30.00	143.17	2.09	11.02	77.42	N
310	Wufeng	7.60	224.70	17.00	28.00	30.75	149.62	2.09	11.02	76.88	N
311	Adapzari	7.40	48.00	71.00	25.00	30.25	67.22	1.21	11.38	96.71	N
312	Adapzari	7.40	21.60	81.00	1.00	2.89	17.30	1.21	11.38	55.32	N
313	Wufeng	7.60	217.20	67.00	17.00	21.13	129.31	2.09	11.02	68.74	N
314	Wufeng	7.60	183.50	35.00	37.00	43.93	144.62	2.09	11.02	90.99	N

Sr. No. 1–220 used in Formulation; Sr. No. 221–314 used in Validation.

2

Un-differenced precise point positioning model using triple GNSS constellations

Akram Afifi[1]* and Ahmed El-Rabbany[1]

*Corresponding author: Akram Afifi, Department of Civil Engineering, Ryerson University, Toronto, Ontario, Canada
E-mail: akram.afifi@ryerson.ca
Reviewing editor: Shuanggen Jin, Shanghai Astronomical Observatory, Chinese Academy of Sciences, China

Abstract: This paper introduces a dual-frequency precise point positioning (PPP) model, which combines the observations of three different global navigation satellite system (GNSS) constellations, namely GPS, Galileo, and BeiDou. A drawback of a single GNSS system such as GPS, however, is the availability of sufficient number of visible satellites in urban areas. Combining GNSS observations offers more visible satellites to users, which in turn is expected to enhance the satellite geometry and the overall positioning solution. However, combining several GNSS observables introduces additional biases, which require rigorous modeling, including the GNSS time offsets and hardware delays. In this paper, un-differenced ionosphere-free linear combination PPP model is developed. The additional biases of the GPS, Galileo, and BeiDou combination are accounted for through the introduction of a new unknown parameter, which is identified as the inter-system bias, in the PPP mathematical model. Natural Resources Canada's GPSPace PPP software is modified to enable a combined GPS, Galileo, and BeiDou PPP solution and to handle the newly introduced biases. A total of four data-sets collected at four different IGS stations are processed to verify the developed PPP model. Precise satellite orbit and clock products from the International GNSS Service Multi-GNSS Experiment (IGS-MGEX) network are used to correct the GPS, Galileo, and BeiDou measurements. It is shown that the un-differenced GPS-only post-processed PPP solution indicates that the model is capable of obtaining a sub-decimeter-level accuracy. However, the solution takes about 20 min

ABOUT THE AUTHOR

Akram Afifi obtained his PhD degree in Geomatics Engineering from the Department of Civil Engineering, Ryerosn University, Canada. He is currently a postdoctoral fellow at Ryerson University, Toronto, Canada. He is acting as an adjunct professor with active membership of the graduate school at Ryerson University. Afifi's areas of interest include Satellite Navigation, Geodesy, Land survey, and Hydrographic Surveying. He published several papers and posters at various journals, conferences, and professional events. He acted as a chair of the Canadian Institute of Geomatics student affair committee from 2013 to 2014. Afifi received numerous awards in recognition of his academic achievements, including three awards from the AOLS in the annual general meeting and Dennis Mock leadership award from Ryerson University.

PUBLIC INTEREST STATEMENT

This paper introduces a dual-frequency precise point positioning (PPP) model, which combines the observations of three different global navigation satellite system (GNSS) constellations, namely GPS, Galileo, and BeiDou. A drawback of a single GNSS system such as GPS, however, is the availability of sufficient number of visible satellites in urban areas. Combining GNSS observations offers more visible satellites to users, which in turn is expected to enhance the satellite geometry and the overall positioning solution. It is shown that the un-differenced GPS-only post-processed PPP solution indicates that the model is capable of obtaining a sub-decimeter-level accuracy. However, the solution takes about 20 min to converge to decimeter-level precision. The convergence time of the combined GNSS post-processed PPP solutions takes about 15 min to reach the decimeter-level precision, which represent a 25% improvement in comparison with the GPS-only post-processed PPP solution.

to converge to decimeter-level precision. The convergence time of the combined GNSS post-processed PPP solutions takes about 15 min to reach the decimeter-level precision, which represent a 25% improvement in comparison with the GPS-only post-processed PPP solution.

Subjects: Aerospace Engineering; Civil, Environmental and Geotechnical Engineering; Earth Sciences

Keywords: PPP; GPS; Galileo; BeiDou

1. Introduction

Precise point positioning (PPP) has proven to be capable of providing positioning accuracy at the sub-decimeter and decimeter levels in static and kinematic modes, respectively. PPP accuracy and convergence time are controlled by the ability to mitigate all potential error sources in the system. Several comprehensive studies have been published on the accuracy and convergence time of un-differenced combined GPS/Galileo PPP model (see, e.g. Afifi & El-Rabbany, 2015; Collins, Bisnath, Lahaye, & Héroux, 2010; Colombo, Sutter, & Evans, 2004; Ge, Gendt, Rothacher, Shi, & Liu, 2008; Hofmann-Wellenhof, Lichtenegger, & Wasle, 2008; Kouba & Héroux, 2001; Leick, 2004; Zumberge, Heflin, Jefferson, Watkins, & Webb, 1997). PPP relies essentially on the availability and use of precise satellite products, namely orbital and clock corrections. At present, the Multi-global navigation satellite systems (GNSS) Experiment (MGEX) of the International GNSS Service (IGS) provides the precise satellite orbital and clock corrections for all the GNSS (Montenbruck et al., 2014).

Unfortunately, the use of a single constellation limits the number of visible satellites, especially in urban areas, which affects the PPP solution. Recently, a number of researchers showed that combining GPS and Galileo observations in PPP solution enhances the positioning convergence and precision in comparison with the GPS-only PPP solution (Afifi & El-Rabbany, 2016a; Melgard, Tegedor, de Jong, Lapucha, & Lachapelle, 2013). At present, the IGS-MEGX network provides the GNSS users with precise clock and orbit products to all currently available satellite systems (Montenbruck et al., 2014). This makes it possible to obtain a PPP solution by combining the observations of two or more GNSS constellations. This research focuses on combining the GPS, Galileo, and BeiDou observations in a PPP model.

Presently, there exist four operational GNSS. These include the US global positioning system (GPS), the Russian global navigation satellite system (GLONASS), the European Galileo system, and the Chinese BeiDou system. Combing the measurements of multiple systems can significantly improve the availability of a navigation solution, especially in urban areas (Afifi & EL-Rabbany, 2016b). GPS satellites transmit signals on three different frequencies, which are controlled by the GPS time frame (GPST). Currently, the GPS users can receive the modernized civil L2C and L5 signals. On the other hand, Galileo satellite constellation foresees 27 operational and three spare satellites positioned in three nearly circular medium earth orbits (MEO). Galileo system transmits six signals on different frequencies using the Galileo time system (GST). Unlike GLONASS satellite system, Galileo and GPS have partial frequency overlaps, which simplify the dual-system integration. In addition, GPS and Galileo operators have agreed to measure and broadcast a GPS to Galileo time offset (GGTO) parameter, in order to facilitate the interchangeable mode (Melgard et al., 2013). BeiDou navigation satellite system, being developed independently by China, is pacing steadily forward toward completing the constellation. China has indicated a plan to complete the second generation of Beidou satellite system by expanding the regional service into global coverage. Beidou system transmits three signals on different frequencies using the BeiDou time frame (BDT). The BeiDou-2 system is proposed to consist of 30 medium Earth orbiting satellites and five geostationary satellites (BeiDou, 2015; ESA, 2015; Hofmann-Wellenhof et al., 2008; IAC, 2015).

In this paper, a triple GNSS constellation (GPS, Galileo, and BeiDou) PPP model is developed. Four combinations are considered in the PPP modeling namely; GPS/Galileo, GPS/BeiDou, Galileo/BeiDou, and GPS/Galileo/BeiDou. All the combined PPP models results are compared with the GPS-only PPP model results. In the developed model GPS L1/L2, Galileo E1/E5a, and BeiDou B1/B2 signals are used in a dual-frequency ionosphere-free linear combination. Precise satellite corrections from the International GNSS Service multi-GNSS experiment (IGS-MEGX) network are used to account for GPS, Galileo and BeiDou satellite orbit and clock errors (Montenbruck et al., 2014). As these products are presently referenced to the GPS time and since we use mixed GNSS receivers that also use GPS time as a reference, the GGTO and the GPS to BeiDou time offset are canceled out in our model. The inter-system bias is treated as an additional unknown parameter. The hydrostatic component of the tropospheric zenith path delay is modeled through the Hopfield model, while the wet component is considered as an additional unknown parameter (Hofmann-Wellenhof et al., 2008; Hopfield, 1972). All remaining errors and biases are accounted for using existing models as shown in Kouba (2009). The inter-system bias parameter was found to be essentially constant over the one-hour observation time span and was receiver dependent. The positioning results of the developed combined GPS/Galileo, GPS/BeiDou, Galileo/BeiDou, and GPS/Galileo/BeiDou PPP models showed a sub-decimeter accuracy level and 25% convergence time improvement in comparison with the GPS-only PPP results.

2. Un-differenced PPP models

PPP has been carried out using dual-frequency ionosphere-free linear combinations of carrier-phase and pseudorange GPS measurements. Equations (1)–(6) show the ionosphere free linear combination of GPS, Galileo, and BeiDou observations (Afifi & El-Rabbany, 2016a).

$$P_{G_{IF}} = \rho_G + c[dt_{rG} - dt^s] + c[\alpha d_{P1} - \beta d_{P2}]_r + c[\alpha d_{P1} - \beta d_{P2}]^s + T_G + \varepsilon_{PG_{IF}} \tag{1}$$

$$P_{E_{IF}} = \rho_E + c[dt_{rG} - GGTO - dt^s] + c[\alpha d_{E1} - \beta d_{E5a}]_r + c[\alpha d_{E1} - \beta d_{E5a2}]^s + T_E + \varepsilon_{E_{IF}} \tag{2}$$

$$P_{B_{IF}} = \rho_B + c[dt_{rG} - GB - dt^s] + c[\alpha d_{B1} - \beta d_{B2}]_r + c[\alpha d_{B1} - \beta d_{B2}]^s + T_B + \varepsilon_{B_{IF}} \tag{3}$$

$$\Phi_{G_{IF}} = \rho_G + c[dt_{rG} - dt^s] + c[\alpha \delta_{L1} - \beta \delta_{L2}]_r + c[\alpha \delta_{L1} - \beta \delta_{L2}]^s + T_G + N_{G_{IF}} + \phi_{r0_{G_{IF}}} + \phi^s_{0_{G_{IF}}} + \varepsilon_{\Phi G_{IF}} \tag{4}$$

$$\Phi_{E_{IF}} = \rho_E + c[dt_{rG} - GGTO - dt^s] + c[\alpha \delta_{E1} - \beta \delta_{E5a}]_r + c[\alpha \delta_{E1} - \beta \delta_{E5a}]^s + T_E + N_{E_{IF}} + \phi_{r0_{E_{IF}}} + \phi^s_{0_{E_{IF}}} + \varepsilon_{\Phi E_{IF}} \tag{5}$$

$$\Phi_{B_{IF}} = \rho_B + c[dt_{rG} - GB - dt^s] + c[\alpha \delta_{B1} - \beta \delta_{B2}]_r + c[\alpha \delta_{B1} - \beta \delta_{B2}]^s + T_B + N_{B_{IF}} + \phi_{r0_{B_{IF}}} + \phi^s_{0_{B_{IF}}} + \varepsilon_{\Phi B_{IF}} \tag{6}$$

where the subscripts G, E, and B refer to the GPS, Galileo, and BeiDou satellite systems, respectively; $P_{G_{IF}}$, $P_{E_{IF}}$ and $P_{B_{IF}}$ are the ionosphere-free pseudoranges in meters for GPS, Galileo, and BeiDou systems, respectively; $\Phi_{G_{IF}}$, $\Phi_{E_{IF}}$ and $\Phi_{B_{IF}}$ are the ionosphere-free carrier phase measurements in meters for GPS, Galileo, and BeiDou systems, respectively; GGTO is the GPS to Galileo time offset; GB is the GPS to BeiDou time offset; ρ is the true geometric range from receiver at reception time to satellite at transmission time in meter; dt_r, dt^s are the clock errors in seconds for the receiver at signal reception time and the satellite at signal transmission time, respectively; d_{P1}, d_{P2}, d_{E1}, d_{E5a}, d_{B1}, d_{B2} are frequency-dependent code hardware delays for the receiver at reception time in seconds; d^s_{P1}, d^s_{P2}, d^s_{E1}, d^s_{E5a}, d^s_{B1}, d^s_{B2a} are frequency-dependent code hardware delays for the satellite at transmission time in seconds; δ_{L1r}, δ_{L2r}, δ_{E1r}, δ_{E5ar}, δ_{B1r}, δ_{B2r} are frequency-dependent carrier-phase hardware delays for the receiver at reception time in seconds; δ^s_{L1}, δ^s_{L2}, δ^s_{E1}, δ^s_{E5a}, δ^s_{B1}, δ^s_{B2} are frequency-dependent carrier-phase hardware delays for the satellite at transmission time in seconds; T is the tropospheric delay in meter; $N_{G_{IF}}$, $N_{E_{IF}}$, $N_{B_{IF}}$ are the ionosphere-free linear combinations of the ambiguity parameters for both GPS, Galileo, and BeiDou carrier-phase measurements in meters, respectively; $\phi_{r0_{G_{IF}}}$, $\phi^s_{0_{G_{IF}}}$, $\phi_{r0_{E_{IF}}}$, $\phi^s_{0_{E_{IF}}}$, $\phi_{r0_{B_{IF}}}$, $\phi^s_{0_{B_{IF}}}$ are ionosphere-free linear combinations of frequency-dependent initial fractional phase biases in the receiver and satellite channels for both GPS, Galileo, and BeiDou in meters, respectively;

c is the speed of light in vacuum in meter per second; $\varepsilon_{P_{IF}}, \varepsilon_{E_{IF}}, \varepsilon_{\Phi G_{IF}}, \varepsilon_{\Phi E_{IF}}, \varepsilon_{B_{IF}}, \varepsilon_{\Phi B_{IF}}$ are the ionosphere-free linear combinations of the relevant noise and un-modeled errors in meter; $\alpha_G, \beta_G, \alpha_E, \beta_E, \alpha_B, \beta_B$ are the ionosphere-free linear combination coefficients for GPS, Galileo, and BeiDou which are given, respectively, by: $\alpha_G = \frac{f_1^2}{f_1^2 - f_2^2}, \beta_G = \frac{f_2^2}{f_1^2 - f_2^2}, \alpha_E = \frac{f_{E1}^2}{f_{E1}^2 - f_{E5a}^2}, \beta_E = \frac{f_{E5a}^2}{f_{E1}^2 - f_{E5a}^2}, \alpha_B = \frac{f_{B1}^2}{f_{B1}^2 - f_{B2}^2}, \beta_B = \frac{f_{B2}^2}{f_{B1}^2 - f_{B2}^2}.$

where f_1 and f_2 are GPS L$_1$ and L$_2$ signals frequencies; f_{E1} and f_{E5a} are Galileo E$_1$ and E$_{5a}$ signals frequencies; f_{B1} and f_{B2} are BeiDou B$_1$ and B$_2$ signals frequencies.

$$N_{G_{IF}} = \alpha_G \lambda_1 N_1 - \beta_G \lambda_2 N_2 \tag{7}$$

$$N_{E_{IF}} = \alpha_E \lambda_{E1} N_{E1} - \beta_E \lambda_{E5a} N_{E5a} \tag{8}$$

$$N_{B_{IF}} = \alpha_B \lambda_{B1} N_{B1} - \beta_B \lambda_{B2} N_{B2} \tag{9}$$

where λ_1 and λ_2 are the GPS L1 and L2 signals wavelengths in meters; λ_{E1} and λ_{E5a} are the Galileo E1 and E5a signals wavelengths in meters; λ_{B1} and λ_{B2} are the BeiDou B1 and B2 signals wavelengths in meters; N_1, N_2 are the integer ambiguity parameters of GPS signals L1 and L2, respectively; N_{E1}, N_{E5a} are the integer ambiguity parameters of Galileo signals E1 and E5a, respectively; N_{B1}, N_{B2} are the integer ambiguity parameters of BeiDou signals B1 and B2, respectively.

Precise orbit and satellite clock corrections of IGS-MGEX networks are produced for both GPS/Galileo observations and are referred to GPS time. IGS precise GPS satellite clock correction includes the effect of the ionosphere-free linear combination of the satellite hardware delays of L1/L2 P(Y) code, while the Galileo counterpart includes the effect of the ionosphere-free linear combination of the satellite hardware delays of the Galileo E1/E5a pilot code. In addition, BeiDou satellite clock correction includes the effect of the ionosphere-free linear combination of the satellite hardware delays of B1/B2 code (Montenbruck et al., 2014). By applying the precise clock products for both GPS/Galileo/BeiDou observations, Equations (1)–(6) will take the following form:

$$P_{G_{IF}} = \rho_G + c[dt_{rG} - dt_{prec}^s] + c[\alpha d_{P1} - \beta d_{P2}]_r + T_G + \varepsilon_{PG_{IF}} \tag{10}$$

$$P_{E_{IF}} = \rho_E + c[dt_{rG} - dt_{prec}^s] + c[\alpha d_{E1} - \beta d_{E5a}]_r + T_E + \varepsilon_{E_{IF}} \tag{11}$$

$$P_{B_{IF}} = \rho_B + c[dt_{rG} - dt_{prec}^s] + c[\alpha d_{B1} - \beta d_{B2}]_r + T_B + \varepsilon_{B_{IF}} \tag{12}$$

$$\begin{aligned}\Phi_{G_{IF}} = \rho_G + cdt_{rG} - c[dt_{prec}^s + [\alpha d_{P1} - \beta d_{P2}]^s] + c[\alpha \delta_{L1} - \beta \delta_{L2}]_r - c[\alpha \delta_{L1} - \beta \delta_{L2}]^s + T_G + N_{G_{IF}} \\ + \phi_{r0_{G_{IF}}} + \phi_{0_{G_{IF}}}^s + \varepsilon_{\Phi G_{IF}}\end{aligned} \tag{13}$$

$$\begin{aligned}\Phi_{E_{IF}} = \rho_E + cdt_{rG} - c[dt_{prec}^s + [\alpha d_{E1} - \beta d_{E5a}]^s] + c[\alpha \delta_{E1} - \beta \delta_{E5a}]_r - c[\alpha \delta_{E1} - \beta \delta_{E5a}]^s \\ + T_E + N_{E_{IF}} + \phi_{r0_{E_{IF}}} + \phi_{0_{E_{IF}}}^s + \varepsilon_{\Phi E_{IF}}\end{aligned} \tag{14}$$

$$\begin{aligned}\Phi_{B_{IF}} = \rho_B + cdt_{rG} - c[dt_{prec}^s + [\alpha d_{B1} - \beta d_{B2}]^s] + c[\alpha \delta_{B1} - \beta \delta_{B2}]_r - c[\alpha \delta_{B1} - \beta \delta_{B2}]^s \\ + T_B + N_{B_{IF}} + \phi_{r0_{B_{IF}}} + \phi_{0_{B_{IF}}}^s + \varepsilon_{\Phi B_{IF}}\end{aligned} \tag{15}$$

For simplicity, the receiver and satellite hardware delays will take the following forms:

$$b_{r_P} = c[\alpha d_{P1} - \beta d_{P2}]_r \qquad\qquad b_P^s = c[\alpha d_{P1} - \beta d_{P2}]^s$$

$$b_{r_E} = c[\alpha d_{E1} - \beta d_{E5a}]_r \qquad\qquad b_E^s = c[\alpha d_{E1} - \beta d_{E5a}]^s$$

$$b_{r_B} = c[\alpha d_{B1} - \beta d_{B2}]_r \qquad\qquad b_B^s = c[\alpha d_{B1} - \beta d_{B2}]^s$$

$$b_{r_\Phi} = c[\alpha \delta_{L1} - \beta \delta_{L2}]_r + \phi_{r0_{G_{IF}}} \qquad\qquad b_\Phi^s = c[\alpha \delta_{L1} - \beta \delta_{L2}]^s + \phi_{0_{G_{IF}}}^s$$

$$b_{r_{E\Phi}} = c[\alpha \delta_{E1} - \beta \delta_{E5a}]_r + \phi_{r0_{E_{IF}}} \qquad\qquad b_{E\Phi}^s = c[\alpha \delta_{E1} - \beta \delta_{E5a}]^s + \phi_{0_{E_{IF}}}^s$$

$$b_{r_{B\Phi}} = c[\alpha \delta_{B1} - \beta \delta_{B2}]_r + \phi_{r0_{B_{IF}}} \qquad\qquad b_{B\Phi}^s = c[\alpha \delta_{B1} - \beta \delta_{B2}]^s + \phi_{0_{B_{IF}}}^s$$

In the combined GPS/Galileo un-differenced PPP model, the GPS receiver clock error is lumped with the GPS receiver differential code biases. In order to maintain consistency in the estimation of a common receiver clock offset, this convention is used when combining the ionosphere-free linear combination of GPS L1/L2, Galileo E1/E5a, and BeiDou B1/B2 observations in PPP solution. This, however, introduces an additional bias in the Galileo ionosphere-free PPP mathematical model, which represents the difference in the receiver differential code biases of both systems. Such an additional bias is commonly known as the inter-system bias, which is referred to as *ISB* in this paper. In our PPP mode, the Hopfield tropospheric correction model along with the Vienna mapping function are used to account for the hydrostatic component of the tropospheric delay (Boehm & Schuh, 2004; Hopfield, 1972). Other corrections are also applied, including the effect of ocean loading (Bos & Scherneck, 2011; IERS, 2010), Earth tide (Kouba, 2009), carrier-phase windup (Leick, 2004; Wu, Wu, Hajj, Bertiger, & Lichten, 1993), Sagnac (Kaplan & Heagarty, 2006), relativity (Hofmann-Wellenhof et al., 2008), and satellite and receiver antenna phase-center variations (Dow, Neilan, & Rizos, 2009). The noise terms are modeled stochastically using an exponential model, as described in Afifi and El-Rabbany (2015). With the above consideration, the GPS/Galileo ionosphere-free linear combinations of both pseudorange and carrier phase can be written as:

$$P_{G_{IF}} = \rho_G + \tilde{d}t_{rG} - dt_{prec}^s + T_G + \varepsilon_{PG_{IF}} \tag{16}$$

$$P_{E_{IF}} = \rho_E + \tilde{d}t_{rG} - dt_{prec}^s + ISB_{GE} + T_E + \varepsilon_{E_{IF}} \tag{17}$$

$$\Phi_{G_{IF}} = \rho_G + \tilde{d}t_{rG} - dt_{prec}^s + T_G + \tilde{N}_{G_{IF}} + \varepsilon_{\Phi G_{IF}} \tag{18}$$

$$\Phi_{E_{IF}} = \rho_E + \tilde{d}t_{rG} - dt_{prec}^s + T_E + \tilde{N}_{E_{IF}} + ISB_{GE} + \varepsilon_{\Phi E_{IF}} \tag{19}$$

where $\tilde{d}t_{rG}$ represents the sum of the receiver clock error and receiver hardware delay $\tilde{d}t_{rG} = cdt_{rG} + b_{r_P}$; *ISB* is the inter system bias as follows $ISB_{GE} = b_{r_E} - b_{r_P}$; $\tilde{N}_{G_{IF}}$ and $\tilde{N}_{E_{IF}}$ are given by:

$$\tilde{N}_{G_{IF}} = N_{G_{IF}} + b_{r_\Phi} + b_{r_P} - b_\Phi^s - b_P^s \tag{20}$$

$$\tilde{N}_{E_{IF}} = N_{E_{IF}} + b_{r_{E\Phi}} + b_{r_P} - b_{E\Phi}^s - b_E^s \tag{21}$$

In case of combining GPS and BeiDou observations in a PPP model ionosphere-free linear combinations of both pseudorange and carrier phase can be written as:

$$P_{G_{IF}} = \rho_G + \tilde{d}t_{rG} - dt_{prec}^s + T_G + \varepsilon_{PG_{IF}} \tag{22}$$

$$P_{B_{IF}} = \rho_B + \tilde{d}t_{rG} - dt_{prec}^s + ISB_{GB} + T_B + \varepsilon_{B_{IF}} \tag{23}$$

$$\Phi_{G_{IF}} = \rho_G + \tilde{d}t_{rG} - dt_{prec}^s + T_G + \tilde{N}_{G_{IF}} + \varepsilon_{\Phi G_{IF}} \tag{24}$$

$$\Phi_{B_{IF}} = \rho_B + \tilde{dt}_{rG} - dt^s_{prec} + T_B + \tilde{N}_{B_{IF}} + ISB_{GB} + \varepsilon_{\Phi B_{IF}} \tag{25}$$

where $\tilde{dt}_{rG}$ represents the sum of the receiver clock error and receiver hardware delay $\tilde{dt}_{rG} = cdt_{rG} + b_{r_P}$; ISB is the inter system bias as follows $ISB_{GB} = b_{r_B} - b_{r_P}$; $\tilde{N}_{G_{IF}}$ and $\tilde{N}_{B_{IF}}$ are given by:

$$\tilde{N}_{G_{IF}} = N_{G_{IF}} + b_{r_\Phi} + b_{r_P} - b^s_\Phi - b^s_P \tag{26}$$

$$\tilde{N}_{B_{IF}} = N_{B_{IF}} + b_{r_{B\Phi}} + b_{r_P} - b^s_{B\Phi} - b^s_B \tag{27}$$

In the combined Galileo and BeiDou un-differenced PPP model, the Galileo receiver clock error is lumped with the Galileo receiver differential code biases. In order to maintain consistency in the estimation of a common receiver clock offset, this convention is used when combining the iono-sphere-free linear combination of Galileo E1/E5a and BeiDou B1/B2 observations. This, however, in-troduces an additional bias in the BeiDou ionosphere-free PPP mathematical model, which represents the difference in the receiver differential code biases of both systems. As a result, the Galileo and BeiDou combined PPP model ionosphere-free linear combinations of both pseudorange and carrier phase can be written as:

$$P_{E_{IF}} = \rho_E + \tilde{dt}_{rE} - dt^s_{prec} + T_E + \varepsilon_{PE_{IF}} \tag{28}$$

$$P_{B_{IF}} = \rho_B + \tilde{dt}_{rE} - dt^s_{prec} + ISB_{EB} + T_B + \varepsilon_{B_{IF}} \tag{29}$$

$$\Phi_{E_{IF}} = \rho_E + \tilde{dt}_{rE} - dt^s_{prec} + T_E + \tilde{N}_{E_{IF}} + \varepsilon_{\Phi E_{IF}} \tag{30}$$

$$\Phi_{B_{IF}} = \rho_B + \tilde{dt}_{rG} - dt^s_{prec} + T_B + \tilde{N}_{B_{IF}} + ISB_{EB} + \varepsilon_{\Phi B_{IF}} \tag{31}$$

where $\tilde{dt}_{rE}$ represents the sum of the receiver clock error and receiver hardware delay $\tilde{dt}_{rE} = cdt_{rE} + b_{r_E}$; ISB is the inter system bias as follows $ISB_{EB} = b_{r_B} - b_{r_E}$; $\tilde{N}_{E_{IF}}$ and $\tilde{N}_{B_{IF}}$ are given by:

$$\tilde{N}_{E_{IF}} = N_{E_{IF}} + b_{r_{E\Phi}} + b_{r_E} - b^s_{E\Phi} - b^s_E \tag{32}$$

$$\tilde{N}_{B_{IF}} = N_{B_{IF}} + b_{r_{B\Phi}} + b_{r_P} - b^s_{B\Phi} - b^s_B \tag{33}$$

When using the combined GPS/Galileo or GPS/BeiDou or Galileo/BeiDou un-differenced PPP model, the ambiguity parameters lose their integer nature as they are contaminated by receiver and satel-lite hardware delays. It should be pointed out that the number of unknown parameters in the com-bined PPP model equals the number of visible satellites from any system plus six parameters, while the number of equations equals double the number of the visible satellites. This means that the re-dundancy equals $n_G + n_E - 6$. In other words, at least six mixed satellites are needed for the solution to exist. In comparison with the GPS-only un-differenced scenario, which requires a minimum of five satellites for the solution to exist, the addition of Galileo or BeiDou satellites increases the redun-dancy by $n_E - 1$. In other words, we need a minimum of two satellites from any GNSS system in order to contribute to the solution.

3. Least-squares estimation technique
Under the assumption that the observations are uncorrelated and the errors are normally distrib-uted with zero mean, the covariance matrix of the un-differenced observations takes the form of a diagonal matrix. The elements along the diagonal line represent the variances of the code and car-rier phase measurements. In our solution, we consider the ratio between the standard deviation of

the code and carrier-phase measurements to be 100. The general linearized form for the above observation equations around the initial (approximate) vector $\boldsymbol{u}^0$ and observables $\boldsymbol{l}$ can be written in a compact form as:

$$f(\boldsymbol{u}, \boldsymbol{l}) \approx \boldsymbol{A}\Delta\boldsymbol{u} - \boldsymbol{w} - \boldsymbol{r} \approx 0 \tag{34}$$

where $\boldsymbol{u}$ is the vector of unknown parameters; $\boldsymbol{A}$ is the design matrix, which includes the partial derivatives of the observation equations with respect to the unknown parameters $\boldsymbol{u}$; $\Delta\boldsymbol{u}$ is the unknown vector of corrections to the approximate parameters $\boldsymbol{u}^0$, i.e. $\boldsymbol{u} = \boldsymbol{u}^0 + \Delta\boldsymbol{u}$; $\boldsymbol{w}$ is the misclosure vector and $\boldsymbol{r}$ is the vector of residuals. The sequential least-squares solution for the unknown parameters $\Delta\boldsymbol{u}_i$ at an epoch i can be obtained from Vanicek and Krakiwsky (1986):

$$\Delta\boldsymbol{u}_i = \Delta\boldsymbol{u}_{i-1} + \boldsymbol{M}_{i-1}^{-1}\boldsymbol{A}_i^T(\boldsymbol{C}_{l_i} + \boldsymbol{A}_i\boldsymbol{M}_{i-1}^{-1}\boldsymbol{A}_i^T)^{-1}[\boldsymbol{w}_i - \boldsymbol{A}_i\Delta\boldsymbol{u}_{i-1}] \tag{35}$$

$$\boldsymbol{M}_i^{-1} = \boldsymbol{M}_{i-1}^{-1} - \boldsymbol{M}_{i-1}^{-1}\boldsymbol{A}_i^T(\boldsymbol{C}_{l_i} + \boldsymbol{A}_i\boldsymbol{M}_{i-1}^{-1}\boldsymbol{A}^T)^{-1}\boldsymbol{A}_i\boldsymbol{M}_{i-1}^{-1} \tag{36}$$

$$\boldsymbol{C}_{\Delta u_i} = \boldsymbol{M}_i^{-1} = \boldsymbol{M}_{i-1}^{-1} - \boldsymbol{M}_{i-1}^{-1}\boldsymbol{A}_i^T(\boldsymbol{C}_{l_i} + \boldsymbol{A}_i\boldsymbol{M}_{i-1}^{-1}\boldsymbol{A}^T)^{-1}\boldsymbol{A}_i\boldsymbol{M}_{i-1}^{-1} \tag{37}$$

where $\Delta\boldsymbol{u}_{i-1}$ is the least-squares solution for the estimated parameters at epoch $i-1$; $\boldsymbol{M}$ is the matrix of the normal equations; $\boldsymbol{C}_l$ and $\boldsymbol{C}_{\Delta u}$ are the covariance matrices of the observations and unknown parameters, respectively. It should be pointed out that the usual batch least-squares adjustment should be used in the first epoch, i.e. for $i = 1$. The batch solution for the estimated parameters and the inverse of the normal equation matrix are given, respectively, by Vanicek and Krakiwsky (1986):

$$\Delta\boldsymbol{u}_1 = [\boldsymbol{C}_{x^0}^{-1} + \boldsymbol{A}_1^T\boldsymbol{C}_{l_1}^{-1}\boldsymbol{A}_1]^{-1}\boldsymbol{A}_1^T\boldsymbol{C}_{l_1}^{-1}\boldsymbol{w}_1 \tag{38}$$

$$\boldsymbol{M}_1^{-1} = [\boldsymbol{C}_{x^0}^{-1} + \boldsymbol{A}_1^T\boldsymbol{C}_{l_1}^{-1}\boldsymbol{A}_1]^{-1} \tag{39}$$

where $\boldsymbol{C}_x^0$ is a priori covariance matrix for the approximate values of the unknown parameters.

In case of the combined GPS/Galileo PPP model, the design matrix A and the vector of corrections to the unknown parameters Δx take the following forms:

$$A = \begin{bmatrix} \left(\frac{x_0-X^{1_G}}{\rho_0^{1_G}}\right) & \left(\frac{y_0-Y^{1_G}}{\rho_0^{1_G}}\right) & \left(\frac{z_0-Z^{1_G}}{\rho_0^{1_G}}\right) & 1 & m_f^{1_G} & 0 & 0 & \cdots & 0 & 0 & \cdots & 0 \\ \left(\frac{x_0-X^{1_G}}{\rho_0^{1_G}}\right) & \left(\frac{y_0-Y^{1_G}}{\rho_0^{1_G}}\right) & \left(\frac{z_0-Z^{1_G}}{\rho_0^{1_G}}\right) & 1 & m_f^{1_G} & 0 & 1 & \cdots & 0 & 0 & \cdots & 0 \\ \vdots & \vdots & \vdots & \vdots & \vdots & \vdots & \vdots & \vdots & \vdots & \vdots & \ddots & \vdots \\ \left(\frac{x_0-X^{n_G}}{\rho_0^{n_G}}\right) & \left(\frac{y_0-Y^{n_G}}{\rho_0^{n_G}}\right) & \left(\frac{z_0-Z^{n_G}}{\rho_0^{n_G}}\right) & 1 & m_f^{1_{nG}} & 0 & 0 & \cdots & 0 & 0 & \cdots & 0 \\ \left(\frac{x_0-X^{n_G}}{\rho_0^{n_G}}\right) & \left(\frac{y_0-Y^{n_G}}{\rho_0^{n_G}}\right) & \left(\frac{z_0-Z^{n_G}}{\rho_0^{n_G}}\right) & 1 & m_f^{1_{nG}} & 0 & 0 & \cdots & 1 & 0 & \cdots & 0 \\ \left(\frac{x_0-X^{1_E}}{\rho_0^{1_E}}\right) & \left(\frac{y_0-Y^{1_E}}{\rho_0^{1_E}}\right) & \left(\frac{z_0-Z^{1_E}}{\rho_0^{1_E}}\right) & 1 & m_f^{1_E} & 1 & 0 & \cdots & 0 & 0 & \cdots & 0 \\ \left(\frac{x_0-X^{1_E}}{\rho_0^{1_E}}\right) & \left(\frac{y_0-Y^{1_E}}{\rho_0^{1_E}}\right) & \left(\frac{z_0-Z^{1_E}}{\rho_0^{1_E}}\right) & 1 & m_f^{1_E} & 1 & 0 & \cdots & 0 & 1 & \cdots & 0 \\ \vdots & \vdots & \vdots & \vdots & \vdots & \vdots & \vdots & \vdots & \vdots & \vdots & \ddots & \vdots \\ \left(\frac{x_0-X^{n_E}}{\rho_0^{n_E}}\right) & \left(\frac{y_0-Y^{n_E}}{\rho_0^{n_E}}\right) & \left(\frac{z_0-Z^{n_E}}{\rho_0^{n_E}}\right) & 1 & m_f^{1_{nE}} & 1 & 0 & \cdots & 0 & 0 & \cdots & 0 \\ \left(\frac{x_0-X^{n_E}}{\rho_0^{n_E}}\right) & \left(\frac{y_0-Y^{n_E}}{\rho_0^{n_E}}\right) & \left(\frac{z_0-Z^{n_E}}{\rho_0^{n_E}}\right) & 1 & m_f^{1_{nE}} & 1 & 0 & \cdots & 0 & 0 & \cdots & 1 \end{bmatrix}_{2n \times (n+6)} \quad \Delta x = \begin{bmatrix} \Delta x \\ \Delta y \\ \Delta z \\ c\,dt_{r_G} \\ zpd_w \\ ISB_{GE} \\ \tilde{N}_G^1 \\ \vdots \\ \tilde{N}_G^{n_G} \\ \tilde{N}_E^1 \\ \vdots \\ \tilde{N}_E^{n_E} \end{bmatrix}_{n+6} \tag{40}$$

where n_G refers to the number of visible GPS satellites; n_E refers to the number of visible Galileo satellites; $n = n_G + n_E$ is the total number of the observed satellites for both GPS/Galileo systems; x_0, y_0, and z_0 are the approximate receiver coordinates; $X^{j_G}, Y^{j_G}, Z^{j_G}, j = 1, 2, \ldots, n_G$ are the known GPS satellite coordinates; $X^{k_E}, Y^{k_E}, Z^{k_E}, k = 1, 2, \ldots, n_E$ are the known Galileo satellite coordinates; ρ_0 is the approximate receiver–satellite range. The unknown parameters in the above system are the corrections to the receiver coordinates, Δx, Δy, and Δz, the wet component of the tropospheric zenith path delay zpd_w, the inter-system bias ISB, and the non-integer ambiguity parameters $\widetilde{N}$.

In case of the combined GPS/BeiDou PPP model, the design matrix A and the vector of corrections to the unknown parameters Δx take the following forms:

$$
A = \begin{bmatrix}
\left(\frac{x_0-X^{1_G}}{\rho_0^{1_G}}\right) & \left(\frac{y_0-Y^{1_G}}{\rho_0^{1_G}}\right) & \left(\frac{z_0-Z^{1_G}}{\rho_0^{1_G}}\right) & 1 & m_f^{1_G} & 0 & 0 & \cdots & 0 & 0 & \cdots & 0 \\
\left(\frac{x_0-X^{1_G}}{\rho_0^{1_G}}\right) & \left(\frac{y_0-Y^{1_G}}{\rho_0^{1_G}}\right) & \left(\frac{z_0-Z^{1_G}}{\rho_0^{1_G}}\right) & 1 & m_f^{1_G} & 0 & 1 & \cdots & 0 & 0 & \cdots & 0 \\
\vdots & \vdots & \vdots & \vdots & \vdots & \vdots & \vdots & \ddots & \vdots \\
\left(\frac{x_0-X^{n_G}}{\rho_0^{n_G}}\right) & \left(\frac{y_0-Y^{n_G}}{\rho_0^{n_G}}\right) & \left(\frac{z_0-Z^{n_G}}{\rho_0^{n_G}}\right) & 1 & m_f^{1_{nG}} & 0 & 0 & \cdots & 0 & 0 & \cdots & 0 \\
\left(\frac{x_0-X^{n_G}}{\rho_0^{n_G}}\right) & \left(\frac{y_0-Y^{n_G}}{\rho_0^{n_G}}\right) & \left(\frac{z_0-Z^{n_G}}{\rho_0^{n_G}}\right) & 1 & m_f^{1_{nG}} & 0 & 0 & \cdots & 1 & 0 & \cdots & 0 \\
\left(\frac{x_0-X^{1_B}}{\rho_0^{1_B}}\right) & \left(\frac{y_0-Y^{1_B}}{\rho_0^{1_B}}\right) & \left(\frac{z_0-Z^{1_B}}{\rho_0^{1_B}}\right) & 1 & m_f^{1_B} & 1 & 0 & \cdots & 0 & 0 & \cdots & 0 \\
\left(\frac{x_0-X^{1_B}}{\rho_0^{1_B}}\right) & \left(\frac{y_0-Y^{1_B}}{\rho_0^{1_B}}\right) & \left(\frac{z_0-Z^{1_B}}{\rho_0^{1_B}}\right) & 1 & m_f^{1_B} & 1 & 0 & \cdots & 0 & 1 & \cdots & 0 \\
\vdots & \vdots & \vdots & \vdots & \vdots & \vdots & \vdots & \vdots & \vdots & \ddots & \vdots \\
\left(\frac{x_0-X^{n_B}}{\rho_0^{n}}\right) & \left(\frac{y_0-Y^{n_B}}{\rho_0^{n}}\right) & \left(\frac{z_0-Z^{n_B}}{\rho_0^{n}}\right) & 1 & m_f^{1_{nB}} & 1 & 0 & \cdots & 0 & 0 & \cdots & 0 \\
\left(\frac{x_0-X^{n_B}}{\rho_0^{n_B}}\right) & \left(\frac{y_0-Y^{n_B}}{\rho_0^{n_B}}\right) & \left(\frac{z_0-Z^{n_B}}{\rho_0^{n_B}}\right) & 1 & m_f^{1_{nB}} & 1 & 0 & \cdots & 0 & 0 & \cdots & 1
\end{bmatrix}_{2n \times (n+6)}
\qquad
\Delta x = \begin{bmatrix}
\Delta x \\ \Delta y \\ \Delta z \\ c\,dt_{r_G} \\ zpd_w \\ ISB_{GB} \\ \widetilde{N}_G^1 \\ \vdots \\ \widetilde{N}_G^{n_G} \\ \widetilde{N}_B^1 \\ \vdots \\ \widetilde{N}_B^{n_B}
\end{bmatrix}_{n+6}
\qquad (41)
$$

where n_G refers to the number of visible GPS satellites; n_B refers to the number of visible BeiDou satellites; $n = n_G + n_B$ is the total number of the observed satellites for both GPS/BeiDou systems; x_0, y_0, and z_0 are the approximate receiver coordinates; $X^{j_G}, Y^{j_G}, Z^{j_G}, j = 1, 2, \ldots, n_G$ are the known GPS satellite coordinates; $X^{k_B}, Y^{k_B}, Z^{k_B}, k = 1, 2, \ldots, n_B$ are the known BeiDou satellite coordinates; ρ_0 is the approximate receiver–satellite range. The unknown parameters in the above system are the corrections to the receiver coordinates, Δx, Δy, and Δz, the wet component of the tropospheric zenith path delay zpd_w, the inter-system bias ISB_{GB}, and the non-integer ambiguity parameters $\widetilde{N}$.

In case of the combined Galileo/BeiDou PPP model, the design matrix A and the vector of corrections to the unknown parameters Δx take the following forms:

$$
A = \begin{bmatrix}
\left(\frac{x_0-X^{1_E}}{\rho_0^{1_E}}\right) & \left(\frac{y_0-Y^{1_E}}{\rho_0^{1_E}}\right) & \left(\frac{z_0-Z^{1_E}}{\rho_0^{1_E}}\right) & 1 & m_f^{1_E} & 0 & 0 & \cdots & 0 & 0 & \cdots & 0 \\
\left(\frac{x_0-X^{1_E}}{\rho_0^{1_E}}\right) & \left(\frac{y_0-Y^{1_E}}{\rho_0^{1_E}}\right) & \left(\frac{z_0-Z^{1_E}}{\rho_0^{1_E}}\right) & 1 & m_f^{1_E} & 0 & 1 & \cdots & 0 & 0 & \cdots & 0 \\
\vdots & \vdots & \vdots & \vdots & \vdots & \vdots & \vdots & \vdots & \vdots & \ddots & \vdots \\
\left(\frac{x_0-X^{n_E}}{\rho_0^{n_E}}\right) & \left(\frac{y_0-Y^{n_E}}{\rho_0^{n_E}}\right) & \left(\frac{z_0-Z^{n_E}}{\rho_0^{n_E}}\right) & 1 & m_f^{1_{nE}} & 0 & 0 & \cdots & 0 & 0 & \cdots & 0 \\
\left(\frac{x_0-X^{n_E}}{\rho_0^{n_E}}\right) & \left(\frac{y_0-Y^{n_E}}{\rho_0^{n_E}}\right) & \left(\frac{z_0-Z^{n_E}}{\rho_0^{n_E}}\right) & 1 & m_f^{1_{nE}} & 0 & 0 & \cdots & 1 & 0 & \cdots & 0 \\
\left(\frac{x_0-X^{1_B}}{\rho_0^{1_B}}\right) & \left(\frac{y_0-Y^{1_B}}{\rho_0^{1_B}}\right) & \left(\frac{z_0-Z^{1_B}}{\rho_0^{1_B}}\right) & 1 & m_f^{1_B} & 1 & 0 & \cdots & 0 & 0 & \cdots & 0 \\
\left(\frac{x_0-X^{1_B}}{\rho_0^{1_B}}\right) & \left(\frac{y_0-Y^{1_B}}{\rho_0^{1_B}}\right) & \left(\frac{z_0-Z^{1_B}}{\rho_0^{1_B}}\right) & 1 & m_f^{1_B} & 1 & 0 & \cdots & 0 & 1 & \cdots & 0 \\
\vdots & \vdots & \vdots & \vdots & \vdots & \vdots & \vdots & \vdots & \vdots & \ddots & \vdots \\
\left(\frac{x_0-X^{n_B}}{\rho_0^{n}}\right) & \left(\frac{y_0-Y^{n_B}}{\rho_0^{n}}\right) & \left(\frac{z_0-Z^{n_B}}{\rho_0^{n}}\right) & 1 & m_f^{1_{nB}} & 1 & 0 & \cdots & 0 & 0 & \cdots & 0 \\
\left(\frac{x_0-X^{n_B}}{\rho_0^{n_B}}\right) & \left(\frac{y_0-Y^{n_B}}{\rho_0^{n_B}}\right) & \left(\frac{z_0-Z^{n_B}}{\rho_0^{n_B}}\right) & 1 & m_f^{1_{nB}} & 1 & 0 & \cdots & 0 & 0 & \cdots & 1
\end{bmatrix}_{2n \times (n+6)}
\qquad
\Delta x = \begin{bmatrix}
\Delta x \\ \Delta y \\ \Delta z \\ c\,dt_{r_G} \\ zpd_w \\ ISB_{GB} \\ \widetilde{N}_G^1 \\ \vdots \\ \widetilde{N}_G^{n_G} \\ \widetilde{N}_B^1 \\ \vdots \\ \widetilde{N}_B^{n_B}
\end{bmatrix}_{n+6}
\qquad (42)
$$

where n_E refers to the number of visible Galileo satellites; n_B refers to the number of visible BeiDou satellites; $n = n_G + n_B$ is the total number of the observed satellites for both Galileo/BeiDou systems; x_0, y_0, and z_0 are the approximate receiver coordinates; $X^{j_E}, Y^{j_E}, Z^{j_E}, j = 1, 2, \ldots, n_E$ are the known

Galileo satellite coordinates; $X^{k_B}, Y^{k_B}, Z^{k_B}, k = 1, 2, \ldots, n_B$ are the known BeiDou satellite coordinates; ρ_0 is the approximate receiver–satellite range. The unknown parameters in the above system are the corrections to the receiver coordinates, Δx, Δy, and Δz, the wet component of the tropospheric zenith path delay zpd_w, the inter-system bias ISB_{EB}, and the non-integer ambiguity parameters $\tilde{N}$.

4. Results and discussion

To verify the developed combined PPP models, GPS, Galileo, and BeiDou observations at four globally distributed stations (Figure 1) were selected from the IGS tracking network (Dow et al., 2009). Those stations are occupied by GNSS receivers, which are capable of simultaneously tracking the GNSS constellations. Only one hour of observations with maximum possible number of Galileo and BeiDou satellites of each data-set is considered in our analysis. All data-sets have an interval of 30 s.

The positioning results for stations DLF1 are presented below. Similar results are obtained for the other stations. However, a summary of the convergence times and the three-dimensional PPP solution standard deviations are presented below for all stations. Natural Resources Canada's GPSPace PPP software was modified to handle data from GPS, Galileo, and BeiDou systems, which enables a combined PPP solution as detailed above (Afifi & El-Rabbany, 2016b). In addition to the combined PPP model, we also obtained the solutions of the un-differenced ionosphere-free GPS-only which is

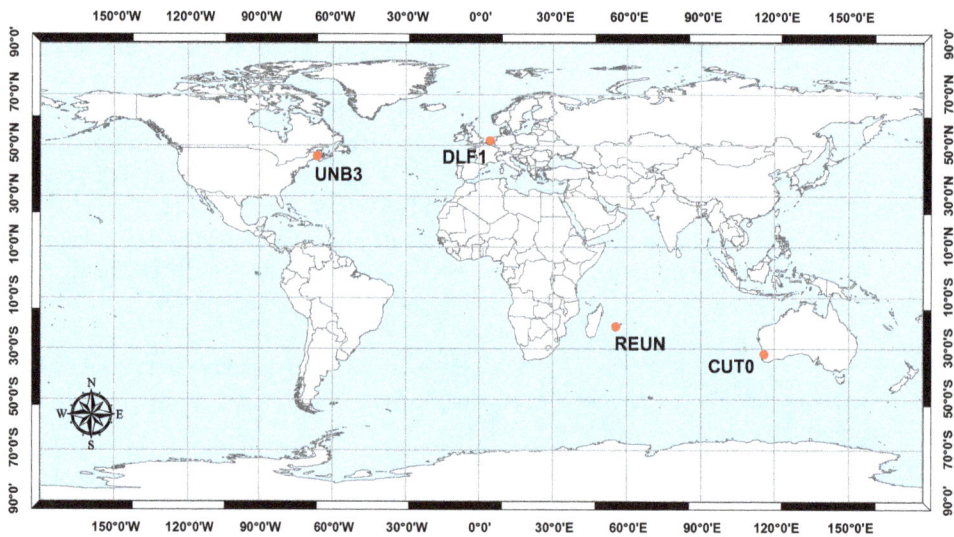

Figure 1. Analysis stations.

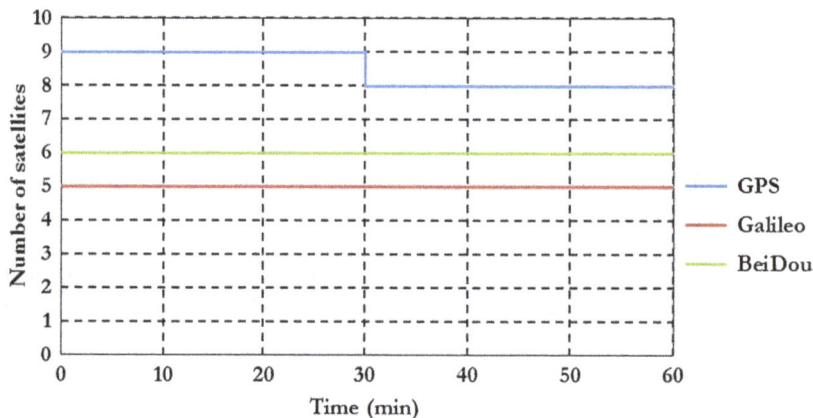

Figure 2. DLF1 station GNSS satellite availability.

used to assess the performance of the newly developed PPP model. Figure 2 summarizes the satellite availability during the analysis time (one hour) for each system at DLF1 station.

As shown in Figure 2, the GPS system offers eight visible satellites for one hour, however by adding the Galileo system the number of visible satellites will be 13 satellites. In case of combining GPS and BeiDou the number of visible satellites will be 14 satellites, however by combining the three satellite systems the number of visible satellites will be 19 satellites. Figure 3 shows the positioning results in

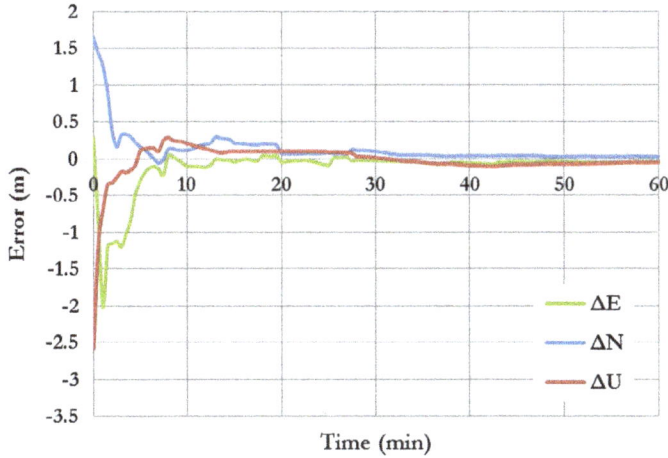

Figure 3. The positioning results of the GPS-only PPP model.

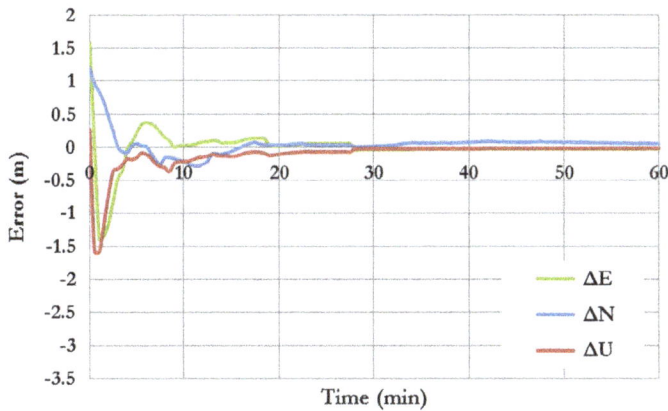

Figure 4. The positioning results of the GPS/Galileo PPP model.

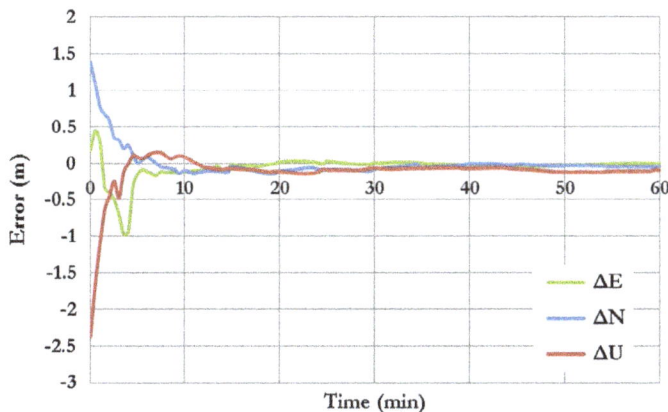

Figure 5. The positioning results of the GPS/BeiDou PPP model.

the East, North, and Up directions, respectively, for the GPS-only PPP model. As can be seen, the un-differenced GPS-only PPP solution indicates that the model is capable of obtaining a sub-decimeter-level accuracy. However, the solution takes about 20 min to converge to decimeter-level accuracy.

Figure 4 shows the positioning results combined GPS/Galileo PPP model. As shown in Figure 4, the positioning results of the combined GPS/Galileo traditional PPP model have a convergence time of 15 min to reach the decimeter-level of accuracy.

Figure 5 shows the combined GPS/BeiDou PPP model positioning results. As shown in Figure 5 the convergence time of the combined GPS/BeiDou PPP model is similar to the combined GPS/Galileo PPP model which is 15 min to reach the decimeter level of accuracy.

Figure 6 shows the combined Galileo/BeiDou PPP model positioning results. Similar to the previous combined PPP models the convergence time of the combined Galileo/BeiDou PPP model is 15 min to reach the decimeter level of accuracy.

Figure 7 shows the combined GPS/Galileo/BeiDou PPP model positioning results. As shown in Figure 7 the convergence time of the combined GPS/Galileo/BeiDou PPP model has a convergence time of 15 min to reach the decimeter level of accuracy.

Figure 8 summarizes the convergence times for all combined PPP models, which confirm the PPP solution consistency at all stations.

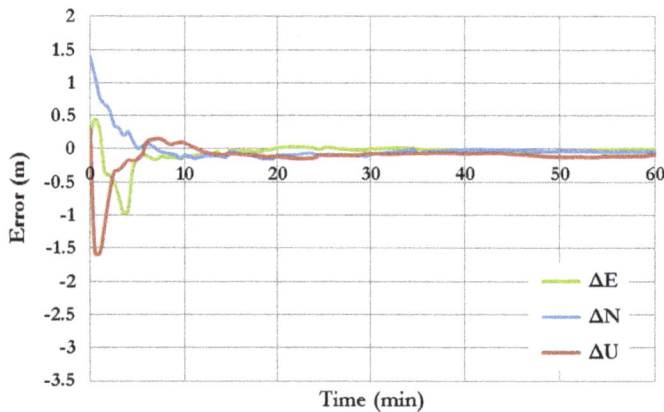

Figure 6. The positioning results of the Galileo/BeiDou PPP model.

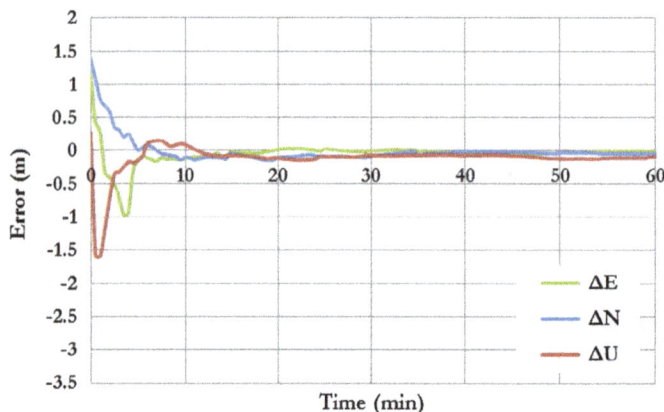

Figure 7. The positioning results of the GPS/Galileo/BeiDou PPP model.

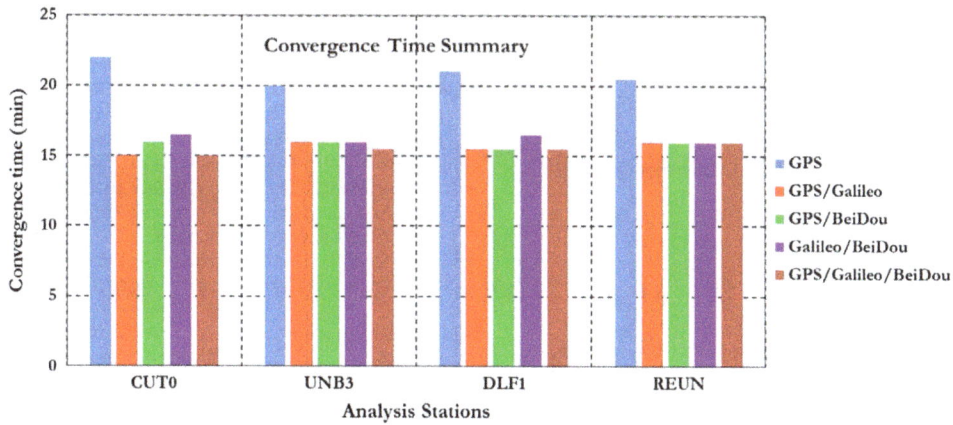

Figure 8. Summary of convergence times of all stations and PPP models.

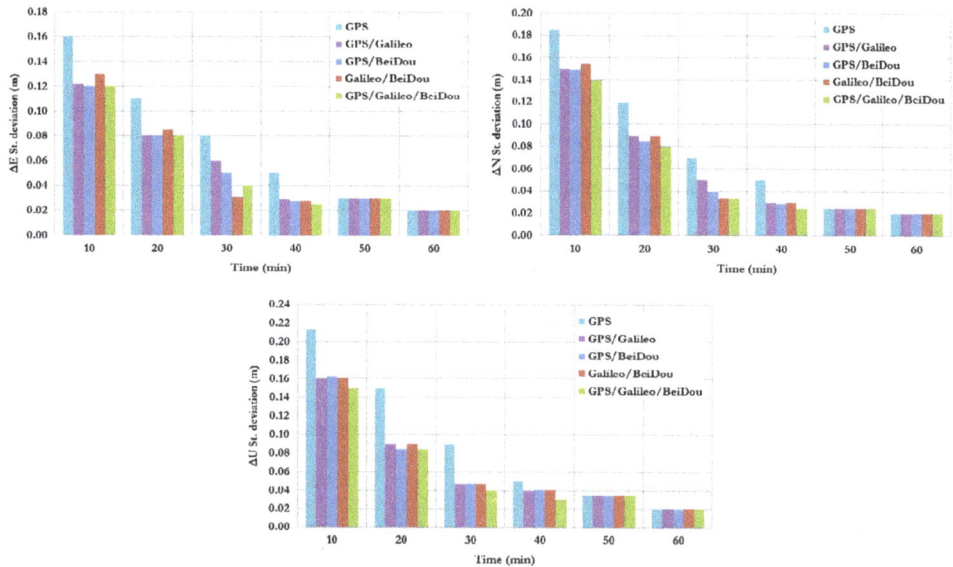

Figure 9. Summary of positioning standard deviations in East, North, and Up directions of all PPP models.

To further assess the performance of the various PPP models, the solution output is sampled every 10 min and the standard deviation of the computed station coordinates is calculated for each sample. Figure 9 shows the position standard deviations in the East, North, and Up directions, respectively. Examining the standard deviations of the combined PPP models is almost comparable to the GPS-only PPP model. As the number of epochs, and consequently the number of measurements, increases the performance of the various models tends to be comparable.

5. Conclusions

This paper presented a PPP model, which combines GPS/Galileo, GPS/BeiDou, Galileo/BeiDou, and GPS/Galileo/BeiDou observations in the un-differenced mode. The developed PPP model accounts for the combined effects of the different GNSS time offsets and hardware delays through the introduction of a new unknown parameter, the inter-system bias, in the PPP mathematical model. Four combinations are considered in the PPP modeling namely; GPS/Galileo, GPS/BeiDou, Galileo/BeiDou, and GPS/Galileo/BeiDou. All the combined PPP models results are compared with the GPS-only PPP model results. In the developed model GPS L1/L2, Galileo E1/E5a, and BeiDou B1/B2 are used in a dual-frequency ionosphere-free linear combination. It has been shown that the positioning results of the GPS-only and GPS/Galileo/BeiDou PPP are comparable and are at the sub-decimeter-level accuracy. However, the convergence time of the combined PPP models improved by about 25% in comparison with the GPS-only PPP.

Acknowledgments

This research was partially supported by the Natural Sciences and Engineering Research Council (NSERC) of Canada. The authors would like to thank the International GNSS service-Multi-GNSS Experiment (IGS-MEGX) network.

Funding

The authors received no direct funding for this research.

Author details

Akram Afifi[1]
E-mail: akram.afifi@ryerson.ca
Ahmed El-Rabbany[1]
E-mail: rabbaany@ryerson.ca
[1] Department of Civil Engineering, Ryerson University, Toronto, Ontario, Canada.

References

Afifi, A., & El-Rabbany, A. (2015). Performance analysis of several GPS/Galileo precise point positioning models. *Sensors, 15*, 14701–14726. doi:10.3390/s150614701

Afifi, A., & El-Rabbany, A. (2016a). Precise point positioning using triple GNSS constellations in various modes. *Sensors, 16*, 779. doi:10.3390/s16060779

Afifi, A., & El-Rabbany, A. (2016b). Improved between-satellite single-difference precise point positioning model using triple GNSS constellations: GPS, Galileo, and BeiDou. *Positioning, 7*, 63–74. doi:10.4236/pos.2016.72006

BeiDou. (2015). *BeiDou navigation satellite system*. Retrieved June 28, 2015, from http://en.beidou.gov.cn/

Boehm, J., & Schuh, H. (2004). Vienna mapping functions in VLBI analyses. *Geophysical Research Letters, 31*, 1601–1604. doi:10.1029/2003gl018984

Bos, M. S., & Scherneck, H.-G. (2011). *Ocean tide loading provider*. Retrieved December 1, 2014, from http://holt.oso.chalmers.se/loading/

Collins, P., Bisnath, S., Lahaye, F., & Héroux, P. (2010). Undifferenced GPS ambiguity resolution using the decoupled clock model and ambiguity datum fixing. *Navigation, 57*, 123–135. http://dx.doi.org/10.1002/navi.2010.57.issue-2

Colombo, O. L., Sutter, A. W., & Evans, A. G. (2004, September 21–24). *Evaluation of precise, kinematic GPS point positioning*. ION GNSS 17th International Technical Meeting of the Satellite Division, Long Beach, CA.

Dow, J. M., Neilan, R. E., & Rizos, C. (2009). The International GNSS service in a changing landscape of global navigation satellite systems. *Journal of Geodesy, 83*, 191–198.

doi:10.1007/s00190-008-0300-3

ESA. (2015). *European space agency*. Retrieved June 28, 2015, from http://www.esa.int/ESA

Ge, M., Gendt, G., Rothacher, M., Shi, C., & Liu, J. (2008). Resolution of GPS carrier-phase ambiguities in precise point positioning (PPP) with daily observations. *Journal of Geodesy, 82*, 401. doi:10.1007/s00190-007-0208-3

Hofmann-Wellenhof, B., Lichtenegger, H., & Wasle, E. (2008). *GNSS global navigation satellite systems: GPS, Glonass, Galileo & more* (p. 501). Wien: Springer.

Hopfield, H. S. (1972). Tropospheric refraction effects on satellite range measurements. *APL Technical Digest, 11*, 11–19.

IAC. (2015). *Fedral space agency, the Information-analytical centre*. Retrieved June 28, 2015, from https://glonass-iac.ru/en/

IERS. (2010). *International earth rotation and reference system services conventions (2010)* (IERS Technical Note 36). Retrieved from http://www.iers.org/IERS/EN/Publications/TechnicalNotes/tn36.html/

Kaplan, E., & Heagarty, C. (2006). *Understanding GPS principles and applications* (650 p.). Boston, MA: Artech House.

Kouba, J. (2009). *A guide to using international GNSS service (IGS) products*. Retrieved from http://igscb.jpl.nasa.gov/igscb/resource/pubs/UsingIGSProductsVer21.pdf

Kouba, J., & Héroux, P. (2001). Precise point positioning using IGS orbit and clock products. *GPS Solutions, 5*, 12–28. http://dx.doi.org/10.1007/PL00012883

Leick, A. (2004). *GPS satellite surveying* (3rd ed., p. 435). New York, NY: Wiley.

Melgard, T., Tegedor, J., de Jong, K., Lapucha, D., & Lachapelle, G. (2013, September 16–20). Interchangeable integration of GPS and Galileo by using a common system clock in PPP. In *ION GNSS+*, Nashville TN: Institute of Navigation.

Montenbruck, O., Steigenberger, P., Khachikyan, R., Weber, G., Langley, R. B., Mervart, L., & Hugentobler, U. (2014). IGS-MGEX: Preparing the ground for multi-constellation GNSS science. *Inside GNSS, 9*, 42–49.

Vanicek, P., & Krakiwsky, E. J. (1986). *Geodesy: The concepts* (2nd ed.). Amsterdam: North-Holland.

Wu, J. T., Wu, S. C., Hajj, G. A., Bertiger, W. I., & Lichten, S. M. (1993). Effects of antenna orientation on GPS carrier phase. *Manuscripta Geodetica, 18*, 91–98.

Zumberge, J. F., Heflin, M. B., Jefferson, D. C., Watkins, M. M., & Webb, F. H. (1997). Precise point positioning for the efficient and robust analysis of GPS data from large networks. *Journal of Geophysical Research: Solid Earth, 102*, 5005–5017. http://dx.doi.org/10.1029/96JB03860

3

Petrology and crustal inheritance of the Cloudy Bay Volcanics as derived from a fluvial conglomerate, Papuan Peninsula (Papua New Guinea): An example of geological inquiry in the absence of *in situ* outcrop

Robert J. Holm[1,2]* and Benny Poke[3]

*Corresponding author. Robert J. Holm, Frogtech Geoscience, 2 King Street, Deakin West ACT 2600, Australia; Geosciences, College of Science & Engineering, James Cook University, Townsville, Queensland 4811, Australia

E-mail: rholm@frogtech.com.au

Reviewing editior: Chris Harris, University of Cape Town, South Africa

Abstract: In regions of enhanced weathering and erosion, such as Papua New Guinea, our ability to examine a complete geological record can become compromised by the absence of *in situ* outcrops. In this study, we provide an example of the insights that can be gained from investigations of secondary deposits. We sampled matrix material and clasts derived from an isolated conglomerate outcrop within a landscape dominated by lowland tropical forest of the southeast Papuan Peninsula, and mapped as belonging to the Cloudy Bay Volcanics. Nine variations of volcanic rock types were identified that range from basalts to trachyandesites. Major and trace element geochemistry characterize the volcanic arc assemblage as shoshonites and provide evidence for differential magma evolution pathways with a subset of samples marked by heavy REE- and Y-depletion, indicative of high-pressure magma fractionation. Zircon U–Pb dating of the individual volcanic clasts indicates activity of the Cloudy Bay Volcanics was largely constrained to the latest Miocene, between ca. 7 and 5 Ma. Of the analyzed zircons, the majority are xenocrystic

ABOUT THE AUTHORS

Robert J. Holm is a senior geoscientist with Frogtech Geoscience and an adjunct lecturer in Geosciences at James Cook University. Robert is an early career geoscientist interested in multidisciplinary and innovative approaches to solve earth science problems, and holds expertise in the research areas of tectonics, igneous petrology, structural geology, geochronology, and metallogeneis. He obtained his PhD from James Cook University, investigating magmatic arcs and porphyry systems of Papua New Guinea to gain new insights into the late Cenozoic tectono-magmatic evolution and subduction histories at the northern Australian plate boundary. Robert continues his research into the South West Pacific and is working toward developing an integrated approach to geosciences in the region and developing collaborations between regional universities, government agencies, together with the minerals, and oil and gas industries to benefit our understanding of the regional geology.

PUBLIC INTEREST STATEMENT

Plate tectonics have shaped the Earth and give rise to natural hazards such as earthquakes and volcanoes. To understand how the Earth changes through time we observe these processes, such as volcanism and magmatism in the geological rock record, and investigate how changes in these processes reflect large-scale changes related to plate tectonics. However, Earth surface processes such as weathering and erosion that shape the landscape will often obscure or remove the rocks that are required to investigate the geological record for a specific region. In this study, we implement an innovative methodology of sampling conglomerates, or river deposits, eroded from ancient volcanic activity in Papua New Guinea. By investigating individual rock clasts within these deposits, we explore ancient volcanic activity and reveal what insights we can gain into the wider formation and evolution of the region up to the present day.

zircons that provide insight into the provenance of the Papuan Peninsula with potentially significant implications for South West Pacific tectonics. Additional Hf-isotope
analysis of the primary igneous zircons suggests a relatively unradiogenic crustal
component contributed to magma compositions, which cannot be readily explained
by current regional tectonic paradigms.

**Subjects: Earth Sciences; Tectonics; Mineralogy & Petrology; Volcanology; Sedimentology &
Stratigraphy**

**Keywords: Papua New Guinea; volcanic; conglomerate; shoshonite; U–Pb geochronology;
zircon provenance; Hf isotope**

1. Introduction
In regions of enhanced weathering and erosion our ability to examine the geological record can
become compromised by earth surface processes. These processes can include, for example, deep
weathering profiles, obstruction beneath cover, or removal via erosion, and are typical of tropical
climates and areas of active tectonics and uplift. As a result, these areas are often characterized by
gaps in our data-sets and difficulties in reconciling the regional geological history. Investigations
into the petrology of first-generation conglomerate clasts, such as presented in this study, can offer
many insights into the nature of rocks exposed in their source catchment at the time of deposition
(e.g. Lamminen, Andersen, & Nystuen, 2015; Samuel, Be'eri-Shlevin, Azer, Whitehouse, & Moussa,
2011; Schott & Johnson, 2001). The power of this methodology is derived in part from sedimentary
processes where clastic detrital material records the bedrock geology that is representative of a
larger watershed or geological terrain. For example, parameters such as age, texture, or composition for a group of conglomerate clasts can be useful in identifying not only the source of the detritus
and associated sedimentary pathways (e.g. detrital zircon provenance; Gehrels, 2014), but also provide us with detailed insights into the nature of terranes that have been lost to erosion, burial, or
tectonic dismemberment (e.g. Graham & Korsch, 1990; Hidaka, Shimizu, & Adachi, 2002; Lamminen
et al., 2015; Samuel et al., 2011; Schott & Johnson, 2001; Wandres et al., 2004).

In this study, rather than surface processes obscuring geological information, we use fluvial and
colluvial surface processes, and the resulting secondary deposits, to complement existing data-sets
and contribute to our understanding of a geological terrane. We report the results of an investigation into an outcrop of conglomerate within an area previously mapped as part of the Cloudy Bay
Volcanics of the southeast Papuan Peninsula of Papua New Guinea (Figure 1). The chosen sample
location on the southern coast of the Papuan Peninsula provides an example of a terrain where exploring the geological history of the region through traditional ground-based geological mapping
techniques is extremely difficult, if not impossible. The terrain is characterized by dense lowland
tropical forest with regional swamp forest and mangroves, with moderately high rainfall (750–1,200-
mm precipitation in the driest quarter; Shearman, Ash, Mackey, Bryan, & Lokes, 2009; Shearman &
Bryan, 2011). Such conditions are characteristic of tropical regions, such as Papua New Guinea,
where dense vegetation and thick soil profiles result in a low density of informative outcrops. In this
example, the sampled assemblage of clasts and matrix material were examined via zircon U–Pb
geochronology, zircon Lu–Hf isotope analysis, and whole-rock major and trace element geochemical
investigations. The results from this work demonstrate that in areas where suitable *in situ* outcrops
may be absent or compromised by earth surface processes, we can still gain valuable insight into the
wider geological history of a region through targeted and innovative sampling methodologies. The
investigation presented here is part of a wider examination of the geology of southeast Papua New
Guinea, in which, we aim to demonstrate the regional tectonic insights that can be gained from such
sampling of secondary deposits.

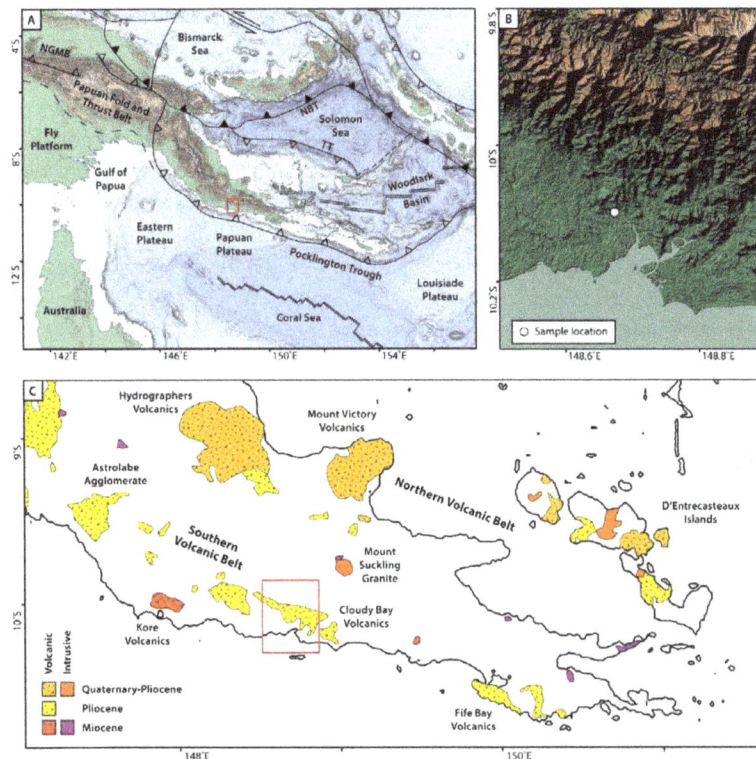

Figure 1. Topography, bathymetry, and major tectonic boundaries of southeast Papua New Guinea and the Papuan Peninsula region. (A) tectonic elements of the southeast Papua New Guinea region (modified from Holm et al., 2016); NBT, New Britain trench; NGMB, New Guinea Mobile Belt; TT, Trobriand trough. Topography and bathymetry after Amante and Eakins (2009). (B) landscape morphology of the Cloudy Bay area and position of the sample site in the lowland rainforest. (C) Miocene to Quaternary volcanic and intrusive rocks of southeast Papua New Guinea (modified from Australian BMR, Australian Bureau of Mineral Resources, 1972; Smith & Milsom, 1984). Inset shows location of B).

2. Geologic setting and samples

2.1. Geological setting

The Papuan Peninsula forms the eastern extent of the Papua New Guinea mainland, between approximately 146 and 151°E (Figure 1). There are two main components that comprise the geological basement of the Papuan Peninsula. These are a core of moderate to high-grade metamorphic rocks, that form the Owen Stanley Metamorphic Complex that transitions into the Milne Terrane to the east, and the overlying Papuan Ultramafic Belt; an obducted sheet of ultramafic rocks and associated mid-ocean ridge-type basalts (Baldwin, Fitzgerald, & Webb, 2012; Davies, 2012; Davies & Smith, 1971; Smith, 2013a).

The Owen Stanley Metamorphic Complex forms the main spine of the Owen Stanley Ranges and the Papuan Peninsula. Two major rock units form the Owen Stanley Metamorphic Complex. The Kagi Metamorphics are primarily composed of pelitic and psammitic sediments derived from felsic volcanism, with minor intercalated volcanics, that have been folded and metamorphosed to greenschist facies (Davies, 2012; Pieters, 1978). The Emo Metamorphics outcrop northeast of, and overlie the Kagi Metamorphics, forming a 1–2-km-thick carapace, which dips shallowly to the north and northeast. The Emo Metamorphics mainly comprise metabasite derived from low-K tholeiitic basalt, dolerite, and gabbro, together with minor volcaniclastic sediments, and metamorphosed to greenschist and blueschist metamorphic facies (Davies, 2012; Pieters, 1978). The protolith of the Emo Metamorphics is interpreted as a supra-subduction extensional back arc setting (Smith, 2013a; Worthing & Crawford, 1996). The Owen Stanley Metamorphic Complex is interpreted to be of middle

Cretaceous age from U–Pb dating of zircon of likely volcanic origin (Aptian–Albian; 120–107 Ma [Kopi, Findlay, & Williams, 2000]), and from preserved macrofossils (Aptian–Cenomanian [Dow, Smit, & Page, 1974]).

The Milne Terrane occupies the equivalent structural domain to the Owen Stanley Metamorphics in the southeast of the Papuan Peninsula and comprises the Goropu Metabasalt and Kutu Volcanics (Smith, 2013a; Worthing & Crawford, 1996). The Goropu Metabasalt consists of low-grade N-MORB-type metabasalts with minor metamorphosed limestone and calcareous schist. Submarine basaltic volcanoes and interbedded lenses of pelagic limestone with minor terrigenous sediments form the interpreted protolith for the Goropu Metabasalt (Smith, 2013a; Smith & Davies, 1976). The Kutu Volcanics are interpreted as the unmetamorphosed continuation of the Goropu Metabasalt comprising dominantly basaltic lava with minor gabbro and ultramafics, agglomerate, tuffaceous and calcareous sediments, and limestone (Smith, 2013a; Smith & Davies, 1976). The age of the Milne Terrane is constrained by microfossils and is interpreted to have been deposited from the Upper Cretaceous (Maestrictian), and potentially as young as Eocene in the southeast of the terrane (Smith & Davies, 1976).

The Papuan Ultramafic Belt occupies the northeast side of the Papuan Peninsula and is juxtaposed above the Owen Stanley Metamorphic Complex along the Owen Stanley Fault (Baldwin et al., 2012; Davies, 2012). The Papuan Ultramafic Belt is interpreted as an ophiolite complex comprising oceanic crust and lithospheric mantle (Davies & Jaques, 1984; Davies & Smith, 1971), which is interpreted as late Cretaceous in age (Davies, 2012; Davies & Smith, 1971). Obduction of the Papuan Ultramafic Belt, and metamorphism of the Owen Stanley Metamorphic Complex, is interpreted at 58.3 ± 0.4 Ma, derived from the cooling age of amphibole within the high-grade metamorphic contact between the two terranes (Davies, 2012; Lus, McDougall, & Davies, 2004).

The Papuan Peninsula was subsequently intruded by a major episode of subduction-related volcanism, which commenced during the middle Miocene and has continued to the present day (e.g. Jakeš & Smith, 1970; Smith, 1972, 1982, 2013b; Smith & Milsom, 1984). This volcanic province is marked by a transition from early submarine–subaerial activity, to entirely subaerial volcanism during the Pliocene and Quaternary, reflecting the emergence of eastern Papua New Guinea during the latter part of the Cenozoic (Davies, 2012).

The Cloudy Bay Volcanics are located some 140 km east-southeast from Port Moresby and extend a further 90 km east along the Papuan Peninsula (Figure 1). The volcanics cover approximately 770 km^2 and form rolling terrain with very low relief, and extend up to the foothills of the Milne Terrane at an elevation of ca. 250 m. The estimated thickness of the volcanics is 500 m, and they dip at a shallow angle to the south (Pieters, 1978). Mapping of the Cloudy Bay Volcanics suggests it mainly comprises basalt, andesitic pyroclastics and lava, and tuffaceous sandstone (Pieters, 1978; Smith, 1976). The tuff-dominated members form rolling terrain with very low relief and moderately spaced meandering streams, whereas the lava and pyroclastic members form small cones, and are associated with greater relief and dendritic drainage patterns (Pieters, 1978).

2.2. Samples
In this study, we sampled a single isolated conglomerate outcrop (Figure 2; 10.102°S 148.673°E) within a landscape dominated by lowland rainforest and swamp, and mapped as part of the Cloudy Bay Volcanics (Figure 1). The outcrop comprised a clast-supported conglomerate and was sampled for clasts of different rock types, together with samples of the matrix. Clasts within the outcrop were predominantly cobble and boulder-sized up to approximately 50 cm in diameter, typically sub-rounded to well-rounded, and elongated to spherical in shape (Figure 2). The matrix material was dominantly made up of clays but also contained recognizable, euhedral plagioclase and pyroxene grains, indicating the detritus was immature and not well-traveled. From the isolated nature of the outcrop it is difficult to establish its context within the Cloudy Bay Volcanics, however, it likely represents a secondary deposit derived from fluvial transport of an eroding volcanic landscape.

Figure 2. Conglomerate outcrop from which samples were collected for this study. Outcrop location is shown in inset of Figure 1. Lens cap used for scale in lower center of the photograph is 46 mm diameter.

In the field, an effort was made to collect a representative variety of clast types that reflected the compositional variation within the outcrop, but were also sufficiently large to conduct analytical procedures. In total, nine clasts types were identified and sampled, in addition to the matrix. Given the weathered and degraded appearance of the clasts in the field (see supplementary material for sample photographs), rock descriptions and petrography were carried out on the least altered rock material following processing (Figure 3; Table 1). All the sampled clasts comprise mafic-intermediate volcanic rock types, which range in composition from basalt to trachyandesite. Most samples are porphyritic in texture with a very fine-grained to aphanitic matrix; samples 103265 and 103267 comprise augite phenocrysts up to 5 mm in length, in a very fine-grained matrix; the remaining samples predominantly consist of plagioclase- and augite-phyric lavas where the phenocryst size is <2 mm. Magnetite (<0.5 mm) and biotite (typically <2 mm) phenocrysts are also present in various samples (Table 1); apatite crystals up to 1 mm are present in sample 103265. The matrix of the selected volcanic samples comprises similar mineralogy, made up of variable amounts of plagioclase, augite, magnetite, apatite, and biotite. Sample 102368c is textually distinct within the sample suite being very fine grained to aphanitic in nature, although augite, plagioclase, magnetite, and apatite microphenocrysts can be identified.

3. Methods

Samples of both clasts and matrix material were collected from the conglomerate outcrop shown in Figure 2. Rock clasts were washed to remove any contamination from matrix material and soil, cut into workable sections, and weathering rinds were removed with diamond implanted grinders. The least weathered material from each clast was selected for petrographic thin sections and geochemical analysis. The amount of rock material sampled for geochemical analysis ranged from 110 g up to 240 g (average of 170 g) depending on the size and weathering state of the clast. The remainder of each sample, including weathered sections free from matrix material and soil contamination, was subsequently utilized for mineral separation. Matrix material predominantly comprised clay with lithic pebbles and remnant minerals. The matrix sample was dried over several days at 100°C before

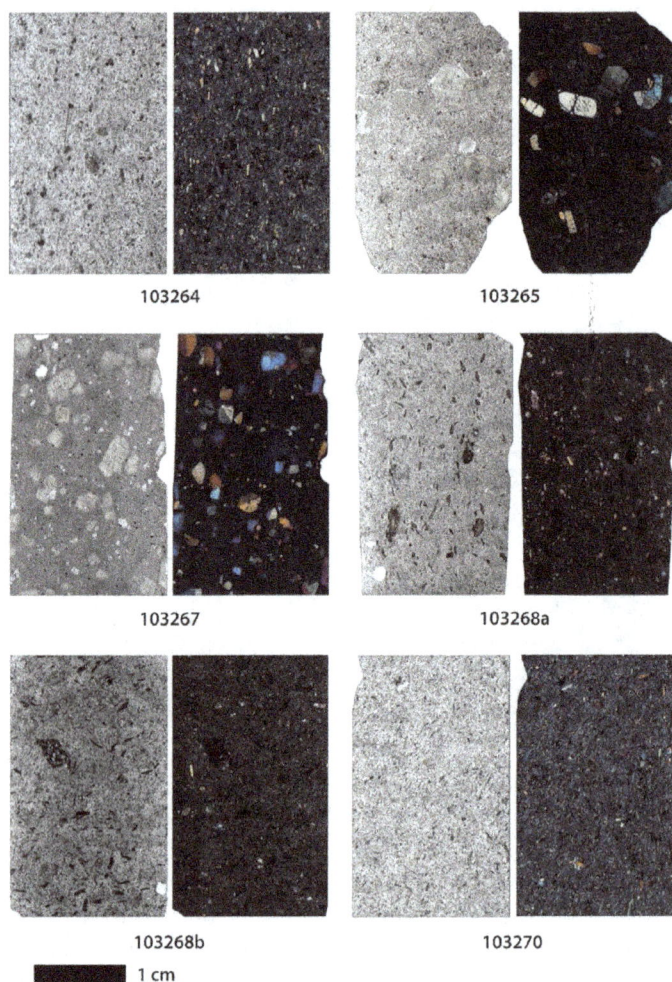

Figure 3. Petrographic mosaic photographs for selected samples used in this study (additional samples are included in the supplementary material). For each sample, the image on the left is observed under plane-polarized light, and the right under cross-polarized light. See Table 1 for sample descriptions.

undergoing mineral separation as a bulk sample. The matrix material was not amenable to petrographic thin section preparation.

All geochemical analyses were carried out by the Bureau Veritas Mineral Laboratories in Vancouver, Canada. Whole rock samples were crushed, split, and pulverized to a 200-mesh grain size, and mixed with $LiBO_2/Li_2B_4O_7$ flux. The cooled bead was dissolved in ACS grade nitric acid and analyzed by a combination of ICP-ES and ICP-MS methods. Additional volatile elements (Mo, Cu, Pb, Zn, Ni, As) were analyzed via aqua regia digestion in combination with ICP-ES and ICP-MS instrumentation. In-house standards (SO-19 and DS10) were used to measure analytical uncertainty. The maximum errors for SiO_2, MgO, and K_2O were 1.0, 1.0, and 3.9%, respectively. Maximum relative errors for representative trace elements Nb, Zr, Y, Nd, and Sm are 3.8, 5.3, 8.2, 8.6, and 11.3%, respectively. The relative errors for other trace elements were similar in magnitude based on a comparison between the measured and accepted trace element concentrations. Loss on ignition (LOI) was determined by igniting a 1-g sample split to 1,000°C for one hour, cooled and then measuring the weight loss.

Mineral separation to extract zircon crystals was carried out at James Cook University (JCU) in a standard process. Samples were crushed and milled to 500 µm, and separation was carried out via the use of a Wilfley table (smaller samples were hand washed to remove the clay fraction), and a

Table 1. Sample descriptions				
Sample	**TAS Classification**	**Phenocrysts**	**Matrix**	**Additional Notes**
103264	basaltic trachyandesite	Aug + Mag + Ap + Bt	Pl + Mag + Ap	Mag + Bt + Aug aggregates
103265	basalt	Aug + Ap + Mag	Pl + Ap + Mag + Bt	Hem rims on Ap
103267	trachybasalt	Aug + Mag	Pl + Aug + Mag + Ap	2 Aug phenocryst generations
103268a	basaltic trachyandesite	Aug + Pl + Mag + Bt	Pl + Mag + Ap	Pl + Aug + Bt + Mag + Ap aggregates
				Pl phenocryst breakdown
103268b	trachyandesite	Aug + Pl + Mag	Pl + Aug + Mag + Ap	Pl + Aug + Bt + Mag + Ap aggregates
				Pl phenocryst breakdown
103268c	basalt	no phenocrysts	Aug + Pl + Mag + Ap	Aug xenocryst
103268d	basaltic trachyandesite	Aug + Pl + Mag	Pl + Mag + Ap	Pl + Aug + Bt + Mag + Ap aggregates
				Pl phenocryst breakdown
103269	trachyandesite	Pl + Aug + Mag	Pl + Aug + Mag + Ap	Aug + Mag + Bt + Ap aggregates
				minor Pl phenocryst breakdown
103270	basaltic trachyandesite	Aug + Pl + Mag + Bt	Pl + Mag + Aug + Ap	alignment of phenocrysts and matrix
				Aug + Mag + Bt + Ap aggregates

Notes: Mineral codes: Aug, augite; Mag, magnetite; Ap, apatite; Bt, biotite; Pl, plagioclase; Hem, hematite.

combination of heavy liquid density separation and magnetic separation. Zircon crystals were hand-picked under a binocular microscope and mounted in epoxy, before being polished and carbon-coated. Cathodoluminescence (CL) images of all zircon crystals were obtained to study internal zonation and structures using a Jeol JSM5410LV scanning electron microscope equipped with a Robinson CL detector, housed at the Advanced Analytical Centre (AAC), JCU (Figure. 4; additionally see supplementary material for complete CL images).

All U–Pb dating work was completed at the AAC, JCU. U–Pb dating of zircons was conducted via Coherent GeolasPro 193 nm ArF Excimer laser ablation system connected to a Bruker 820-ICP-MS following the methodology outlined in Holm, Spandler, and Richards (2013, 2015). All zircons were analyzed with a beam spot diameter of 44 μm and selection of analytical sample spots was guided by CL images. Data reduction was carried out using the Glitter software (Van Achterbergh, Ryan, Jackson, & Griffin, 2001). Drift in instrumental measurements was corrected following analysis of drift trends in the raw data using measured values for the GJ1 primary zircon standard (608.5 ± 0.4 Ma; Jackson, Pearson, Griffin, & Belousova, 2004). Secondary zircon standards Temora 2 (416.8 ± 0.3 Ma; Black et al., 2004) and AusZ2 (38.8963 ± 0.0044 Ma; Kennedy, Wotzlaw, Schaltegger, Crowley, & Schmitz, 2014) were used for verification of GJ1 following drift correction (see supplementary material). For quantification of U and Th concentration in zircon samples, analysis of the NIST SRM 612 reference glass was conducted throughout every analytical session at regular intervals, with ^{29}Si used as the internal standard assuming perfect zircon stoichiometry. Background corrected analytical count rates, calculated isotopic ratios and 1σ uncertainties were exported for further processing and data reduction.

Age regression and data presentation for all samples was carried out using Isoplot (Ludwig, 2009). Correction for initial Th/U disequilibrium during zircon crystallization related to the exclusion of ^{230}Th due to isotope fractionation, and resulting in a deficit of measured ^{206}Pb as a ^{230}Th decay product (Parrish, 1990; Schärer, 1984), was applied to all analyses of Cenozoic age. Correction of $^{206}Pb/^{238}U$

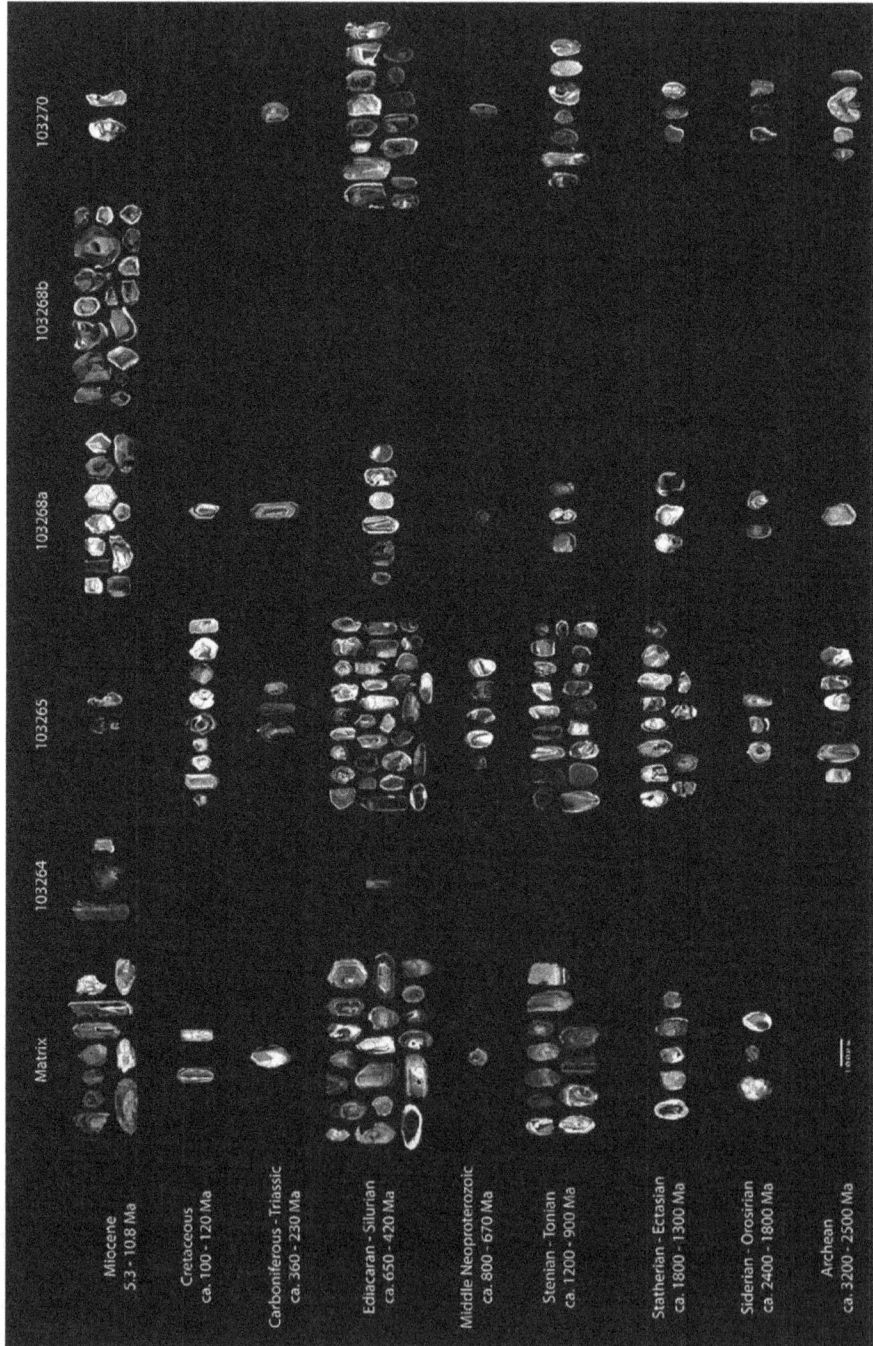

Figure 4. Cathodoluminescence for all concordant zircon grains, and grains interpreted to belong to concordant igneous populations in this study. Grains are arranged by sample and interpreted age to illustrate distribution of zircon ages and grain morphology.

Note: Red circles denote laser ablation spot locations.

ages for the [206]Pb deficit utilized Th and U concentrations for zircon determined during LA-ICP-MS analysis; equivalent concentrations in the melt were determined from bulk rock analysis for samples where zircons were derived from conglomerate clasts. A melt Th/U ratio of 3 ± 0.3 was assumed for zircons extracted from the matrix samples. Uncertainties associated with the correction were propagated into errors on the corrected ages according to Crowley, Schoene, and Bowring (2007). The effect of common Pb was taken into account by the use of Tera–Wasserburg Concordia plots (Jackson et al., 2004; Tera & Wasserburg, 1972). Spot ages were corrected for common Pb by utilizing the Age7Corr algorithms in Isoplot, with the isotopic common-Pb composition modeled from Stacey and Kramers (1975). Weighted mean [206]Pb/[238]U age calculations were carried out using Isoplot. All errors for Cenozoic zircons were propagated at 2σ level and reported at 2σ and 95% confidence for concordia and weighted averages, respectively.

Isotopic data derived from detrital and xenocrystic zircon grains (greater than Miocene in age) were discriminated based on age. Where grain ages were in excess of 1,000 Ma, the [207]Pb/[206]Pb age was preferred and assessed for discordance between the [207]Pb/[206]Pb and [206]Pb/[238]U age systems; while grains below 1,000 Ma in age were reported according to the [206]Pb/[238]U age and assessed for discordance between [207]Pb/[206]Pb and [206]Pb/[238]U and [206]Pb/[238]U and [207]Pb/[235]U age systems. A 20% discordance threshold was used as the cut-off limit beyond which analyses were excluded from further data reduction. The preferred inherited ages taken forward for analysis were a combination of [207]Pb/[206]Pb and [206]Pb/[238]U ages and all errors were propagated and reported at a 1σ level; these were plotted using the cumulative probability plot and histogram function of Isoplot. A similar methodology was used for Miocene ages using a 30% discordance cut-off to provide for the increased level of uncertainty in young U–Pb ages.

Selected samples that featured a prevalence of zircons crystals underwent additional *in situ* analysis for zircon Hf isotopes. Laser ablation analyses of zircons for Lu–Hf isotope ratios were carried out at the Advanced Analytical Centre, JCU, using a GeoLas193-nmArF laser and a Thermo-Scientific Neptune multicollector ICP-MS following the setup outlined in Næraa et al. (2012) and Kemp et al. (2009). Suitable zircon crystals were selected on the basis of size and U–Pb dating results, and ablation was carried out at a repetition rate of 4 Hz and a spot size of 60 μm. All [176]Hf/[177]Hf ratios for standard and sample zircons were normalized to measurements of the Mud Tank reference zircon (average measured 176Hf/177Hf ratio during this study was 0.282495 [n = 18], normalized to solution value of 0.282507) and compared with the FC1 secondary zircon standard (average 176Hf/177Hf value for this study is 0.282177, with reference to the solution value of 0.282167 ± 10; Kemp et al., 2009). Epsilon Hf values for the data were calculated following the procedure of Holm et al. (2015).

4. Results

4.1. U–Pb geochronology

All samples underwent mineral separation procedures; however, the zircon yield was highly variable for the selected sample suite. Both the volcanic nature of the rock types, and the sample size are considered contributing factors to inconsistent zircon yields. Of the nine clast samples studied, only five samples yielded zircons, and the number of zircons grains returned from the samples varied significantly (Table 2).

Results from the U–Pb zircon dating returned ages ranging from the latest Miocene–earliest Pliocene up to the Mesoarchean. These will be presented as two distinct sets of results. The first comprises ages ranging from 5.26 ± 0.27 Ma up to 14.67 ± 0.64 Ma (2σ error), broadly spanning the middle–late Miocene. These zircon crystals exhibit morphology and CL textures that are predominantly euhedral and prismatic with oscillatory zoning (Figure 4). The Th/U ratio among this suite of Miocene zircons ranges from 0.44 to 2.88 with an average of 1.41. Corresponding Tera–Wasserburg concordia and weighted average plots for these data are shown in Figure 5 and results are outlined in Table 2; we will cite the concordia ages of the samples. Sample 103264 yielded an age of 5.99 ± 0.31 Ma (n = 4), while samples 103268a and 103268b returned similar ages of 5.77 ± 0.21 Ma

(n = 7) and 5.71 ± 0.13 Ma (n = 12), respectively. Coupled with the euhedral, oscillatory-zoning CL textures (Figure 4; also see supplementary material), the relatively high U/Th ratios indicate these zircon U–Pb dating results mostly likely reflect igneous crystallization ages (e.g. Ahrens, Cherry, & Erlank, 1967; Corfu, Hanchar, Hoskin, & Kinny, 2003; Heaman, Bowins, & Crocket, 1990; Hoskin & Schaltegger, 2003). The matrix did produce a concordant zircon population that yielded an age of 8.22 ± 0.19 Ma (n = 6), however, younger zircon ages within the same sample suggest this is an in-herited magmatic age. Results for all samples passing the 30% discordance cut-off within this age suite form a distinct concentration of ages between approximately 5 and 7 Ma, and peaking at 6 Ma, with isolated ages extending back to 11 Ma (Figure 5).

Of the 267 zircon grains analyzed for this study, 185 of these (passing 20% discordance) yielded ages older than Miocene, and indeed older than Cenozoic (Figures 4 and 5), with the youngest pre-Miocene age at 99.8 ± 1.7 Ma (1σ error). Significant age populations within this data-set include a Cretaceous population that exhibit euhedral and prismatic grain morphology (Figure 4), and spans ca. 100–120 Ma (12 grains [6%]); a minor population of euhedral and prismatic zircons of Carboniferous–Triassic age (ca. 360–230 Ma) that comprises just 5 grains (3%); a broad population from the Ediacaran to the Silurian (ca. 650–420 Ma; 72 grains [39%]), with two peaks at approxi-mately 460 Ma and 580 Ma. These zircon grains range from euhedral and prismatic with distinct oscillatory zoning to rounded grains with patchy and/or diffuse CL textures. Subsequent older zircon populations are largely comprised of rounded zircon grains with a range of CL textures (Figure 4); these include a middle Neoproterozoic population (ca. 670–800 Ma) of just 8 grains [4%]); a Stenian to Tonian Population (ca. 1200–900 Ma; 38 grains [21%]), with a significant peak at ~990 Ma, and a smaller peak at ~1120 Ma; minor populations extend back into the Mesoarchean, with peaks at ~1450, ~1630, and ~1890 Ma. The distribution of ages did not vary significantly between the zircon grains separated from the clasts and those from the matrix (Figure 5).

4.2. Major and trace elements

Major and trace element composition of clast samples are given in Table 3. Loss on ignition (LOI) values are generally moderate (2.0–3.5%), with the exception of sample 103262 at 8.2%. Geochemical compositions indicate the sampled rocks are alkaline, according to the classification of Irvine and Baragar (1971), and form a continuum with previous analyses of the Cloudy Bay Volcanics and the regional Fife Bay Volcanics (Figure 6(a); Smith, 1976). In comparison, representative analyses of the Northern Volcanic Belt of the Papuan Peninsula (see Figure 1; Smith, 1982) are distinct from the Cloudy Bay and Fife Bay Volcanics in that they are subalkaline in composition. Rock types show high variation and include basalt, trachybasalt, basaltic trachyandesite, and trachyandesite composi-tions (Figure 6; Le Maitre, 2002). The majority of these analyses also belong to the shoshonite series (Figure 6(b)) and are predominantly distributed across both the absarokite and shoshonite composi-tional fields of Peccerillo and Taylor (1976).

Major element variation diagrams (Figure 7) have been used to show geochemical trends for clasts derived from the Cloudy Bay Volcanics, together with previous analyses of the Cloudy Bay Volcanics and the Fife Bay Volcanics (Smith, 1976). SiO_2 contents of samples from this study (nor-malized for volatile-free compositions) vary from 48 to 56 wt.% (45 to 55 wt.% as measured), and MgO vary from approximately 8.5 to 3 wt.%. This range is similar to previous analyses of the Cloudy Bay Volcanics (48 wt.% < SiO_2 < 61 wt.%; 1 wt.% < MgO < 9 wt.%) and generally reflect more evolved compositions when compared to the Fife Bay Volcanics (47 wt.% < SiO_2 < 56 wt.%; 3 wt.% < MgO < 14 wt.%). The major elements Fe_2O_3 and CaO, together with MnO exhibit a good positive correlation with MgO (or negative with SiO_2), whereas Al_2O_3, K_2O, and TiO_2 show negative correlations with MgO. A similar, but weaker, negative correlation is evident between Na_2O and MgO. The major element data presented here, together with previous results from the Cloudy Bay Volcanics and the Fife Bay Volcanics, appear to form a compositional continuum.

Table 2. Zircon U–Pb geochronology results														
Sample	Number of analyses	Miocene ages[a]	> Miocene ages[b]	Youngest age[a]	1σ Error	Concordia age	2σ Error	MSWD	Probability of fit	Weighted average	95% Confidence	MSWD	Probability of fit	N
Matrix	63	7	42	5.3	0.1	8.22[c]	0.2	1.5	0.19	8.18[c]	0.22	1.3	0.26	6
103264	6	4	1	6.0	0.1	5.99	0.3	0.07	0.93	6.09	0.13	0.2	0.89	4
103265	103	2	90	5.5	0.1									
103268a	36	11	18	5.7	0.1	5.77	0.2	0.56	0.73	5.92	0.15	2.2	0.043	7
103268b	19	17	0	5.4	0.1	5.71	0.1	1.2	0.28	5.82	0.06	1.5	0.14	12
103270	40	2	34	5.3	0.1									

[a]Ages passing 30% discordance cut-off.
[b]Ages greater than Miocene passing 20% discordance cut-off.
[c]Representative of an inherited magmatic age.

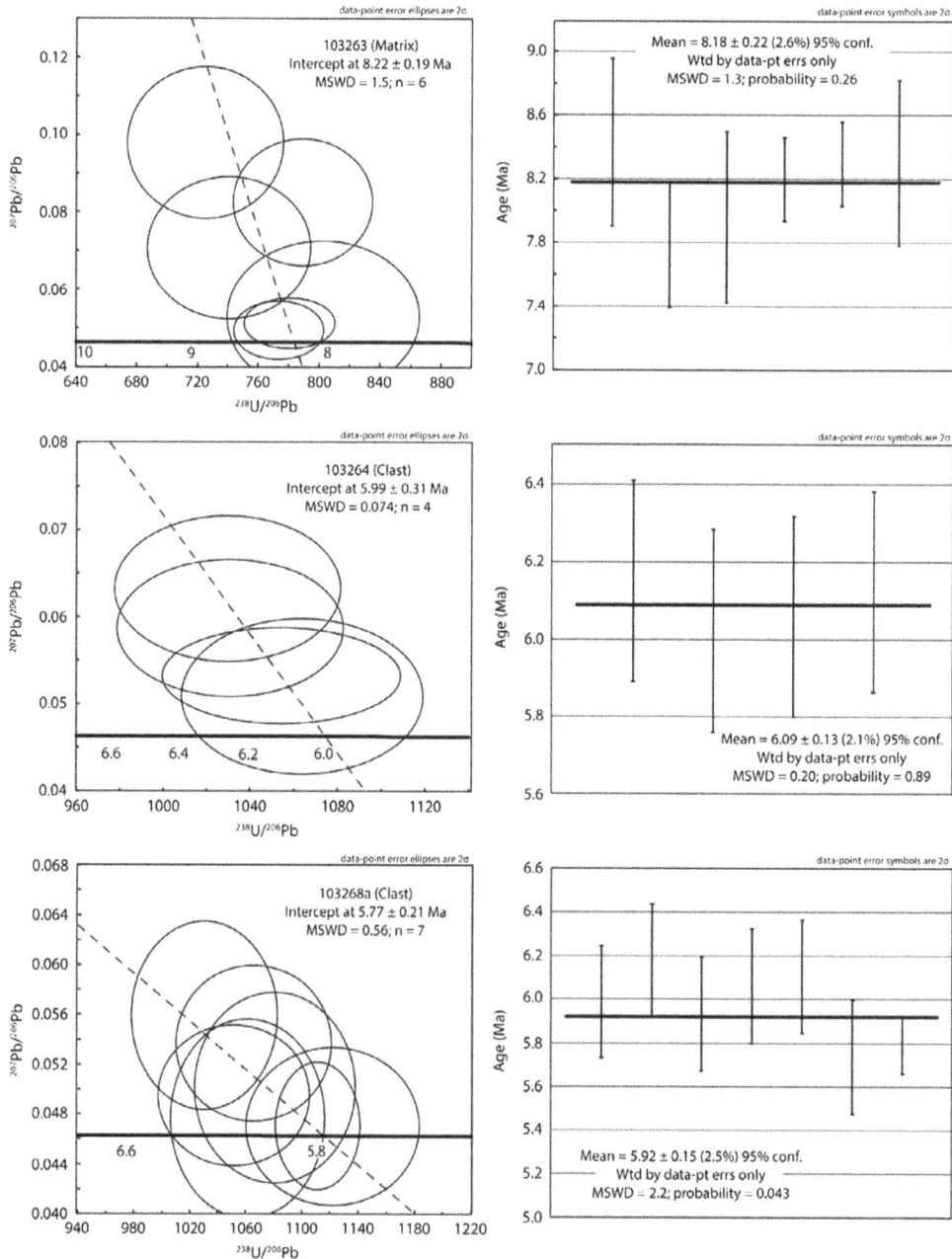

Figure 5. Zircon U–Pb dating results for the Cloudy Bay Volcanics. Tera–Wasserburg concordia and weighted average are constructed from zircon isotopic compositions and U–Pb calculated ages, respectively (detailed isotopic data in supplementary material). Tera–Wasserburg plots are corrected for initial Th disequilibrium; weighted average plots are corrected for initial Th disequilibrium and $^{207}Pb/^{206}Pb$ common Pb. All error bars, data point error ellipses and calculated errors are 2σ and 95% confidence for concordia and weighted averages, respectively. Probability–density histograms are shown for concordant grains, with Miocene ages <30% discordance and greater than >Miocene (pre-Cenozoic) ages <20% discordance.

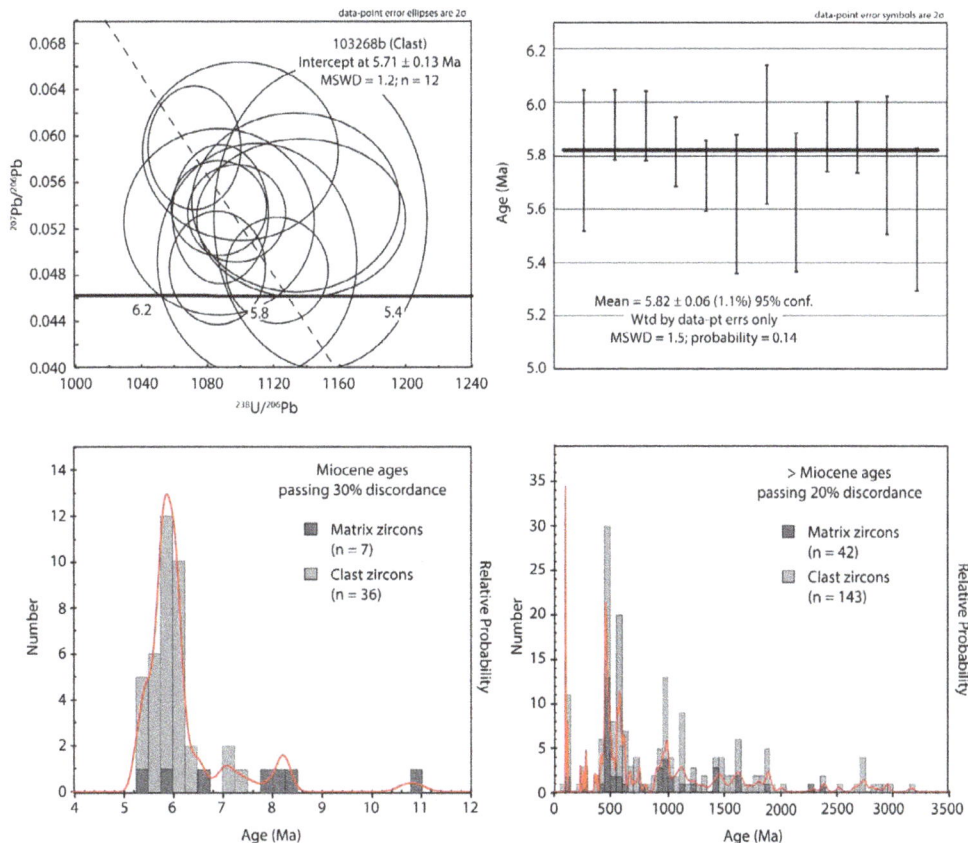

Figure 5. (*Continued*)

Trace element data are presented in Figure 8 by way of normalized multi-element plots and *X–Y* plots. All samples from the Cloudy Bay Volcanics exhibit subduction-related geochemical affinities (Figure 8(a)) with negative Nb, Ta, and Ti anomalies, and relative enrichments in large-ion lithophile elements (LILE), Th, U, Pb, and Sr. On comparison with previous analyses (Figure 8(b)), the results presented herein correlate well with prior results of the Cloudy Bay Volcanics from Smith (1976). Weathering and alteration has affected the samples to varying extents; we will therefore, not focus on the LILE (e.g. K, Rb, Cs) or Sr in detail as these elements are easily mobilized during alteration and will instead focus more on the less mobile HFSE (Nb, Ti, Zr, Hf), REE, and Th (e.g. Floyd & Winchester, 1978). All samples exhibit light REE-enriched patterns (Figure 8(c)) that are typical of evolved sub-duction-related magmas. The sample suite generally appears to form a compositional continuum with different degrees of light REE enrichment. This is supported by, for example, Figure 8(d) that demonstrates a positive correlation between the slope of the REE trend and Nb, and similarly Zr concentrations (Figure 8(f)). There are minor differences, however, where samples 103265, 103267, 103268c, and 103269 are generally characterized by lower light REE abundances in comparison to samples 103264, 103268a, 103268b, 103268d, and 103270. The latter suite of samples also exhibits minor negative Eu anomalies and REE trends more consistent with heavy REE depletion (Figure 8(c)). Differences in the two sample suites are also evident in plots of Y (Figure 8(e) and (g)), where two different correlation trends are apparent. That is, samples 103264, 103268a, 103268b, 103268d, and 103270 are marked by heavy REE-depleted trends, and exhibit a relative Y-depletion, compared to the remainder of the samples. These characteristics of variable Y contents also distinguish previous analyses of the Cloudy Bay Volcanics from the Fife Bay Volcanics (Figure 8(g)).

Table 3. Representative major and trace element data

Sample	103264	103265	103267	103268a	103268b	103268c	103268d	103269	103270
SiO_2	51.39	45.01	47.19	52.31	52.66	48.15	51.37	54.59	52.21
Al_2O_3	14.76	14.45	13.9	16.48	15.83	15.25	15.56	17.19	14.31
Fe_2O_3	8.5	11.1	9.87	8.97	8.92	10.5	9.37	7.42	8.34
MgO	4.83	7.34	8.35	2.97	3.23	5.62	3.94	2.86	4.63
CaO	8.11	8.51	10.56	5.3	5.16	10.1	6.74	5.8	7.93
Na_2O	3.45	1.44	2.05	2.76	2.25	3.35	3.49	3.77	3.99
K_2O	3.04	1.17	3.66	5.11	6.66	1.09	2.99	4.14	2.95
TiO_2	1.2	1.02	0.92	1.26	1.21	1.1	1.3	0.74	1.16
P_2O_5	0.93	0.91	0.76	0.91	0.86	0.77	1.01	0.54	0.89
MnO	0.15	0.21	0.16	0.11	0.13	0.15	0.13	0.08	0.15
Cr_2O_3	0.01	0.068	0.038	0.019	0.017	0.004	0.01	0.013	0.009
LOI	2.9	8.2	2	3.3	2.6	3.4	3.4	2.5	2.8
Total	99.54	99.7	99.59	99.68	99.73	99.7	99.55	99.77	99.55
Sc	21	32	27	22	21	28	22	15	20
V	229	253	263	239	107	309	198	149	223
Co	22.6	45.5	36	28.7	23.2	31.7	23.7	17.2	22.2
Ni	59	107.3	61.5	67.4	19.2	58.9	48.8	28.6	32.4
Cu	78.6	116.5	120.3	75.1	28.1	110.2	53.9	51.2	72.6
Zn	80	51	66	80	20	75	51	37	75
Ga	18.5	14.9	15.4	20.6	18.8	17.1	19.7	17	18.3
As	1.3	1	1	1	6.2	0.5	1.5	2.1	1.8
Rb	35.7	34	324.5	184.4	273.1	81.7	36.3	197.5	64.9
Sr	1855	680.5	1476	1288.7	1158.4	900	1905.4	927.5	1890.3
Y	28.8	20.7	14.3	19.3	16.7	19.6	23.1	24.1	27.5
Zr	399.9	85.7	81.9	224.9	212.2	89.4	407.3	133.7	384.9
Nb	12.7	3	3.1	7.8	7.1	3.4	13.8	4.5	12.6
Mo	0.2	0.3	0.7	0.5	0.3	1.1	0.3	0.8	0.3
Sn	6	2	1	3	4	1	6	1	6
Cs	29.2	1	7.6	3.1	1.7	57.7	1.2	3.4	63.4
Ba	1950	2018	780	1681	1560	2511	1961	918	2032
La	50.8	16.7	13.8	34.6	31.6	11.6	49.7	19.7	49.6
Ce	107	36.3	27.8	71.7	70	24.4	103.2	34.4	104.5

(Continued)

Petrology and crustal inheritance of the Cloudy Bay Volcanics as derived...

Table 3. (Continued)

Sample	103264	103265	103267	103268a	103268b	103268c	103268d	103269	103270
Pr	14.26	5.07	3.94	9.78	8.8	3.83	13.64	5.61	14.09
Nd	58.2	22.5	17.8	39.8	34.9	17.7	53.6	22.9	57.3
Sm	11.35	4.84	3.74	7.88	6.97	4.08	9.99	4.87	10.89
Eu	2.8	1.57	1.23	2.06	1.78	1.36	2.57	1.57	2.76
Gd	9.22	4.89	3.47	6.5	5.71	4.16	7.89	4.96	8.83
Tb	1.16	0.68	0.48	0.83	0.73	0.63	1.03	0.69	1.15
Dy	5.93	4.08	2.8	4.24	3.64	3.61	4.81	3.9	5.61
Ho	1	0.77	0.5	0.71	0.64	0.73	0.85	0.82	0.98
Er	2.59	2.14	1.39	1.87	1.62	2.02	2.14	2.26	2.64
Tm	0.35	0.31	0.19	0.25	0.21	0.28	0.29	0.3	0.33
Yb	2.11	1.87	1.13	1.5	1.27	1.81	1.73	1.93	2.07
Lu	0.32	0.28	0.17	0.23	0.19	0.29	0.25	0.3	0.33
Hf	11.2	2.6	2.4	6.1	5.9	2.6	11.5	3.5	10.8
Ta	0.7	0.2	0.2	0.4	0.5	0.3	0.9	0.3	0.8
W	0.9	<0.5	0.6	0.6	0.8	<0.5	2	2.6	1.5
Tl	0.2	0.3	0.6	0.2	<0.1	0.2	<0.1	<0.1	0.5
Pb	2.9	5.6	12	6.5	5.7	2	6.6	5.8	4.2
Th	24.8	4.3	4.1	16	14.9	2.6	26.3	6.9	25.3
U	7.2	0.8	1.6	4.6	3.4	1	7.7	2.4	7.3

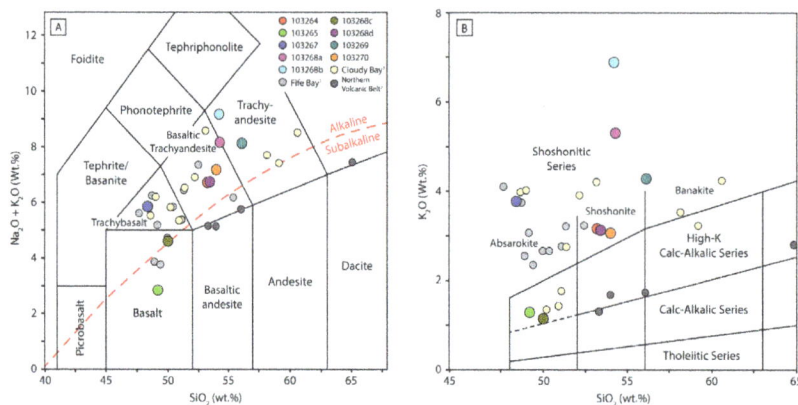

Figure 6. (A) Total alkali versus silica classification diagram (TAS; Le Maitre, 2002) with alkaline–subalkaline curve of Irvine and Baragar (1971), and **(B)** K₂O vs. SiO₂ diagram of Peccerillo and Taylor (1976). Additional data from previous analyses of the Cloudy Bay Volcanics, Fife Bay Volcanics, and the Northern Volcanic Belt is from ¹Smith (1976) and ²Smith (1982). Data is normalized and plotted on a volatile-free basis.

4.3. Lu–Hf isotopes

Selected zircons from samples 103263 (matrix), 103264, 103268a, and 103268b were analyzed for Lu–Hf isotopic ratios. The zircon crystals selected for this additional analysis were chosen to provide a representative range of ages across the sample suite as established by U–Pb dating. The results are reported in Table 4 and illustrated in Figure 8(h). All εHf values fall within a range between +11.1 and +6.9. However, there are distinct ranges for εHf values that correlate with the zircon U–Pb ages. Although only a single analysis, the oldest zircon grain at ca. 10.8 Ma yielded the second most mantle-like εHf value of +10.7; a second cluster comprising three analyses ranging between 8.3 and 8.0 Ma in age yielded an average εHf value of +10.2. The majority of analyzed zircons fall within the age range of 6.4–5.5 Ma and provided a range of εHf values between +9.8 and +6.9, with a corresponding average εHf value of +7.9.

5. Discussion

5.1. Petrology of the Cloudy Bay volcanics derived from a conglomerate

Up to nine variations of volcanic rocks were identified as clasts derived from a secondary conglomerate deposit and interpreted as an erosional product of the Cloudy Bay Volcanics. The volcanic rocks exhibit variable textures ranging from volcanic glass to porphyritic lavas and represent a compositional continuum from basalt to trachyandesite. The geochemical results derived from the conglomerate clasts outlined above are consistent with previous studies of regional late Miocene to Recent arc-type volcanic activity, which comprises a variety of high-K calc-alkaline rocks and volcanic–plutonic shoshonite suites (Jakeš & Smith, 1970; Smith, 1972, 1982, 2013b; Smith & Milsom, 1984).

The geochemical investigations of the Cloudy Bay Volcanics presented herein support previous interpretations that the volcanics are derived from partial melting of subduction-modified mantle in a volcanic arc setting (Figure 8). As the context of the outcrop is not conclusively constrained within the Cloudy Bay Volcanics and does suffer from a moderate degree of weathering and alteration, we will not focus here on detailed petrogenesis of the volcanics. Instead, we will emphasize the geochemical indicators relevant to wider scale studies of tectonics and crustal processes. We find that the sample suite generally reflects a compositional continuum, but with minor distinctions that suggest the potential for different magma evolution pathways. We interpret that the volcanic rock clasts can be divided into two sample suites on the basis of REE trends correlated with differential Y contents. We refer to samples 103265, 103267, 103268c, and 103269 as Suite 1, comprising basalts, a trachybasalt, and one trachyandesite. Suite 2 includes samples 103264, 103268a, 103268b, 103268d, and 103270, and comprises basaltic trachyandesites and trachyandesites. Although subtle, differences in REE trends of Suite 2 rocks, in comparison with Suite 1, provide evidence of an

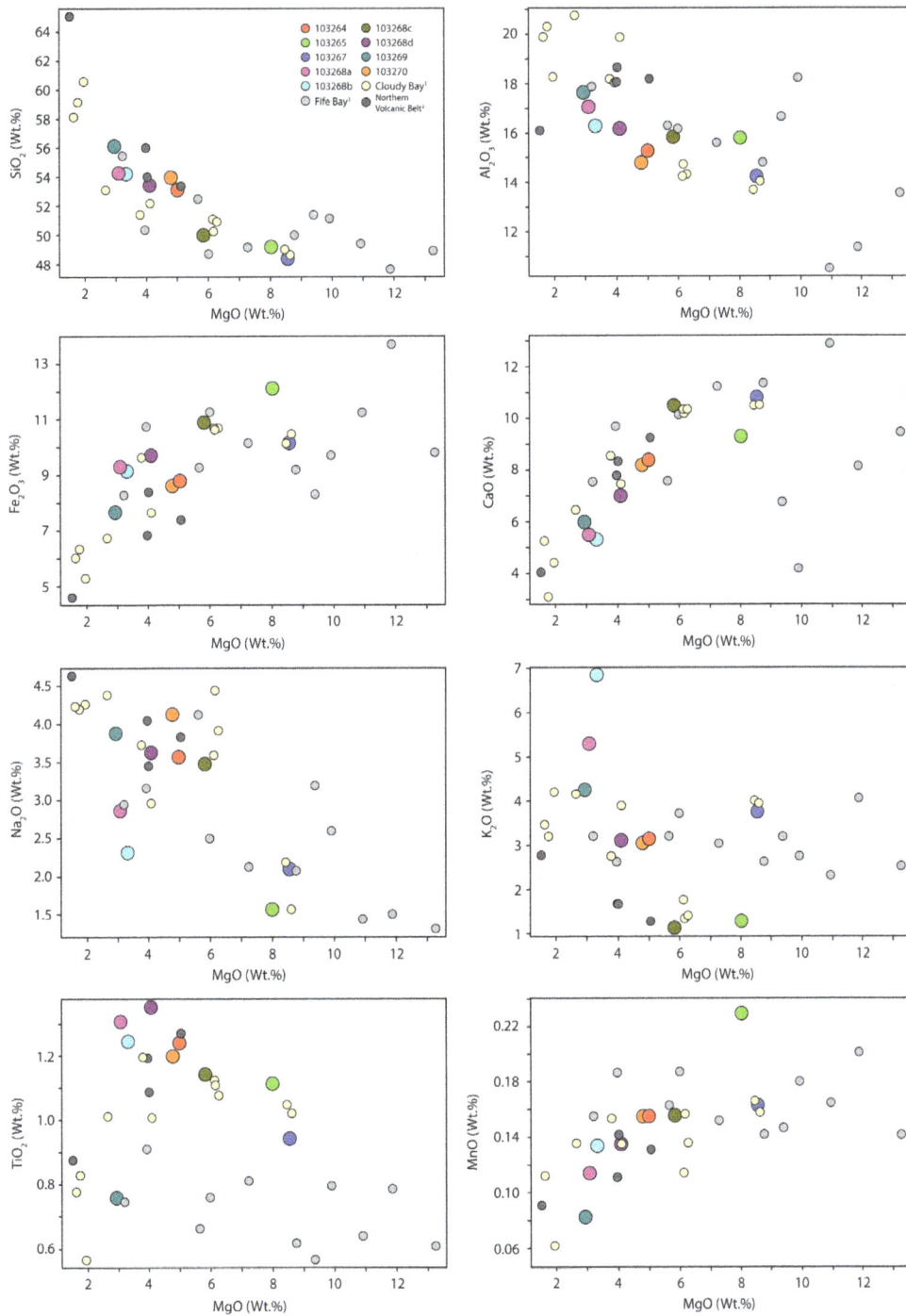

Figure 7 Major element variation diagrams. Additional data from previous analyses of the Cloudy Bay Volcanics, Fife Bay Volcanics, and the Northern Volcanic Belt is from [1]Smith (1976) and [2]Smith (1982). Data are normalized and plotted on a volatile-free basis.

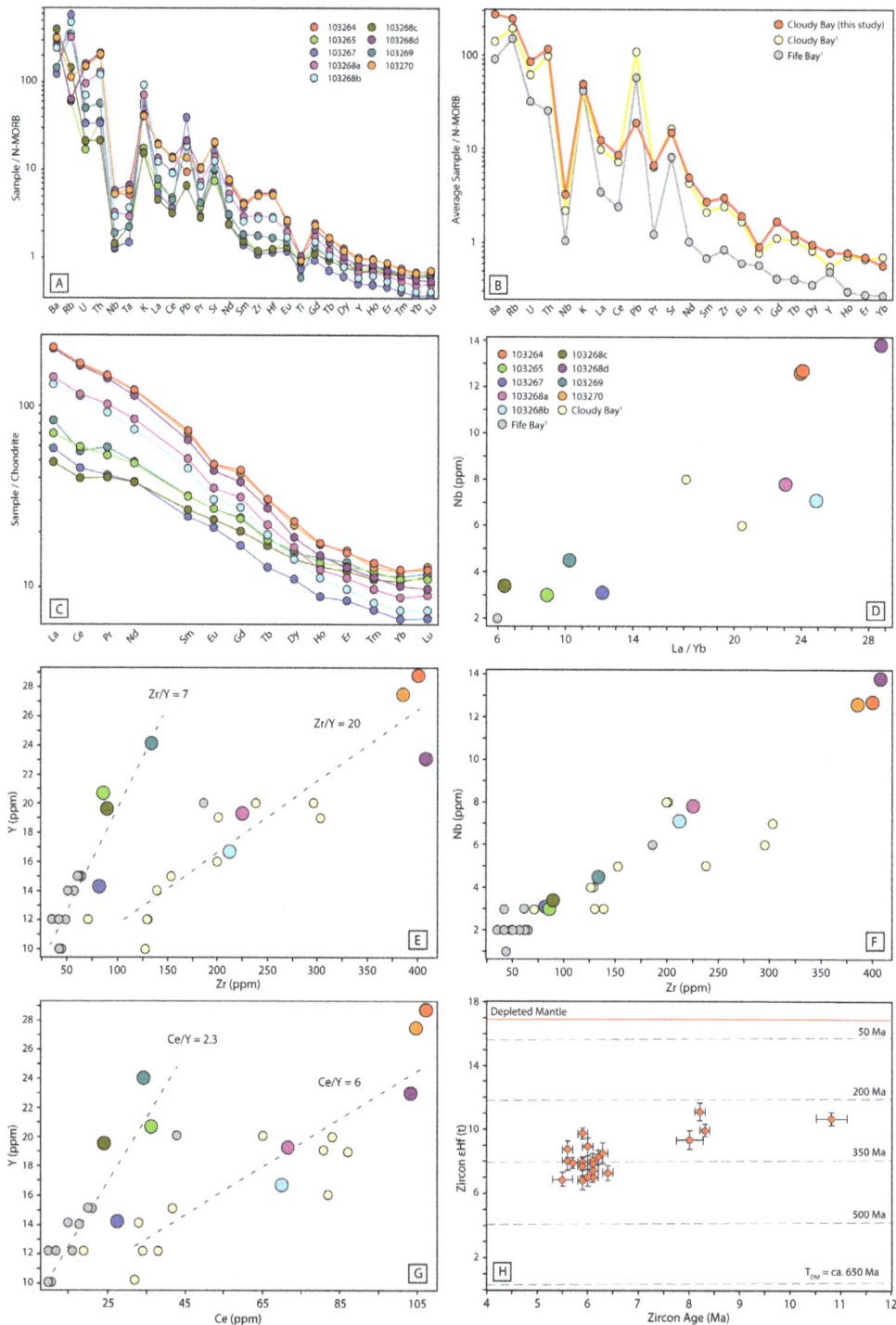

Figure 8. (A) N-MORB normalized multi-element plot; (B) N-MORB normalized multi-element plot for average compositions for samples from this study, and previous analyses from Cloudy Bay and Fife Bay ([1]Smith, 1976); (C) C1 chondrite normalized REE plots for the Cloudy Bay Volcanics. N-MORB and C1 chondrite normalizations are from Sun and McDonough (1989). (D)–(G) Trace element X–Y scatter plots, and (H) Zircon εHf (t) values for zircons of the Cloudy Bay Volcanics; crosses are 1σ errors.

Table 4. Lu–Hf isotope data.

Sample	Spot ID	U–Pb Spot Age (Ma)	±1σ	$^{176}Hf/^{177}Hf$	±1σ	$^{176}Lu/^{177}Hf$	±1σ	εHf (t = age)	±1σ
103263	21	10.8	0.3	0.283081	0.000012	0.001971	0.000007	10.7	0.4
	23	8.3	0.1	0.283062	0.000011	0.001206	0.000010	10.0	0.4
	27	8.2	0.1	0.253095	0.000016	0.006811	0.000091	11.1	0.6
	31	8.0	0.3	0.283045	0.000016	0.002608	0.000074	9.4	0.6
103264	2	6.1	0.1	0.282991	0.000014	0.001816	0.000014	7.4	0.5
	5	6.1	0.1	0.283010	0.000012	0.001316	0.000025	8.1	0.4
103268a	1	6.0	0.1	0.282979	0.000015	0.001840	0.000034	7.0	0.5
	15	6.2	0.1	0.283015	0.000013	0.001303	0.000029	8.3	0.5
	19	5.9	0.1	0.283000	0.000017	0.001765	0.000025	7.7	0.6
	20	6.1	0.1	0.282980	0.000009	0.001678	0.000016	7.0	0.3
	27	6.1	0.1	0.283003	0.000008	0.001458	0.000022	7.8	0.3
	33	6.4	0.1	0.282987	0.000013	0.001780	0.000015	7.3	0.4
	36	6.3	0.1	0.283022	0.000019	0.002961	0.000064	8.5	0.7
103268b	2	5.9	0.1	0.283007	0.000009	0.001611	0.000014	8.0	0.3
	3	5.9	0.1	0.283058	0.000010	0.001513	0.000014	9.8	0.3
	5	5.7	0.1	0.283005	0.000010	0.001193	0.000008	7.9	0.3
	6	6.0	0.1	0.283035	0.000015	0.001620	0.000007	9.0	0.5
	7	5.6	0.1	0.283010	0.000017	0.001693	0.000019	8.1	0.6
	9	5.6	0.1	0.283030	0.000014	0.001944	0.000030	8.8	0.5
	10	5.9	0.1	0.283075	0.000018	0.001771	0.000030	6.9	0.6
	18	5.5	0.2	0.283077	0.000012	0.001517	0.000074	6.9	0.4

increased level of fractionation, with more pronounced light REE enrichment and negative Eu anomalies. These characteristics are attributed to plagioclase fractionation, and typical of high-K calc-alkaline magmas (Gill, 1981). Samples belonging to Suite 2 also exhibit heavy REE- and Y-depletion, which are both typically linked to magma that has undergone fractionation of garnet at high pressures in the upper mantle or lower crust (Chiaradia, Merino, & Spikings, 2009; Macpherson, Dreher, & Thirlwall, 2006). We also observe that in general, Suite 2 samples are also more fractionated in that they are marked by higher relative SiO_2, K_2O, and TiO_2 concentrations compared with Suite 1 samples (Figure 7), and lower MgO, FeO, and CaO concentrations. We therefore interpret that the sample suite represents a geochemical compositional continuum of shoshonitic volcanics, but that different magma evolution pathways are apparent within the sample suite. That is, Suite 2 volcanic rocks have undergone additional magma fractionation under high-pressure conditions.

Zircon U–Pb dating results from the clast samples indicate that activity of the Cloudy Bay Volcanics was largely constrained to the latest Miocene–earliest Pliocene, at between ca. 7 and 5 Ma in age. Zircon dating also provides evidence for magmatism in the area as old as 11 Ma, but it is not clear if this is related to an earlier phase of volcanic activity or is inherited and belongs to a separate suite of magmatism. The Cloudy Bay Volcanics were previously inferred to be of equivalent age to the Fife Bay Volcanics (Figure 1), dated at 12.6 Ma (Smith & Davies, 1976; Smith & Milsom, 1984); high-K basalts of Woodlark Island, dated at 11.2 Ma (Ashley & Flood, 1981; Smith & Milsom, 1984); and plutons and dykes swarms that intrude the eastern Milne Basic Complex, and dated at 16–12 Ma (Smith, 1972; Smith & Milsom, 1984). Given the revised age for the Cloudy Bay Volcanics, this volcanic activity should now be associated with a later generation of regional volcanism, namely andesites and shoshonites north of, and adjacent to the D'Entrecasteaux islands, dated at 5.5 Ma, and a granite intrusive complex at Mt Suckling dated at 6.3 Ma (Smith & Davies, 1976; Smith & Milsom, 1984).

This study also presents the first Lu–Hf isotope results from the Papuan Peninsula. The Hf isotope results are derived from zircons belonging to volcanic rocks of Suite 2 and the outcrop matrix, and as a result we cannot draw any conclusions about differences between the two geochemical suites. Instead, we interpret these results within the context of the regional tectonics. The Hf results display a tight range of εHf values between +9.8 and +6.9 for the majority of the sampled zircons indicating a relatively homogonous magma composition (Figure 8(h)). The εHf values are much less positve than would be anticipated from Miocene–Pliocene mantle-derived basaltic melts, which is interpreted to reflect contamination of the magma by comparavtively unradiogenic crust. The corresponding depleted mantle model age for the same εHf values of ~200–400 Ma does provide comfirmation that older crustal material has contributed to the magma composition, but given the complex mixing and assimilation processes in arc magmatism we place little emphasis on the precise model ages. We do note that the apparent decrease in εHf values and associated model age (Figure 8(h)) over time may reflect either a changing source or an increase in the degree of assimilation and mixing with foreign crustal material.

Potential sources of crustal contamination will arise from either the melt source region in the mantle, or lithospheric crust through which the magma has migrated. Rocks of the Milne Terrane can provide a first pass proxy for the upper crustal material in the vicinity of Cloudy Bay. These are comprised largely of Upper Cretaceous N-MORB-type basaltic volcanics (Smith, 2013a); such primative rocks are unlikely to yield sufficiently unradiogenic crustal signatures to yield the Lu–Hf ratios reported herein, however, this requires further work. An alterative isotopic source that must be considered is the crust of the Eastern and Papuan Plateaus to the south of the Papuan Peninsula. The geology of the plateaus is not clear at present, however, the Queensland Plateau forming the congugate southern margin of the Coral Sea has been previously interpreted to form part of the New England Orogen of Eastern Australia (Mortimer, Hauff, & Calvert, 2008; Shaanan, Rosenbaum, Hoy, & Mortimer, 2018). Although far removed from the context of the present study, previous Hf isotope data from the southern New England Orogen has provided εHf values ranging from +3.4 to +11.3 (Kemp et al., 2009). While this is speculative at this stage we infer that underthrust portions of the

Eastern and Papuan Plateau beneath the Papuan Peninsula related to past convergent events do have the potential to contribute to magma genesis and could go someway to explaining the Lu–Hf isotope ratios of the Cloudy Bay Volcanics.

The newly established timing for activity of the Cloudy Bay Volcanics, the recognition of multiple magma evolution pathways with similar ages, and the interpretation for the involvement of comparatively unradiogenic crust are important contributions in the context of the regional tectonic evolution of Papua New Guinea. The middle Miocene to early Pliocene is a time of tectonic upheaval, marked by middle to late Miocene closure of the Pocklington trough and development of the New Guinea Orogen related to collision of the Australian continent (Cloos et al., 2005; Holm et al., 2015; Webb, Baldwin, & Fitzgerald, 2014), thickened arc crust (Drummond, Collins, & Gibson, 1979; Milsom & Smith, 1975), and interpreted lithospheric delamination beneath the Papuan Highlands from ca. 6 Ma (Cloos et al., 2005; Holm et al., 2015); delamination has similarly been interpreted beneath the Papuan Peninsula (Abers & Roecker, 1991; Eilon, Abers, Gaherty, & Jin, 2015). Early Pliocene regional tectonics are marked by the onset of rifting and sea floor spreading in the Woodlark Basin from at least 5 Ma (Holm, Rosenbaum, & Richards, 2016; Taylor, Goodliffe, & Martinez, 1999; Taylor, Goodliffe, Martinez, & Hey, 1995; Wallace et al., 2014), Pliocene subduction of the Solomon Sea plate at the Trobriand trough (Holm et al., 2016), and late Miocene to Recent formation of the Aure–Moresby fold–thrust belt (Ott & Mann, 2015). An in-depth interpretation of the tectonic context for activity of the Cloudy Bay Volcanics is beyond the scope of this study, however, ongoing work on the tectonic and magmatic evolution of the Papuan Peninsula will aim to provide greater insights into the dynamic regional setting.

5.2. Provenance of zircon xenocrysts
The majority of zircon grains recovered from the sampled clasts and matrix material are of pre-Cenozoic age (Figures 4 and 5) and represent xenocrystic zircon grains inherited from a foreign source. There are two possible mechanisms to explain the presence of inherited zircons within the Cloudy Bay Volcanics. The first appeals to assimilation of older crustal basement into the magma during migration and ascent through the crust. Alternatively, the grains may have been present at the Earth's surface, for example within fluvial systems, and subsequently incorporated into the rocks during subaerial volcanism. Sample petrography does not identify any notable component of xenoliths or xenocrysts that could be considered as surface material during eruption and cooling. We therefore suggest that the xenocrystic zircon grains were introduced to the magma via assimilation of crustal basement (e.g. Buys, Spandler, Holm, & Richards, 2014; Paquette & Le Pennec, 2012; Van Wyck & Williams, 2002).

The geological setting of the Papuan Peninsula at the present day, and similarly throughout the late Cenozoic, is bound by oceanic basins to the north and south (e.g. Hall, 2002; Holm et al., 2015, 2016; Schellart, Lister, & Toy, 2006), and precludes the introduction of zircons from distal crustal sources into the Papuan Peninsula setting leading up to the late Miocene volcanism. Instead, the presence of inherited or detrital zircons within the Cloudy Bay Volcanics requires assimilation of older basement crust (e.g. Buys et al., 2014; Paquette & Le Pennec, 2012; Van Wyck & Williams, 2002). As outlined above, the greater Papuan Peninsula is comprised or, or underlain by the Owen Stanley Metamorphic Complex and Milne Terrane (Davies, 2012; Smith, 2013a), the protolith of which is interpreted to be largely made up of sedimentary detritus derived from middle Cretaceous felsic magmatism (Davies, 2012) and Upper Cretaceous basaltic magmatism (Smith, 2013a). Although only a minor xenocrystic population, the youngest xenocrystic zircons are middle Cretaceous in age. The youngest grain, at ca. 100 Ma, correlates well with the interpreted Albian–Cenomanian depositional age for the protolith of the Owen Stanley Metamorphic Complex (Dow et al., 1974; Kopi et al., 2000). The rocks of the Owen Stanley Metamorphic Complex, therefore, represent the most likely source for the xenocrystic zircons within the Cloudy Bay Volcanics. The abraded and rounded morphology of the majority of pre-Cretaceous zircon grains, however, indicate they are detrital in origin and do not originate from Papuan Peninsula per se (Figure 4). And by proxy, evaluation of the xenocrystic zircons can be interpreted to reflect the detrital provenance of the Owen Stanley

Metamorphic Complex (e.g. Buys et al., 2014; Fergusson, Henderson, & Offler, 2017; Paquette & Le Pennec, 2012; Van Wyck & Williams, 2002).

The xenocryst/detrital zircon record presented here is comparable to established zircon records from the eastern highlands of Papua New Guinea (Van Wyck & Williams, 2002), and recent studies from offshore of northeast Australia from the Solomon Islands (Tapster, Roberts, Petterson, Saunders, & Naden, 2014), New Caledonia (Adams, Cluzel, & Griffin, 2009; Pirard & Spandler, 2017) and Vanuatu (Buys et al., 2014). Similar to these regional studies, the pre-Cretaceous zircon populations within the Cloudy Bay Volcanics are characterized by Ediacaran to Silurian (ca. 650–420 Ma) and Mesoproterozoic to early Neoproterozoic (~1600–900 Ma) populations. Drawing from previous interpretations, these detrital age signatures are consistent with detritus that was likely sourced from recycling of basins on the eastern margin of Gondwana. These sources include the Mossman, Thomson, New England, and Lachlan Orogens of eastern Australia, of which Ordovician and Mesoproterozoic zircons form a significant component (Fergusson et al., 2017; Henderson, 1986; Pell, Williams, & Chivas, 1997; Van Wyck & Williams, 2002), and also the Charters Towers Province, and the Georgetown and Coen Inliers of northern Queensland, where major crust-forming events occurred at ca 1.55 Ga, 420–400 Ma, and 300–284 Ma (Blewett & Black, 1998; Fergusson, Henderson, Fanning, & Withnall, 2007; Fergusson, Henderson, Withnall, & Fanning, 2007; Fergusson et al., 2017; Pell et al., 1997).

An important distinction can, however, be drawn between the established zircon provenance from northeast Australia and related South West Pacific terranes, and the Cloudy Bay Volcanics. Namely, this is the absence of a significant Carboniferous–Permian zircon population that is characteristic of the Kennedy Igneous Association (330–270 Ma; Champion & Bultitude, 2013), and the main phase of the Hunter-Bowen Orogeny in the New England Orogen (Korsch et al., 2009). Regional reconstructions (e.g. Schellart et al., 2006; Tapster et al., 2014; Whattam, Malpas, Ali, & Smith, 2008) imply that during the Cretaceous, and prior to the opening of the Coral Sea, the Papuan Peninsula was located adjacent to northeast Queensland forming a continuation of the extended continental crust. Given the importance of this regional crust-forming event and the prevalence of reworked zircons related this time period in other regional terranes (e.g. Kubor and Bena Bena Blocks, eastern Papuan Highlands [Van Wyck & Williams, 2002]; Espiritu Santo, Vanuatu (Buys et al., 2014); New Caledonia (Adams et al., 2009; Pirard & Spandler, 2017), it is unusual that zircons of this age only form a minor population (3%) of the total xenocrystic zircon yield. This implies that erosion of these Carboniferous–Triassic rocks did not contribute significant volumes of material to form the sedimentary protolith of the Owen Stanley Metamorphic Complex. And furthermore, that either a barrier was in place that prevented transport of sedimentary material to the Owen Stanley terrane prior to break-up of eastern Gondwana and opening of the Coral Sea, or that reconstructions of the Owen Stanley terrane are incorrect. These results are not yet conclusive and ongoing research is underway to expand the scope of investigations into the volcanics and provenance of the Papuan Peninsula, but it does suggest that regional provenance models and tectonic reconstructions for eastern Gondwana may require revision.

6. Conclusions
We present an investigation into the petrology of the Cloudy Bay Volcanics of the southeast Papuan Peninsula, Papua New Guinea. However, this study is distinct in that we sample clasts derived from an isolated conglomerate outcrop within a region comprising dense lowland tropical forest with regional swamp forest and mangroves. Tropical regions such as this are characterized by enhanced weathering and erosion, where our ability to examine the geological record can become compromised by the absence of informative *in situ* geological outcrops. Our findings suggest that the Cloudy Bay Volcanics were largely active during the latest Miocene, between 7 and 5 Ma. The shoshonitic volcanics largely form a geochemical continuum in overall composition but form two distinct suites reflecting different magma evolution pathways. Evidence for this arises from disparate REE trends, with an anomalous volcanic suite marked by heavy REE- and Y-depletion indicative of high-pressure magma fractionation. Inclusion of a considerable number of xenocrystic zircons within the Cloudy

Bay Volcanics provide additional insights into the nature of the Owen Stanley Metamorphic Complex and Milne Terrane, which form the basement of the Papuan Peninsula, and suggests an eastern Gondwana provenance for detrital zircons. The results of this study illustrate that in such areas, characterized by incomplete data-sets and difficulties in reconciling the regional geological history, we can still gain insight into the geology by sampling of secondary deposits.

Acknowledgments

John Wardell and Kelly Heilbronn are thanked for assisting with rock processing and U–Pb analyses, respectively; Yi Hu and staff of the Advanced Analytical Centre (JCU) provided support with analytical work; we also thank Stephanie Mrozek, Chris Harris, and an anonymous reviewer for helpful comments and suggestions on the manuscript.34th International Geological Congress Travel GrantEarly Career Researcher Rising Stars Leadership Program34th International Geological Congress Travel Grant

Funding

Funding for field work was provided by the 34th International Geological Congress Travel Grant Scheme for Early-Career Australian and New Zealand Geoscientists from the Australian Geoscience Council and the Australian Academy of Science. Dulcie Saroa, Nathan Mosusu, and staff of the Mineral Resources Authority are thanked for their assistance with fieldwork and logistics. The James Cook University Early Career Researcher Rising Stars Leadership Program grant provided funding for analytical work.

Author details

Robert J. Holm[1,2]
E-mail: rholm@frogtech.com.au
ORCID ID: http://orcid.org/0000-0001-5470-2612
Benny Poke[3]
E-mail: bpoke@mra.gov.pg
[1] Frogtech Geoscience, 2 King Street, Deakin West ACT 2600, Australia.
[2] Geosciences, College of Science & Engineering, James Cook University, Townsville, Queensland 4811, Australia.
[3] Geological Survey Division, Mineral Resources Authority, PO Box 1906, Port Moresby 121, Papua New Guinea.

References

Abers, G. A., & Roecker, S. W. (1991). Deep structure of an arc-continent collision: Earthquake relocation and inversion for upper mantle P and S wave velocities beneath Papua New Guinea. *Journal of Geophysical Research: Solid Earth, 96*, 6379–6401. https://doi.org/10.1029/91JB00145

Adams, C. J., Cluzel, D., & Griffin, W. L. (2009). Detrital zircon ages and geochemistry of sedimentary rocks in basement Mesozoic terranes and their cover rocks in New Caledonia, and provenances at the Eastern Gondwanaland margin. *Australian Journal of Earth Sciences, 56*, 1023–1047. https://doi.org/10.1080/08120090903246162

Ahrens, L. H., Cherry, R. D., & Erlank, A. J. (1967). Observations on the Th-U relationship in zircons from granitic rocks and from kimberlites. *Geochemica et Cosmochimica Acta, 29*, 711–716.

Amante, C., & Eakins, B. W. (2009). *ETOPO1 1 arc-minute global relief model: Procedures, data sources and analysis.* Silver Spring: NOAA Technical Memorandum NESDIS NGDC-24. National Geophysical Data Center, NOAA.

Ashley, P. M., & Flood, R. H. (1981). Low-K tholeiites and high-K igneous rocks from Woodlark Island, Papua New Guinea. *Journal of the Geological Society of Australia, 28*, 227–240. https://doi.org/10.1080/00167618108729158

Australian Bureau of Mineral Resources. (1972). *Geology of Papua New Guinea, 1:1,000,000 map.* Australian Bureau of Mineral Resources.

Baldwin, S. L., Fitzgerald, P. G., & Webb, L. E. (2012). Tectonics of the New Guinea Region. *Annual Review of Earth and Planetary Sciences, 40*, 495–520. https://doi.org/10.1146/annurev-earth-040809-152540

Black, L. P., Kamo, S. L., Allen, C. M., Davis, D. W., Aleinikoff, J. N., Valley, J. W., ... Foudoulis, C. (2004). Improved [206]Pb/[238]U microprobe geochronology by the monitoring of trace-element-related matrix effect; SHRIMP, ID-TIMS, ELA-ICP-MS and oxygen isotope documentation for a series of zircon standards. *Chemical Geology, 205*, 115–140. https://doi.org/10.1016/j.chemgeo.2004.01.003

Blewett, R. S., & Black, L. P. (1998). Structural and temporal framework of the Coen Region, north Queensland: Implications for major tectonothermal events in east and north Australia. *Australian Journal of Earth Sciences, 45*, 597–609. https://doi.org/10.1080/08120099808728415

Buys, J., Spandler, C., Holm, R. J., & Richards, S. W. (2014). Remnants of ancient Australia in Vanuatu: Implications for crustal evolution in island arcs and tectonic development of the southwest Pacific. *Geology, 42*, 939–942. https://doi.org/10.1130/G36155.1

Champion, D. C., & Bultitude, R. J. (2013). Kennedy igneous association. In P. A. Jell (Ed.), *Geology of Queensland* (pp. 473–514). Brisbane: Geological Survey of Queensland.

Chiaradia, M., Merino, D., & Spikings, R. (2009). Rapid transition to long-lived deep crustal magmatic maturation and the formation of giant porphyry-related mineralization (Yanacocha, Peru). *Earth and Planetary Science Letters, 288*, 505–515. https://doi.org/10.1016/j.epsl.2009.10.012

Cloos, M., Sapiie, B., van Ufford, A. Q., Weiland, R. J., Warren, P. Q., & McMahon, T. P. (2005). *Collisional delamination in New Guinea: The geotectonics of subducting slab breakoff* (Special Paper, p. 400). Boulder: Geological Society of America.

Corfu, F., Hanchar, J. M., Hoskin, P. W. O., & Kinny, P. (2003). Atlas of zircon textures. In J. M. Hanchar & P. W. O. Hoskin (Eds.), *Reviews in mineralogy & geochemistry 53: Zircon.* Chantilly: Mineralogical Society of America.

Crowley, J. L., Schoene, B., & Bowring, S. A. (2007). U-Pb dating of zircon in the Bishop Tuff at the millennial scale. *Geology, 35*, 1123–1126. https://doi.org/10.1130/G24017A.1

Davies, H. L. (2012). The geology of New Guinea—The cordilleran margin of the Australian continent. *Episodes, 35*, 87–102.

Davies, H. L., & Jaques, A. L. (1984). Emplacement of ophiolite in Papua New Guinea. *Geological Society, London, Special Publications, 13*, 341–349.

Davies, H. L., & Smith, I. E. (1971). Geology of Eastern Papua. *Geological Society of America Bulletin, 82*, 3299–3312. https://doi.org/10.1130/0016-7606(1971)82[3299:GOEP]2.0.CO;2

Dow, D. B., Smit, J. A. J., Page, R. W. (1974). *Wau–1:250,000 Geological Series. Explanatory notes to accompany Wau 1:250,000 geological map: Geological Survey of Papua New Guinea.* Explanatory Notes SB/55-14.

Drummond, B. J., Collins, C. D. N., & Gibson, G. (1979). The crustal structure of the Gulf of Papua and north-west Coral Sea. *BMR Australian Journal of Geology and Geophysics, 4*, 341–351.

Eilon, Z., Abers, G. A., Gaherty, J. B., & Jin, G. (2015). Imaging continental breakup using telesiesmic body waves: The Woodlark Rift, Papua New Guinea. *Geochemistry, Geophysics, Geosystems, 16*, 2529–2548. https://doi.org/10.1002/2015GC005835

Fergusson, C. L., Henderson, R. A., Fanning, C. M., & Withnall, I. W. (2007). Detrital zircon ages in Neoproterozoic to Ordovician siliciclastic rocks, northern Australia: Implications for the tectonic history of the East Gondwana continental margin. *Journal of the Geological Society, 164*, 215–225. https://doi.org/10.1144/0016-76492005-136

Fergusson, C. L., Henderson, R. A., & Offler, R. (2017). Late Neoproterozoic to early Mesozoic sedimentary rocks of the Tasmanides, eastern Australia: Provenance switching associated with development of the East Gondwana active margin. In R. Mazumder (Ed.), *Sediment Provenance* (pp. 325–369). Netherlands: Elsevier. https://doi.org/10.1016/B978-0-12-803386-9.00013-7

Fergusson, C. L., Henderson, R. A., Withnall, I. W., & Fanning, C. M. (2007). Structural history of the Greenvale Province, north Queensland: Early Palaeozoic extension and convergence on the Pacific margin of Gondwana. *Australian Journal of Earth Sciences, 54*, 573–595. https://doi.org/10.1080/08120090701188970

Floyd, P. A., & Winchester, J. A. (1978). Identification and discrimination of altered and metamorphosed volcanic rocks using immobile elements. *Chemical Geology, 21*, 291–306. https://doi.org/10.1016/0009-2541(78)90050-5

Gehrels, G. (2014). Detrital zircon U-Pb geochronology applied to tectonics. *Annual Review of Earth and Planetary Sciences, 42*, 127–149. https://doi.org/10.1146/annurev-earth-050212-124012

Gill, J. B. (1981). *Orogenic Andesites and plate tectonics.* Berlin: Springer-Varlag. https://doi.org/10.1007/978-3-642-68012-0

Graham, I. J., & Korsch, R. J. (1990). Age and provenance of granitoid clasts in Moeatoa Conglomerate, Kawhia Syncline, New Zealand. *Journal of the Royal Society of New Zealand, 20*, 25–39. https://doi.org/10.1080/03036758.1990.10426731

Hall, R. (2002). Cenozoic geological and plate tectonic evolution of SE Asia and the SW Pacific: Computer-based reconstructions, model and animations. *Journal of Asian Earth Sciences, 20*, 353–431. https://doi.org/10.1016/S1367-9120(01)00069-4

Heaman, L. M., Bowins, R., & Crocket, J. (1990). The chemical composition of igneous zircon suites: Implications for geochemical tracer studies. *Geochimica et Cosmochimica Acta, 54*, 1597–1607. https://doi.org/10.1016/0016-7037(90)90394-Z

Henderson, R. A. (1986). Geology of the Mt Windsor subprovince—A lower Palaeozoic volcano-sedimentary terrane in the northern Tasman orogenic zone. *Australian Journal of Earth Sciences, 33*, 343–364. https://doi.org/10.1080/08120098608729371

Hidaka, H., Shimizu, H., & Adachi, M. (2002). U-Pb geochronology and REE geochemistry of zircons from Palaeoproterozoic paragneiss clasts in the Mesozoic Kamiaso conglomerate, central Japan: Evidence for an Archean provenance. *Chemical Geology, 187*, 279–293. https://doi.org/10.1016/S0009-2541(02)00058-X

Holm, R. J., Spandler, C., & Richards, S. W. (2013). Melanesian arc far-field response to collision of the Ontong Java Plateau: Geochronology and petrogenesis of the Simuku Igneous Complex, New Britain, Papua New Guinea. *Tectonophysics, 603*, 189–212. https://doi.org/10.1016/j.tecto.2013.05.029

Holm, R. J., Spandler, C., & Richards, S. W. (2015). Continental collision, orogenesis and arc magmatism of the Miocene Maramuni arc, Papua New Guinea. *Gondwana Research, 28*, 1117–1136. https://doi.org/10.1016/j.gr.2014.09.011

Holm, R. J., Rosenbaum, G., & Richards, S. W. (2016). Post 8 Ma reconstruction of Papua New Guinea and Solomon Islands: Microplate tectonics in a convergent plate boundary setting. *Earth-Science Reviews, 156*, 66–81. https://doi.org/10.1016/j.earscirev.2016.03.005

Hoskin, P. W. O., & Schaltegger, U. (2003). The composition of zircon and igneous and metamorphic petrogenesis. In J. M. Hanchar & P. W. O. Hoskin (Eds.), *Reviews in Mineralogy & Geochemistry 53: Zircon.* Chantilly: Mineralogical Society of America.

Irvine, T. N., & Baragar, W. R. A. (1971). A guide to the chemical classification of the common volcanic rocks. *Canadian Journal of Earth Sciences, 8*, 523–548. https://doi.org/10.1139/e71-055

Jackson, S. E., Pearson, N. J., Griffin, W. L., & Belousova, E. E. (2004). The application of laser ablation-inductively coupled plasma-mass spectrometry to *in situ* U-Pb zircon geochronology. *Chemical Geology, 211*, 47–69. https://doi.org/10.1016/j.chemgeo.2004.06.017

Jakeš, P., & Smith, I. E. M. (1970). High potassium calc-alkaline rocks from Cape Nelson, eastern Papua. *Contributions to Mineralogy and Petrology, 28*, 259–271. https://doi.org/10.1007/BF00388948

Kemp, A. I. S., Hawkesworth, C. J., Collins, W. J., Gray, C. M., Blevin, P. L., & EIMF. (2009). Isotopic evidence for rapid continental growth in an extensional accretionary orogen: The Tasmanides, eastern Australia. *Earth and Planetary Science Letters, 284*, 455–466. https://doi.org/10.1016/j.epsl.2009.05.011

Kennedy, A. K., Wotzlaw, J.-F., Schaltegger, U., Crowley, J. L., & Schmitz, M. (2014). Eocene zircon reference material for microanalysis of U-Th-Pb isotopes and trace elements. *The Canadian Mineralogist, 52*, 409–421. https://doi.org/10.3749/canmin.52.3.409

Kopi, G., Findlay, R. H., & Williams, I. (2000). *Age and provenance of the Owen Stanley Metamorphic Complex. East Papuan Composite Terrane, Papua New Guinea:* Geological Survey of Papua New Guinea, Report. (unpublished).

Korsch, R. J., Adams, C. J., Black, L. P., Foster, D. A., Murray, C. G., Foudoulis, C., & Griffin, W. L. (2009). Geochronology and provenance of the Late Paleozoic accretionary wedge and Gympie Terrane, New England Orogen, eastern Australia. *Australian Journal of Earth Sciences, 56*, 655–685. https://doi.org/10.1080/08120090902825776

Lamminen, J., Andersen, T., & Nystuen, J. P. (2015). Provenance and rift basin architecture of the Neoproterozoic Hedmark Basin, South Norway inferred from U-Pb ages and Lu-Hf isotopes of conglomerate clasts and detrital zircons. *Geological Magazine, 152*, 80–105. https://doi.org/10.1017/S0016756814000144

Le Maitre, R. W. (Ed.). (2002). *Igneous rocks: A classification and glossary of terms* (2nd ed., pp. 236). Cambridge: Cambridge University Press.

Ludwig, K. R. (2009). *User's manual for isoplot 3.70: A geochronological toolkit for microsoft excel.* Berkeley: Berkeley Geochronology Center Special Publication No. 4.

Lus, W. Y., McDougall, I., & Davies, H. L. (2004). Age of the metamorphic sole of the Papuan Ultramafic Belt ophiolite, Papua New Guinea. *Tectonophysics, 392,* 85–101. https://doi.org/10.1016/j.tecto.2004.04.009

Macpherson, C. G., Dreher, S. T., & Thirlwall, M. F. (2006). Adakites without slab melting: High pressure differentiation of island arc magma, Mindanao, the Philippines. *Earth and Planetary Science Letters, 243,* 581–593. https://doi.org/10.1016/j.epsl.2005.12.034

Milsom, J., & Smith, I. E. (1975). Southeastern Papua: Generation of thick crust in a tensional environment. *Geology, 3,* 117–120. https://doi.org/10.1130/0091-7613(1975)3<117:SPGOTC>2.0.CO;2

Mortimer, N., Hauff, F., & Calvert, A. T. (2008). Continuation of the New England Orogen, Australia, beneath the Queensland Plateau and Lord Howe Rise. *Australian Journal of Earth Sciences, 55,* 195–209. https://doi.org/10.1080/08120090701689365

Næraa, T., Scherstén, A., Rosing, M. T., Kemp, A. I. S., Hoffmann, J. E., Kokfelt, T. F., & Whitehouse, M. J. (2012). Hafnium isotope evidence for a transition in the dynamics of continental growth 3.2 Gyr ago. *Nature, 485,* 627–630. https://doi.org/10.1038/nature11140

Ott, B., & Mann, P. (2015). Late Miocene to Recent formation of the Aure-Moresby fold-thrust belt and foreland basin as a consequence of Woodlark microplate rotation, Papua New Guinea. *Geochemistry, Geophysics, Geosystems, 16,* 1988–2004. https://doi.org/10.1002/2014GC005668

Paquette, J.-L., & Le Pennec, J.-L. (2012). 3.8 Ga zircons sampled by Neogene ignimbrite eruptions in Central Anatolia. *Geology, 40,* 239–242. https://doi.org/10.1130/G32472.1

Parrish, R. R. (1990). U-Pb dating of monazite and its application to geological problems. *Canadian Journal of Earth Sciences, 27,* 1431–1450. https://doi.org/10.1139/e90-152

Peccerillo, A., & Taylor, S. R. (1976). Geochemistry of Eocene calc-alkaline volcanic rocks from the Kastamonu area, Northern Turkey. *Contributions to Mineralology and Petrology, 58,* 68–81.

Pell, S. D., Williams, I. S., & Chivas, A. R. (1997). The use of protolith zircon-age fingerprints in determining the protosource areas for some Australian dune sands. *Sedimentary Geology, 109,* 233–260. https://doi.org/10.1016/S0037-0738(96)00061-9

Pieters, P. E. (1978). Port Moresby-Kalo-Aroa, Papua New Guinea - 1:250 000 Geological Map Series (p. 55). BMR Australia Explanatory Notes.

Pirard, C., & Spandler, C. (2017). The zircon record of high-pressure metasedimentary rocks of New Caledonia: Implications for regional tectonics of the south-west Pacific. *Gondwana Research, 46,* 79–94. https://doi.org/10.1016/j.gr.2017.03.001

Samuel, M. D., Be'eri-Shlevin, Y., Azer, M. K., Whitehouse, M. J., & Moussa, H. E. (2011). Provenance of conglomerate clasts from the volcano-sedimentary sequence at Wadi Rutig in southern Sinai, Egypt as revealed by SIMS U-Pb dating of zircon. *Gondwana Research, 20,* 450–464. https://doi.org/10.1016/j.gr.2010.11.021

Schärer, U. (1984). The effect of initial [230]Th disequilibrium on young U-Pb ages: The Makalu case, Himalaya. *Earth and Planetary Science Letters, 67,* 191–204. https://doi.org/10.1016/0012-821X(84)90114-6

Schellart, W. P., Lister, G. S., & Toy, V. G. (2006). A late cretaceous and Cenozoic reconstruction of the Southwest Pacific region: Tectonics controlled by subduction and slab rollback processes. *Earth-Science Reviews, 76,* 191–233. https://doi.org/10.1016/j.earscirev.2006.01.002

Schott, R. C., & Johnson, C. M. (2001). Garnet-bearing trondhjemite and other conglomerate clasts from the Gualala basin, California: Sedimentary record of the missing western portion of the Salinian magmatic arc? *Geological Society of America Bulletin, 113,* 870–880. https://doi.org/10.1130/0016-7606(2001)113<0870:GBTAOC>2.0.CO;2

Shaanan, U., Rosenbaum, G., Hoy, D., & Mortimer, N. (2018). Late Paleozoic geology of the Queensland Plateau (offshore northeastern Australia). *Australian Journal of Earth Sciences.* doi: https://doi.org/10.1080/08120099.2018.1426041

Shearman, P., & Bryan, J. (2011). A bioregional analysis of the distribution of rainforest cover, deforestation and degradation in Papua New Guinea. *Austral Ecology, 36,* 9–24. https://doi.org/10.1111/aec.2011.36.issue-1

Shearman, P. L., Ash, J., Mackey, B., Bryan, J. E., & Lokes, B. (2009). Forest conversion and degradation in Papua New Guinea 1972–2002. *Biotropica, 41,* 379–390. https://doi.org/10.1111/btp.2009.41.issue-3

Smith, I. E. (1972). High-potassium intrusives from southeastern Papua. *Contributions to Mineralogy and Petrology, 34,* 167–176. https://doi.org/10.1007/BF00373771

Smith, I. E. M. (1976). *Volcanic rocks from southeastern Papua: The Evolution of volcanism at a plate boundary* (Unpublished PhD Thesis, p. 298). Canberra: Australian National University.

Smith, I. E. (1982). Volcanic evolution in eastern Papua. *Tectonophysics, 87,* 315–333. https://doi.org/10.1016/0040-1951(82)90231-1

Smith, I. E. M. (2013a). The chemical characterization and tectonic significance of ophiolite terrains in southeastern Papua New Guinea. *Tectonics, 32,* 1–12.

Smith, I. E. M. (2013b). High-magnesium andesites: The example of the Papuan Volcanic Arc. In A. Gómez-Tuena, S. M. Straub, & G. F. Zellmer (Eds.), *Orogenic andesites and crustal growth* (p. 385). London: Geological Society, Special Publications.

Smith, I. E., & Davies, H. L. (1976). Geology of the southeast Papuan mainland. *BMR Journal of Australian Geology and Geophysics, 165,* 86p.

Smith, I. E., & Milsom, J. S. (1984). Late Cenozoic volcanism and extension in Eastern Papua. *Geological Society, London, Special Publications, 16,* 163–171. https://doi.org/10.1144/GSL.SP.1984.016.01.12

Stacey, J. S., & Kramers, J. D. (1975). Approximation of terrestrial lead isotope evolution by a two-stage model. *Earth and Planetary Science Letters, 26,* 207–221. https://doi.org/10.1016/0012-821X(75)90088-6

Sun, S., & McDonough, W. F. (1989). Chemical and isotopic systematics of oceanic basalts: Implications for mantle composition and processes. *Geological Society, London, Special Publications, 42,* 313–345. https://doi.org/10.1144/GSL.SP.1989.042.01.19

Tapster, S., Roberts, N. M. W., Petterson, M. G., Saunders, A. D., & Naden, J. (2014). From continent to intra-oceanic arc: Zircon xenocrysts record the crustal evolution of the Solomon island arc. *Geology, 42,* 1087–1090. https://doi.org/10.1130/G36033.1

Taylor, B., Goodliffe, A., Martinez, F., & Hey, R. (1995). Continental rifting and initial sea-floor spreading in the Woodlark basin. *Nature, 374,* 534–537. https://doi.org/10.1038/374534a0

Taylor, B., Goodliffe, A. M., & Martinez, F. (1999). How continents break up: Insights from Papua New Guinea. *Journal of Geophysical Research: Solid Earth, 104,* 7497–7512. https://doi.org/10.1029/1998JB900115

Tera, F., & Wasserburg, G. J. (1972). U-Th-Pb systematics in three Apollo 14 basalts and the problem of initial Pb in lunar rocks. *Earth and Planetary Science Letters, 14*, 281–304. https://doi.org/10.1016/0012-821X(72)90128-8

Van Achterbergh, E., Ryan, C. G., Jackson, S. E., Griffin, W. L. (2001). Appendix. In P. J. Sylvester (Ed.), *Laser Ablation-ICP-Mass Spectrometry in the Earth Sciences: Principle and Applications* (Vol. 29, p. 239). Ottawa, ON: Mineralog. Assoc. Can. (MAC) Short Course Series.

Van Wyck, N., & Williams, I. S. (2002). Age and provenance of basement metasediments from the Kubor and Bena Bena Blocks, central Highlands, Papua New Guinea: Constraints on the tectonic evolution of the northern Australian cratonic margin. *Australian Journal of Earth Sciences, 49*, 565–577. https://doi.org/10.1046/j.1440-0952.2002.00938.x

Wallace, L. M., Ellis, S., Little, T., Tregoning, P., Palmer, N., Rosa, R., ... Kwazi, J. (2014). Continental breakup and UHP rock exhumation in action: GPS results from the Woodlark Rift, Papua New Guinea. *Geochemistry, Geophysics, Geosystems, 15*, 4267–4290. https://doi.org/10.1002/2014GC005458

Wandres, A. M., Bradshaw, J. D., Weaver, S., Maas, R., Ireland, T., & Eby, N. (2004). Provenance analysis using conglomerate clast lithologies: A case study from the Pahau terrane of New Zealand. *Sedimentary Geology, 167*, 57–89. https://doi.org/10.1016/j.sedgeo.2004.02.002

Webb, L. E., Baldwin, S. L., & Fitzgerald, P. G. (2014). The early-middle miocene subduction complex of the louisiade archipelago, southern margin of the woodlark rift. *Geochemistry, Geophysics, Geosystems, 15*, 4024–4046. https://doi.org/10.1002/2014GC005500

Whattam, S. A., Malpas, J., Ali, J. R., & Smith, I. E. M. (2008). New SW Pacific tectonic model: Cyclical intraoceanic magmatic arc construction and near-coeval emplacement along the Australia-Pacific margin in the Cenozoic. *Geochemistry, Geophysics, Geosystems, 9*, Q03021.

Worthing, M. A., & Crawford, A. J. (1996). The igneous geochemistry and tectonic setting of metabasites from the Emo Metamorphics, Papua New Guinea; a record of the evolution and destruction of a backarc basin. *Mineralogy and Petrology, 58*, 79–100. https://doi.org/10.1007/BF01165765

Advantages of unmanned aerial vehicle (UAV) photogrammetry for landscape analysis compared with satellite data

Kotaro Iizuka[1,2]*, Masayuki Itoh[2,3], Satomi Shiodera[2,4], Takashi Matsubara[5], Mark Dohar[6] and Kazuo Watanabe[2,4]

*Corresponding author: Kotaro Iizuka, Center for Spatial Information Science (CSIS), University of Tokyo, Chiba, Japan
E-mail: kotaro_iizuka@cseas.kyoto-u. ac.jp; kiizuka@csis.u-tokyo.ac.jp
Reviewing editor: Anshuman Bhardwaj, Computer Science, Electrical and Space Engineering, Lulea Tekniska Universitet, Sweden

Abstract: This study presents the advantages of detailed landscape analysis by UAV (drone hereafter) photogrammetry compared with satellite remote sensing data. First, satellite data are used for generating a coarse-scale land use/land-cover (LULC) map of the study region using conventional GIS techniques. The Advanced Land Observation Satellite-2 (ALOS-2) Phased Array L-band Synthetic Aperture Radar-2 (PALSAR-2) L-Band backscattering data are processed with a multilayer perceptron (MLP) supervised classification for generating a categorical map. The satellite-derived classification map resulted in eight general land-cover types with a ground resolution of 7.5 m, providing a moderate-resolution representation of the island landscapes. Second, the drone's image data are used to collect ground survey information and microscale information of the local site by implementing a structure from motion (SfM) technique to develop mosaicked orthorectified images of the sites. The orthophoto and digital surface model (DSM) derived from the drone-based data had resolutions of 0.05 m and 0.1 m, respectively. The SAR-based LULC map showed an overall accuracy of 78.1%, and the drone-based LULC map had an overall average accuracy of 92.3%. The subset area of the SAR map was compared with the drone-based map and showed average Kappa statistics of 0.375,

ABOUT THE AUTHOR

Kotaro Iizuka is an assistant professor at the Center for Spatial Information Science, University of Tokyo, Japan. He is specialized in Remote Sensing and GIS (Geographical information System), receiving his PhD degree at Graduate School of Science, Department of Earth Sciences, Chiba University, Japan. His research interests is in developing and finding new ideas/information's relating to spatial information science, which can be processed and used as base knowledge for further decision makings, which can be linked with various situations from local policies up to global perspectives. He mainly deals with, but not limited to, forestry by utilizing optical/radar satellites, UAVs, LiDAR to make advancements in relation with precision forestry.

PUBLIC INTEREST STATEMENT

Every year, new techniques or tools are developed for more advancing researches. Remote sensing field is one of the rapid developing area, which gives advantages when conducting large-scale analysis, such as from using satellite information. On the other hand, in the past few years, the emerging trend of the unmanned aerial systems (UASs) shows it's potential in utilizing at various opportunities to conduct experiments or researches focusing more relating to local-scale analysis. The purpose of this work is to show what can be achieved from utilizing the former satellite data and the latter new technology of UAS, by focusing on conventional approach of image classifications for the landscape analysis at both large/local scale. Satellite analysis showed a promising overview of larger spatial extent, while UAS gave precision in local scale analysis with addition of very detailed topography information.

demonstrating that satellite data cause challenges in correctly delineating the local land environment. The terrain information generated by the SfM method provided a good representation of the topography showing the drastic changes in the environment. The results indicate the usefulness of drone-based landscape analysis for future land-use planning at a local village scale.

Subjects: Environmental Management; Computer Science (General); Geography; Remote Sensing

Keywords: Land use; GIS; UAV; PALSAR-2; structure from motion; remote sensing

1. Introduction

Remote sensing data, such as spaceborne data (satellite data) and airborne data (aircraft observation), in association with GIS techniques are advantageous for observing and analyzing areas from local to global scales to understand environmental conditions, situations, or changes occurring on this planet (Gurney, Foster, & Parkinson, 1993). GIS techniques have shown promising results in various fields such as forestry (Iizuka & Tateishi, 2015), ecology (Apan, 1996; Honnay, Piessens, Landuyt, Hermy, & Gulinck, 2003; Martínez, Ramil, & Chuvieco, 2010), urban planning (Weber, 2003), land-use change (Tang, Wang, & Yao, 2008; Villa, 2012; Wu et al., 2006), and other related fields. Satellite sensors are either passive (optical) or active, such as synthetic aperture radar (SAR), and both types of sensors have advantages and disadvantages. The type of sensor used depends on the objectives of the project and the required details of the analysis and processing. A critical issue for optical sensors is cloud cover, which makes it difficult or impossible to detect and analyze the land surface under the clouds. For SAR sensors, the radar distortions and resulting complex feature interpretations are disadvantages compared to optical sensors (Richards, 2009). The data from many satellites have a low ground resolved distance, which blurs the images and makes it difficult to identify objects compared to airborne imagery. Sanga-Ngoie, Iizuka, and Kobayashi (2012) stated the importance of analyzing detailed land features, and the level of detail can impact the results of environmental assessments. Although fine-resolution satellite imagery such as GeoEye, WorldView, and other high-resolution satellite sensors exist, the collection of visible and analyzable scenes may be less frequent due to cloud cover, providing unsatisfactory results and limiting the analysis. Alternatives are required to overcome these issues so that current land information can be updated for land-use planning and monitoring. Unmanned aerial vehicle (UAV, hereafter also referred to as drone) technology can be used as such an alternative. This approach allows us to enhance and advance the analysis compared to using only satellite imagery, while additional surveying of the sites can be improved compared to conventional ground surveying (Congalton & Green, 2009).

Various research studies have demonstrated the usefulness of implementing UAVs in different fields of study (Flynn & Chapra, 2014; Getzin, Nuske, & Wiegand, 2014; Jaud et al., 2016; Luna & Lobo, 2016). For example, Luna and Lobo (2016) utilized UAVs to assess the planting quality of crops by mapping gaps in the crop canopy to guide replanting. UAV images have been used not only for field investigations but also for terrain analysis based on photogrammetry (Jaud et al., 2016), and such information can be used to understand the local topography. Implementing a drone-based technique for topographic analysis has better quality/accuracy than satellite remote sensing techniques, such as the generation of digital elevation models (DEM) from interferometric SAR data (Crosetto, 2002), which limits the resolution and quality depending on various factors, resulting in discontinuities in obtaining correct land information. Research on integrating satellite information and drone photogrammetry is still developing, and it is important to compare the scale of the information that can be obtained from both data sets.

The objective of this study is to present the advantages of the drone photogrammetry method in collecting more detailed and useful local landscape information than satellite data. Postmining sites in Indonesia were selected as the case study sites because their complex topography and landscape were suitable to compare the results of both SAR- and UAV-based methods. In these sites, massive tin mining activity has degraded the land and has changed the landscape, presenting evidence of how human interactions can change the environment. The results could provide better precision to obtain detailed land information including the microtopography (elevation, slope, etc.), which is required for strategic land-use management and the reclamation of such environmentally degraded lands.

2. Methodology

2.1. Study area
The study area is located at Belitung Island in Indonesia, located between Sumatra and Kalimantan, just east of Bangka Island. The center of the island is located at 2.83° S and 107.916° E. The land area is approximately 4,800 km², and the population is 262,357 according to the 2010 census (http://sp2010.bps.go.id/index.php/navigation/wilayah). The west coast of the island has been developed into tourist resorts, while the majority of the island consists of preserved forested areas (either natural or secondary forest) or plantation areas. On the east side of the island, there are large areas of degraded land due to the tin-mining activity (Erman, 2007; Pöyhönen, 2009). The history of tin mining in Belitung began in the eighteenth century (Heidhues, 1991), and the mining continued in the colonial era; today, an Indonesian government-owned company runs the mines. Due to depressed tin prices in recent years, transitions are seen from industrial to agricultural production, resulting in a reduction in tin production, and different products are now produced such as white pepper. However, there are still individual mining activities that have raised some issues (Pöyhönen, 2009). This is evident by interpreting images of areas where the mining activity ended and where lakes were generated (dark features resembling holes); these features can be observed throughout the island (Figure 1). The focus of this study is on these postmining areas, covering the east side of the island. The postmining areas are identified from the large bare soil area with several hole-like lakes in the surrounding area. Mining concession data are used to confirm the postmining sites and to determine whether the site has been abandoned. The information was collected from the support of the local company. Two test sites were selected for implementing the aerial survey with the drone (Figure 1).

   Site 1 is located in the southcentral area of the island. The site consists of four major land use/land-cover (LULC) types. The first is vegetation, which is broad in structure, from small grasses to shrubs to tall trees. The second is bare soil, where the major bare soil is white soil, which is the sandy tailings from the mining activity, and a small area of brown soil is observed (the path along the north-south direction located on the west side of the site), which is mainly used as a pathway for the locals. Water bodies (small lakes) occur as hole-like shapes, which result from open-pit mining, where the water pumps used for mining can directly flood the holes, but also local precipitation contributes to generating small lakes. This is the same for most of the mining sites throughout the island. Site 2 is located in the eastern part of the island. Similar grass and shrubs can be seen in the vegetative classes, but a plantation area was seen on the west side of the site. The site consists of LULC types similar to Site 1, but with larger homogenous areas of bare soil. White soil is again the result from mining activity, and the brown soil on the west side of the site is used as pathways for accessing the plantation area. Few houses are seen along the roads that cross in the east–west direction near the center of the site.

2.2. Data sets
One main data set is the satellite remote sensing data from the Advanced Land Observation Satellite-2 (ALOS-2) Phased Array L-band Synthetic Aperture Radar-2 (PALSAR-2). Since cloud-free optical data were difficult to obtain, SAR data were utilized. PALSAR-2 is a microwave sensor acquiring L-band (1.2 GHz) wavelength information. Data were obtained for 2 June 2016, specifying a Level 1.1 stripmap mode, high sensitivity (6 m), dual polarized product in single look complex (SLC) format. Both HH (Horizontal transmitted, Horizontal received) and

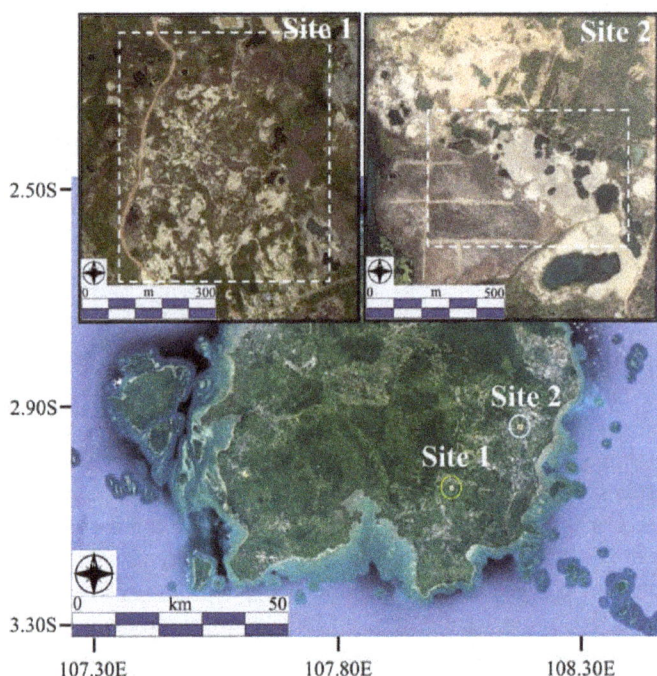

Figure 1. Overview of Belitung Island and the selected sites for implementing the aerial survey with the drone. In the eastern part of the island, mining activities are evident. The dark features are small lakes generated by the mining process (zoomed image). The dotted area within the image represents the survey area for the flight zones. The selected sites are hereafter referred to as Sites 1 and 2. Images are sourced from Google Earth.

HV (Horizontal transmitted, Vertical received) polarized data were used for further analysis. Because the focus is on the mining area of the island, only the images with high coverage over the east part of the island were collected.

The second type of data consisted of drone-based aerial photos acquired at the study site. A DJI Phantom 4 quadcopter (DJI, Shenzhen, China) was used, and a sufficient number of photos were acquired to collect detailed information of the area.

2.3. SAR preprocessing and LULC classification

Preprocessing of the PALSAR-2 data was performed using the software Sentinel-1 toolbox 4.0.2 developed by the European Space Agency (ESA), and Clark Labs' TerrSet software (Clarks Lab, MA, USA) was used for the image classification. Preprocessing of the SAR data was performed using a 7 × 7 Lee Sigma speckle filter to reduce the noise caused by multiple surface objects, referred to as speckles (Lee, Wen, Thomas, Chen, & Chen, 2009). The filtered product was then processed using terrain flattening (slope correction) to flatten the area to reduce distortions to the imagery due to hills and mountains (Small, 2011). A Range Doppler Terrain Correction was applied to the terrain-flattened product to convert from radar coordinates to geographical coordinates. Finally, an additional 3 × 3 Lee Filter was applied to reduce the number of artifacts caused by the radiometric correction. The ground range pixel spacing of the processed data is 7.5 m.

Other than the preprocessed HH and HV bands, the data were processed by extracting texture information derived from the HH and HV data. Using a 3 × 3 window size, the following texture parameters were extracted using the grey-level co-occurrence matrix (GLCM) method (Wen, Zhang, & Deng, 2009): Contrast, Dissimilarity, Homogeneity, Angular Second Moment, Energy, Maximum Probability, Entropy, GLCM Mean, GLCM Variance, and GLCM Correlation. Each texture parameter was generated from both HH and HV bands, and all data except for the GLCM correlation were stacked and used as variables for the classification process. Training data for the land-cover classes were obtained using Google Earth imagery as reference information (Sharma, Tateishi, Hara, & Iizuka, 2016). Through the interpretation of multiple imageries at different time periods, sufficient training samples were collected representing each land-cover class on the island for use in the classification process. Eight classes were used for the classification process: Residential, Forest, Wet Forest (riparian forest), Shrub, Logged, Plantation, Bare land, and Water. Multi-layer perceptron (MLP) neural network supervised classification (Eastman, 2015) was implemented to develop an LULC map of the eastern Belitung Island. The advantage of implementing MLP is the ability to derive patterns and trends from complicated/numerous data, which could be difficult to extract using other techniques; moreover, the method has advantages in characterizing nonlinear trends (Iizuka et al., 2017). The classification map was processed with an additional 3 by 3 mode filter to smooth the classification results.

2.4. Drone observations, orthophoto generation and LULC classification

Field observations were conducted at the two sites selected from the satellite imagery. The Phantom 4 drone flew across the area to collect aerial imagery and video to obtain detailed information on the sites. For the manual flight control, the software DJI GO (DJI Technology Co., Ltd; http://www.dji.com/goapp) was used, and for the automatic flight control and the photo acquisition, the software Altizure (Everest Innovation Technology; https://www.everest-innovation.com/) was used. The aerial photos were processed using Photoscan Pro (Agisoft, St. Petersburg, Russia) to generate a mosaicked orthorectified image of each site. Photoscan outputs a mosaicked image of the area by generating a point cloud from the photos using a structure-from-motion (SfM) technique (Mancini et al., 2013). Based on the estimated camera positions and photos, a 3D polygon mesh is developed by the software (http://www.agisoft.com/pdf/photoscan-pro_1_2_en.pdf). After constructing the geometry, an orthophoto can be generated. Orthorectification refers to the process of correcting the geometry of the images based on the angles and distances of the features from the sensor's position, and the process results in an image with a nadir view (directly vertical), unlike an uncorrected image, which results in inclined features (Zhou, Chen, Kelmelis, & Zhang, 2005). Aerial surveys were performed on 21 July 2016 at Site 1 and on 22 July 2016 at Site 2. Due to drone regulations and climatic conditions, the photos were acquired by maintaining a stable altitude of 140 m for Site 1 (considering the flight limit of 150 m), and 100 m for Site 2 (due to strong wind). Figure 2 shows the flight paths at the sites.

For image classification, the same MLP method was implemented on the drone imagery for developing a categorical map. The RGB composite image (ortho image) was decomposed into individual red (R), green (G) and blue (B) bands. The bands were transformed into an additional color space, hue, lightness and saturation. An additional green-red vegetation index (GRVI) (Motohka, Nasahara, Oguma, & Tsuchida, 2010) was generated, and each band is stacked as variables for the classification process.

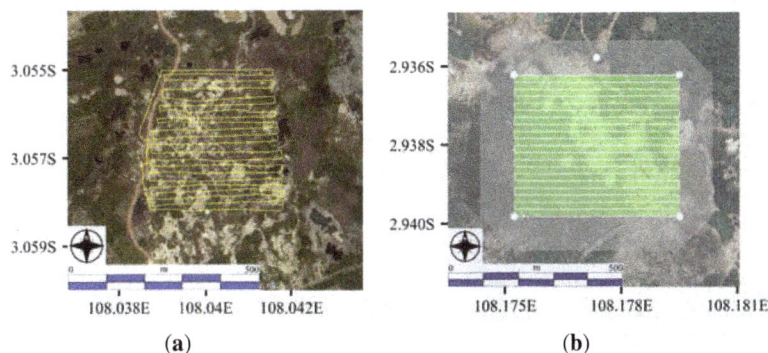

(a)   (b)

Figure 2. Flight paths of the Phantom 4 drone across (a) Site 1 and (b) Site 2. Site 1 was flown using manual flight con-trol and Site 2 was flown using an automated flight plan.

2.5. Digital surface model (DSM) generation

Along with generating a mosaicked orthophoto of the study site, additional data were generated to determine the topography. A digital surface model (DSM), a 3D representation of the terrain's surface, was developed based on the dense point cloud information created in Photoscan. A DSM represents the earth's surface and includes the heights of all objects. For example, the height of the terrain will be shown as the top layer of each object, where this represents the top canopy layer for trees and the roof for a house. The local terrain and the characteristics of each site were examined by further processing to generate slope information from the DSM data. The inverse distance weighting (IDW) interpolation method was implemented to generate the DSM from dense point cloud data.

(1) Evaluation and Comparison of LULC Maps

The two LULC maps generated by SAR- and drone-based methods was first evaluated separately in different ways due to the differences in the spatial resolutions. The SAR-based LULC map was validated by the cross-validation method in the classification process to determine the classification capability of the model. Fifty percent of the data were used for training from the whole training data set, while the other 50% was used for validation. A total of 10,000 samples were used for each LULC class. The conventional accuracy assessment method was applied to the drone-based LULC map to validate the overall accuracy of the categorical map (Congalton & Green, 2009). The ground truth information was collected through the aerial image, and each class was confirmed from the images taken during the ground observation. Selected points were automatically generated using stratified random sampling, and 300 points per class were generated. For the comparison between the two maps, the Kappa index of agreement (KIA) (Pontius, 2000) was computed to quantify the reliability of the SAR map to the drone map.

3. Results

3.1. SAR-based LULC map of east Belitung island

Figure 3 shows the results of the classification using the PALSAR-2 data with a resolution of 7.5 m for the categorical map. An accuracy assessment was performed using 5000 pixels per class, resulting in a model accuracy of 78.1%. The training and testing root mean square (RMS) were 0.2017 and 0.2022, respectively, indicating that there were no issues with overfitting. Table 1 shows the variables used in the neural network model and the accuracy of the categorical map with regard to each variable. The skill measure represents the measured accuracy minus the accuracy expected by chance. For this map, the most influential variable was HV mean, and the least influential variable was HH contrast. Overall, a higher accuracy was obtained when HV-polarized data were used in the model, which is probably due to the high proportion of vegetated land cover in the region, because it is well known that HV data are sensitive to vegetative classes (Iizuka & Tateishi, 2014).

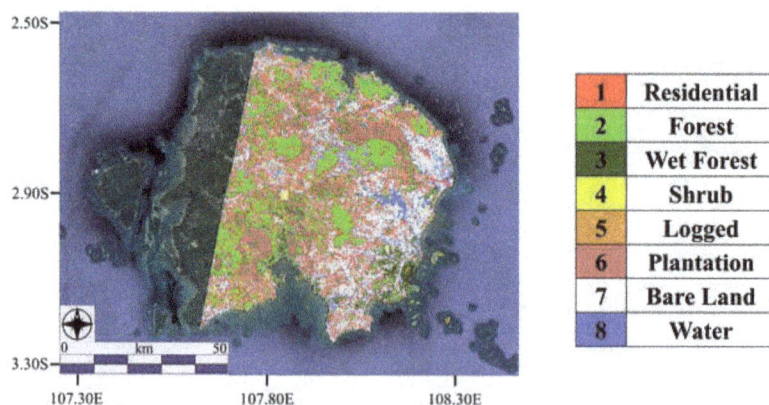

1	Residential
2	Forest
3	Wet Forest
4	Shrub
5	Logged
6	Plantation
7	Bare Land
8	Water

Figure 3. Land-use/land-cover map of Belitung Island based on PALSAR-2 data.

Table 1. Sensitivity of the MLP model to forcing a single independent variable to be constant. Skill measure represents the measured accuracy minus the accuracy expected by chance. If skill measure shows higher decrease from the accuracy, the variable is considered to have higher influence to the model

Model	Accuracy (%)	Skill measure	Order of influence
With all variables	78.11	0.7499	N/A
HH/HV ratio	73.9	0.7017	13
HV variance	66.2	0.6138	9
HV mean	54.79	0.4833	1 (most influential)
HV max	60.23	0.5454	3
HV homogeneity	73.11	0.6927	11
HV entropy	64.52	0.5945	8
HV energy	75.64	0.7216	17
HV dissimilarity	75.44	0.7193	16
HV contrast	77.71	0.7453	20
HV ASM	60.61	0.5498	4
HH variance	59.98	0.5426	2
HH mean	63.67	0.5848	5
HH max	64.46	0.5938	7
HH homogeneity	73.52	0.6974	12
HH entropy	75.9	0.7246	18
HH energy	75.19	0.7165	15
HH dissimilarity	76.55	0.732	19
HH contrast	78.05	0.7492	21 (least influential)
HH ASM	68.91	0.6446	10
HV original (dB)	63.69	0.585	6
HH original (dB)	73.93	0.702	14

3.2. Drone orthophoto and LULC map

At Site 1, the Phantom 4 drone acquired multiple photos over the area shown in Figure 2a. Due to difficulties with network connections, an automated flight course could not be planned for this area; instead, manual flight control was used for the flight paths and photo acquisition, resulting in the unstable tracks seen in the figure. A total of 655 photos were acquired with a minimum of 85% forward and side overlap. Figure 2b shows the flight track for Site 2, which was acquired using automated flight control, and at this site, 517 photos were acquired with 85% forward and side overlap using the Altizure app. Using the Photoscan software, an orthophoto of the site was generated (Figure 4). The total areal coverage was approximate 30 ha for Site 1 and 32 ha for Site 2.

3.3. Interpretation of the topography

Figure 5 shows the DSMs and the slope information generated from the DSM for each site, representing the local topography. The elevation is based on the height from the ellipsoid and on the WGS84 datum. Because the elevation represents DSM data and not digital terrain model (DTM) data, some objects, such as shrubs and trees, are represented with much higher elevations or higher sloped features compared to the surroundings. In addition, the water bodies have a similar rugged slope due to the difficulty in obtaining the actual water elevation through photogrammetry.

(a)

(b)

(c)

(d)

Figure 4. Orthophoto of the two sites generated by using Photoscan Pro, with 655 photos for Site 1 and 517 photos for Site 2. (a) A screen capture of generating the point cloud from the photos at Site 1; the blue marks are the estimated cam-era positions and the black perpendicular line is the axis. (b) The data for Site 2. (c) The final orthophoto generated for Site 1 and (d) for Site 2.

The topography at Site 1 exhibits a trend of higher elevation in the southern part and a gradually decreasing elevation toward the northern part. This trend can be interpreted by looking at the small lakes in the orthophoto (Figure 5a), where the majority of the lakes are located in the northern area at low elevations. The soils were possibly scraped due to mining activity using the water located in the northern area, and the sandy tailings were piled on the southern side, generating a gentle decreasing slope to the north. Ignoring all the vegetated areas and the water bodies, the slope information was extracted only from the bare soils in the image. At Site 1, the average slope was 7.20°, with a standard deviation of 6.50°. The brown- and white-colored soil areas had average slope values (and standard deviations) of 6.40° (5.84) and 7.50° (6.71), respectively.

At Site 2, a general trend of decreasing elevation was observed from the east to the west side of the image. The highest part is in the lower right corner of the image, with a gradual decrease in elevation toward the top left corner. The top right corner of the image is a forested area; therefore, higher elevation data occur in this area. On the west side of the image, there is a stream, and the decreasing elevation in this area continues toward the stream. The area appears similar to flood-plains; however, the gradual elevation trends can be made by similar explanation to Site 1. As in Site 1, the terrain information is extracted from the bare soil areas, which obtained an average slope of 5.71°, with a standard deviation of 5.10°. The slope values in the brown- and white-colored soil areas (Figure 6b) were on average (and standard deviations) 5.41 and 5.97° (4.39 and 5.62°)), respectively.

Overall, the brown soil areas are more leveled because they were pathways for the locals, which includes vehicles. Additionally, similar pathways are seen at the plantations for transportation purposes (but not main asphalt roads).

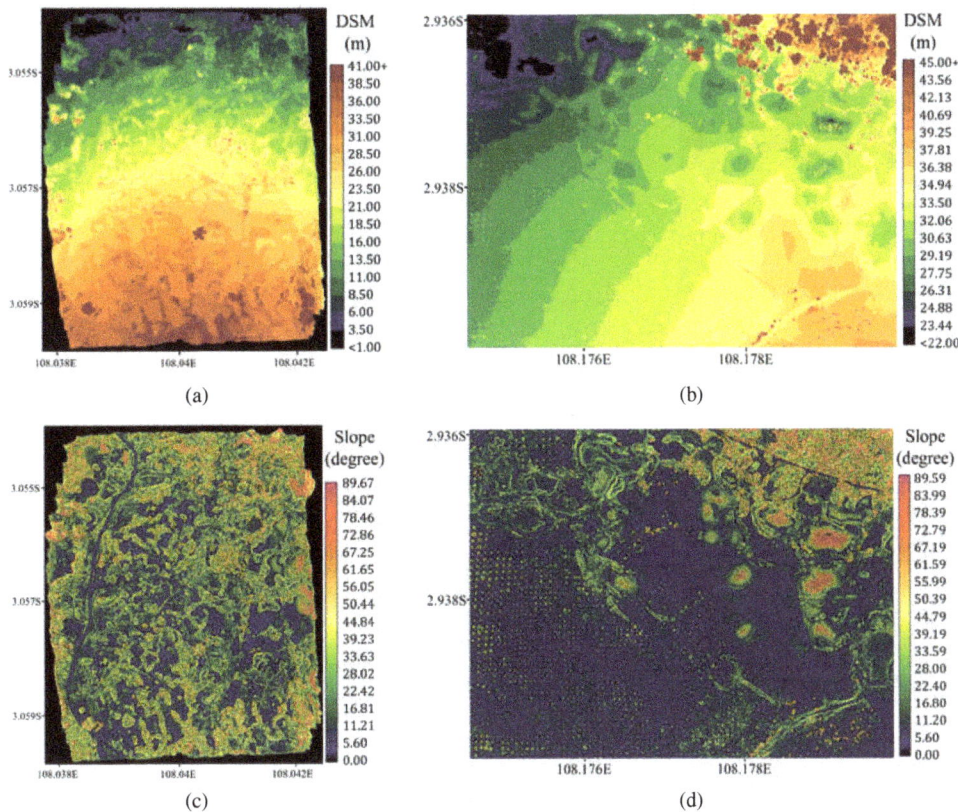

Figure 5. Topographical infor-mation for Site 1 and Site 2. (a) and (b) are the DSMs showing the elevation heights from the ellipsoid for Site 1 and Site 2, and (c) and (d) are the slope maps of Site 1 and Site 2, respectively.

The white bare soil areas result from the human impacts of mining activities. The small-scale open-pit method for mining generates many small hills in the surrounding areas. This is likely one of the causes of the rough topography in the area, especially for Site 1. The composition of the white soil at Site 1 in the highly sloped area is characterized by coarse sand (2.0–0.2 mm): 52.3%, fine sand (0.2–0.02 mm): 42.7%, silt (0.02–0.002 mm): 0.3%, and clay (< 0.002 mm): 4.7%. The soil in the low sloped area has coarse sand: 23.4%, fine sand: 74.1%, silt: 0.9%, and clay: 1.6%. The fine sand has possibly runoff to the low sloped areas. At Site 2, a more homogenous area of bare soil is seen with a smoother topography than Site 1. The environment looks similar to a floodplain considering the surrounding environment. Usually, this kind of environment experiences flood events that transport the sediments and leave the fine-grained sand more inland; however, in the case of this area, the gradual relief of the sandy soils is formed from the past mining activity. The open-pit method has a high water demand. It uses high-pressure water to scrape off the soil, separating the minerals and the sandy tailings, and piling the tailings in one area. As this area is observed to have adequate water capacity at the west side, the mining site would have been owned by a large company, leaving larger volumes of soil in one place than the small-scale activities at Site 1. The composition of the white soil in Site 2 is characterized by coarse sand: 69.6%, fine sand: 29.4%, silt: 0.2%, and clay: 0.8%. Sites 1 and 2 both have low pH values, with an average of 5.72. The soil analysis has been requested to the Japan Soil Association.

A comparison of the two sites from the DSM analysis indicates that Site 2 is flatter than Site 1; simply by considering the flatness and the LULC homogeneity of the area, Site 2 could be a candidate for developments (e.g., residential), because a major cost of developments is from land clearing and leveling. Even though a large water stream is seen in the west area, the elevation at the white soil area can be as high as 10 m from the water, with lower risks of flood events than

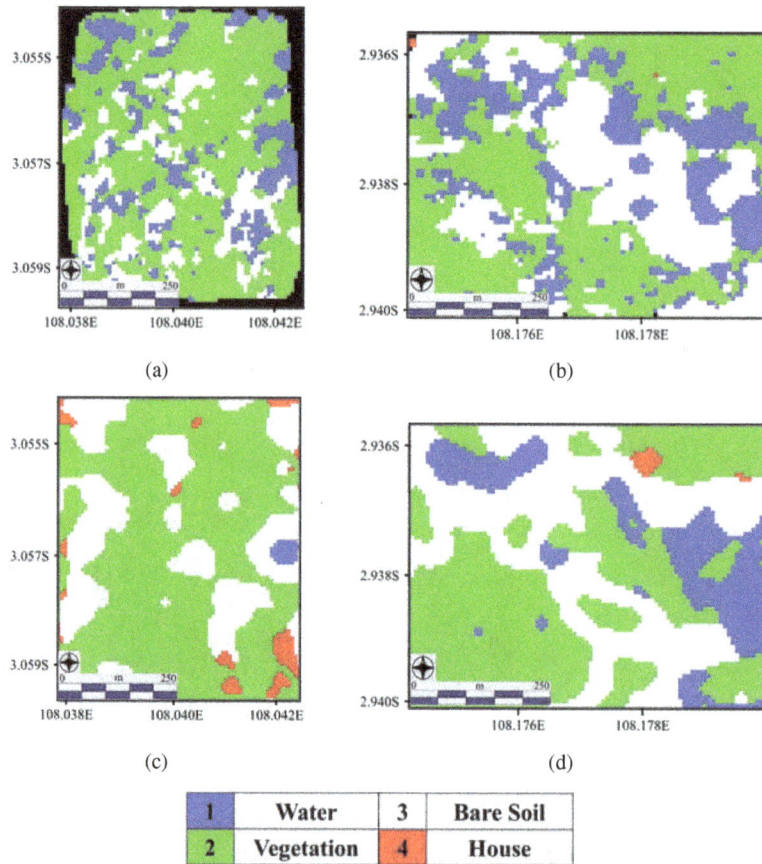

Figure 6. Top images show the LULC maps generated from the drone imagery for (a) Site 1 and (b) Site 2. Bottom images indicate the subset area (enlarged area) of the candidate sites for the LULC maps based on the SAR imagery for (c) Site 1 and (d) Site 2.

at other locations. Moreover, using the topography data, parameters such as the soil volume of the area can be calculated, and further decisions can be made if there are needs for additional ground level raising and volume adjustments. These kinds of decisions can be made from the drone imagery with the SfM method at a local scale, which cannot be easily performed with only SAR interpretation.

(1) SAR-based LULC map comparison to the local-scale drone map

Figure 6 shows the categorical map and the classes from both the SAR and drone analyses at the two sites. The resolution of the SAR- and drone-based LULC maps is 7.5 m and 0.05 m, respectively. The categorical map generated from the drone image resulted in a total of four LULC classes: Water Bodies, Vegetation, Bare Soil and Houses (houses were added manually due to their small number and area). Using the ground truth points collected through the imagery, an accuracy assessment was performed for the drone-based map, and the overall accuracy result was 89.9% and 94.7% for Site 1 and Site 2, respectively. A sample size of 300 points was selected for each LULC class, resulting in a total of 900 points (houses were omitted from the assessment). Most of the error occurs between the water and bare soil classes, with confusion between the murky water and wet soils.

The detectability of the LULC classes for both sources is different; generally, the resolution is what causes the differences. Specifically, the SAR data are being confused due to the scattering mechanism of the microwaves, depending on the land objects, moisture content, topography, etc. for the correct classification (Ji & Wu, 2015; Richards, 2009). To compare the maps, relying on the conventional accuracy assessment does not make sense; therefore, the SAR-based map was quantitatively assessed for its similarity with the drone (reference) map using KIA. To combine both maps, the original LULC classes were merged into a general class. Forest, Wet Forest, Shrub, Logged, and Plantation is now combined into a "Vegetation" class, and Deep Water and Shallow Water are now a "Water" class. Now, the two maps display the four general classes: Water, Vegetation, Bare Soil and Residential. Second, the drone LULC map was resampled to 7.5 m using the nearest neighbor algorithm with a 3 × 3 mode filter. The SAR-based map was compared to the drone map by computing KIA. As a result, the KIA showed a value of 0.42 and 0.33 for Site 1 and Site 2, respectively. KIA can be interpreted where 0 indicates that the agreement of the maps is caused just by chance, a value between 0 and 1 indicates an agreement that is better than chance, and KIA = 1 is a perfect agreement. Therefore, in this case, it could be interpreted that the SAR-based map is on average 37.5% more reliable than an agreement just by chance.

4. Discussion

A classified map of eastern Belitung Island was developed using PALSAR-2 microwave remote sensing data. It was effective in determining and understanding the landscape of the island compared to utilizing optical satellite data, which are often limited due to the frequent cloud cover that obscures the underlying surface. This is a critical problem for determining surface conditions and is especially a problem for determining land features in tropical regions (Asner, 2001). SAR data are not affected by cloud cover, and the major land-cover types could be distinguished, resulting in a coarse representation of the landscape. However, it is difficult to obtain fine resolution information because of the limitation with SAR and its resolution of the imagery (Ji & Wu, 2015; Reigber et al., 2013). It is often difficult to classify land features accurately unless the collection occurs in full polarimetric mode, collecting HH, HV, VV, and VH data (Ji & Wu, 2015). The image classification utilizing PALSAR-2 with the MLP method resulted in a model accuracy of 78.1%. Variables were used based on the information that can be derived from the dual polarized SAR product. Increasing the number of variables improved the accuracy of the classified map, and the most influential variable for the classification was the HV mean data, while the least influential variable was HH contrast. The ranking of the variables depends largely on the LULC class types, and the results may change. The model still has room for improvements; however, from the interpretations of the classified map, distinguishing the major LULC types can be acknowledged to some degree. However, when this at a more local scale, the accuracy could be very limited. As shown by the results, SAR map with an average KIA of 0.375 is not very high to give satisfying results for precise landscape analysis. It is likely that an in-depth analysis of the LULC types at a local scale is erroneous with the space-borne SAR data because the heterogeneity of the land features cannot be discriminated due to the relatively low resolution; thus, alternative approaches are required to detect fine details in the landscape.

To aid in such a problem, the drone information can be utilized as a tool for fine, local-scale landscape analysis. The orthophotos generated from multiple photos (Figure 4) demonstrate the potential for obtaining the detailed information of a landscape. The ground resolution was approximately 0.05 m (depending on the camera sensor and the flying height above the ground, the resolution can be higher), which is much higher than the resolution of any current satellite imagery, allowing for the in-depth analysis of the landscape. Small patches of water bodies or bare soils were easily detected with the drone image and the generated map (Figure 6), while the SAR map tends to show a more generalized (merged) area. The average accuracy was 92.3%, which can represent a better understanding and reliability of the local landscape information. The limitation is that the drone image is captured from a normal RGB camera, which lacks additional information that can usually be collected from multispectral satellite images (e.g., infrared). Therefore, delineating subclasses of LULC types could be difficult. Onboarding different multi-spectral sensors might work for precise classifications (Ahmed et al., 2017).

If topography needs to be considered, an alternative source of topographic data is usually utilized. However, the SfM method can also generate terrain information. Therefore, compared to

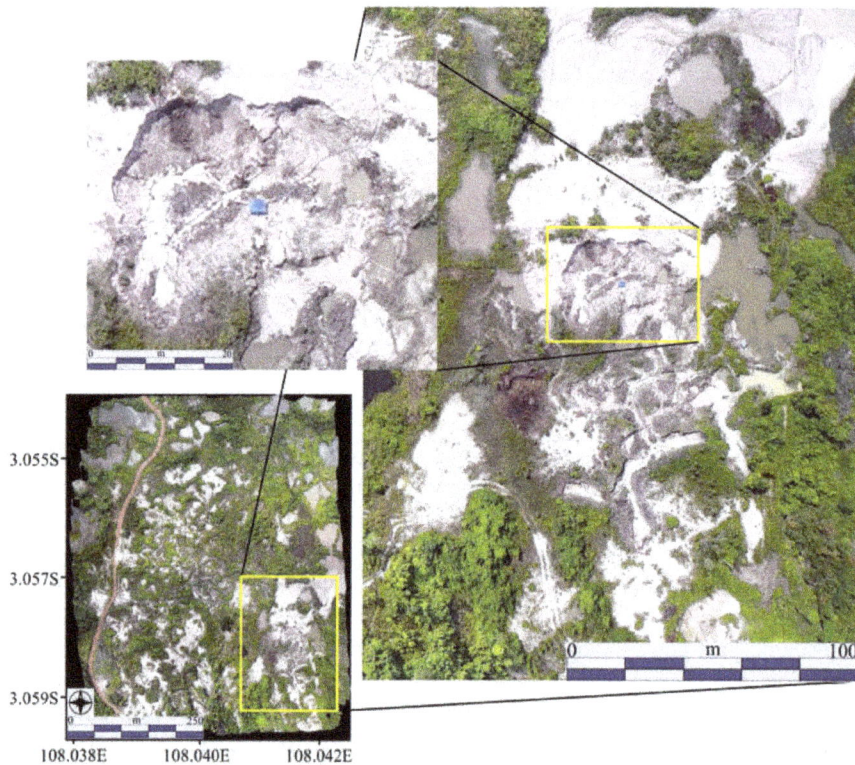

Figure 7. Local worker using water pumps for tin mining. They are usually difficult to recognize *in situ* because they tend to stay away from the main roads. Massive terrain degradation from the activity can be clearly observed using the drone images, which is dif-ficult to detect from the SAR image (at Site 1).

conventional data such as from the 90 m resolution Shuttle Radar Topography Mission (SRTM), Sites 1 and 2 showed an excellent interpretation of the drastic changes in the topography within the sites. This was possible because of the generated DSM resulting in a 0.1 m resolution. At Site 1, local workers (active tin mining) were detected through the analysis, which can be seen in the orthophoto (Figure 7). These individual miners cause issues of frequent land degradation, but since it is at an individual scale, it is often overlooked from satellite observations. They can be identified by the differences in the slope angles suddenly increasing within the LULC class compared to the smoother relief in the surrounding area. Notably, the surrounding slope angles in the smoother area average 6.16°, but the mining pits are 19.34° on average. Clearly, the water pumps used for scraping the sediments are causing such degradation. This cannot be observed in coarse satellite images due to the resolution, while the aerial photos taken from the drone demonstrate one of their potentials.

5. Conclusions

This study presents the advantages of drone photogrammetry in collecting more detailed and useful local landscape information than the satellite imagery. The changes in the local environment on Belitung Island are rapid and dramatic. Many areas face changes from one land-cover type to another depending on the demands. In the case of the post tin mining sites, anthropogenic activity has influenced the environment of the landscape. The SAR data with the use of the MLP method have demonstrated their usefulness in developing a coarse LULC map of the island. To provide a wide view of the large spatial extent, the conventional approach utilizing satellite imagery was promising. However, in regard to the local-scale analysis, detailed observation of the land needs a more in-depth analysis, but the SAR data were not applicable for this. A more precise analysis of the landscape can be performed by the drone using the SfM photogrammetry method. Although the spatial extent of the drone imagery was smaller than the satellite-derived data, the drone-based information provided higher precision and handled a broader area than conventional ground surveying. The SfM method can provide two important types of information: orthoimage and terrain data. To detect the conditions in the post mining sites and to analyze the area prior to landscape-level planning for development, business, or conducting research, it is crucial to understand the local LULC distribution and the terrain characteristics. The results indicate that the method can detect and analyze general LULC classes and characterize the heterogeneous topography using fine-detail photos. This study demonstrates that this technique can be considered for regions where various land-use decisions are required.

Acknowledgements
We would like to express our sincere appreciation to the two anonymous reviewers for the critical reviewing of the work.

Funding
This work was supported by the Japan Society for the Promotion of Science [grant number 15H05625], [grant number 18H02238] and Research Institute for Humanity and Nature [grant number 14200117].

Author details
Kotaro Iizuka[1,2]
E-mail: kotaro_iizuka@cseas.kyoto-u.ac.jp
Masayuki Itoh[2,3]
E-mail: itoma@cseas.kyoto-u.ac.jp
Satomi Shiodera[2,4]
E-mail: sshiodera@gmail.com
Takashi Matsubara[5]
E-mail: matsubara.takashi@obayashi.co.jp
Mark Dohar[6]
E-mail: Mark.Dohar@anj-group.com
Kazuo Watanabe[2,4]
E-mail: isseiw@cseas.kyoto-u.ac.jp
[1] Center for Spatial Information Science (CSIS), University of Tokyo, Chiba, Japan.
[2] Center for Southeast Asian Studies (CSEAS), Kyoto University, Kyoto, Japan.
[3] School of Human Science and Environment, University of Hyogo, Hyogo, Japan.
[4] National Institutes for the Humanities, Research Institute for Humanity and Nature (RIHN), Kyoto, Japan.
[5] Technical Research Institute, Obayashi Corporation, Tokyo, Kiyose, 204-8558, Japan.
[6] PT Austindo Aufwind New Energy, Jakarta, Indonesia.

References
Ahmed, O. S., Shemrock, A., Chabot, D., Dillon, C., Williams, G., Wasson, R., & Franklin, S. (2017). Hierarchical land cover and vegetation classification using multispectral data acquired from an unmanned aerial vehicle. *International Journal of Remote Sensing, 38*, 2037–2052. doi:10.1080/01431161.2017.1294781

Apan, A. A. (1996). Tropical landscape characterization and analysis for forest rehabilitation planning using satellite data and GIS. *Landscape and Urban Planning, 34*, 45–54. doi:10.1016/0169-2046(95)00201-4

Asner, G. P. (2001). Cloud cover in landsat observations of the Brazilian Amazon. *International Journal of Remote Sensing, 22*, 3855–3862. doi:10.1080/01431160010006926

Congalton, R. G., & Green, K. (2009). *Assessing the accuracy of remotely sensed data; principles and practices* (2nd ed. ed.). New York: CRC/Taylor & Francis.

Crosetto, M. (2002). Calibration and validation of SAR interferometry for DEM generation. *ISPRS Journal of Photogrammetry and Remote Sensing, 57*, 213–227. doi:10.1016/S0924-2716(02)00107-7

Eastman, J. R. (2015). *TerrSet tutorial.* Worcester, MA: Clark University.

Erman, E. (2007). *Rethinking of legal and illegal economy: A case study of tin mining in Bangka Island.* Retrieved from http://globetrotter.berkeley.edu/GreenGovernance/papers/Erman2007.pdf

Flynn, K. F., & Chapra, S. C. (2014). Remote sensing of submerged aquatic vegetation in a shallow non-turbid river using an unmanned aerial vehicle. *Remote Sensing, 6*, 12815–12836. doi:10.3390/rs61212815

Getzin, S., Nuske, R. S., & Wiegand, K. (2014). Using unmanned aerial vehicles (UAV) to quantify spatial gap patterns in forests. *Remote Sensing, 6*, 6988–7004. doi:10.3390/rs6086988

Gurney, R. J., Foster, J. L., & Parkinson, C. L. (1993). *Atlas of satellite observations related to global change.* Great Britain: Cambridge University Press.

Heidhues, M. F. S. (1991). Company Island: A note on the history of Belitung, Indonesia. *51*, 1–20. doi:10.2307/3351063

Honnay, O., Piessens, K., Landuyt, W. V., Hermy, M., & Gulinck, H. (2003). Satellite based land use and landscape complexity indices as predictors for regional plant species diversity. *Landscape and Urban Planning, 63*, 241–250. doi:10.1016/S0169-2046(02)00194-9

Iizuka, K., Johnson, B. A., Onishi, A., Magcale-Macandog, D. B., Endo, I., & Bragais, M. (2017). Modeling future urban sprawl and landscape change in the Laguna de Bay Area, Philippines. *Land, 6*, 26. doi:10.3390/land6020026

Iizuka, K., & Tateishi, R. (2014). Simple relationship analysis between l-band backscattering intensity and the stand characteristics of Sugi (*Cryptomeria japonica*) and Hinoki (*Chamaecyparis obtusa*) trees. *Advances in Remote Sensing, 3*, 219–234. doi:10.4236/ars.2014.34015

Iizuka, K., & Tateishi, R. (2015). Estimation of CO_2 sequestration by the forests in Japan by discriminating precise tree age category using remote sensing techniques. *Remote Sensing, 7*, 15082–15113. doi:10.3390/rs71115082

Jaud, M., Passot, S., Le Bivic, R., Delacourt, C., Grandjean, P., & Le Dantec, N. (2016). Assessing the accuracy of high resolution digital surface models computed by PhotoScan® and MicMac® in sub-optimal survey conditions. *Remote Sensing, 8*, 465. doi:10.3390/rs8060465

Ji, K., & Wu, Y. (2015). Scattering mechanism extraction by a modified cloude-pottier decomposition for dual polarization SAR. *Remote Sensing, 7*, 7447–7470. doi:10.3390/rs70607447

Lee, J. S., Wen, J. H., Thomas, L. A., Chen, K. S., & Chen, A. J. (2009). Improved sigma filter for speckle filtering of sar imagery. *IEEE Transactions on Geoscience and Remote Sensing, 47*, 202–213. doi:10.1109/TGRS.2008.2002881

Luna, I., & Lobo, A. (2016). Mapping crop planting quality

in sugarcane from UAV imagery: A pilot study in Nicaragua. *Remote Sensing, 8*, 500. doi:10.3390/rs8060500

Mancini, F., Dubbini, M., Gattelli, M., Stecchi, F., Fabbri, S., & Gabbianelli, G. (2013). Using unmanned aerial vehicles (UAV) for high-resolution reconstruction of topography: The structure from motion approach on coastal environments. *Remote Sensing, 5*, 6880–6898. doi:10.3390/rs5126880

Martínez, S., Ramil, P., & Chuvieco, E. (2010). Monitoring loss of biodiversity in cultural landscapes. *New Methodology Based on Satellite Data. Landscape and Urban Planning, 94*, 127–140. doi:10.1016/j.landurbplan.2009.08.006

Motohka, T., Nasahara, K. N., Oguma, H., & Tsuchida, S. (2010). Applicability of green-red vegetation index for remote sensing of vegetation phenology. *Remote Sensing, 2*, 2369–2387. doi:10.3390/rs2102369

Pontius, R. G. (2000). Quantification error versus location error in comparison of categorical maps. *Photogrammetric Engineering and Remote Sensing, 66*, 1011–1016.

Pöyhönen, P. (2009) *Legal and illegal blurred: Update on tin production for consumer electronics in Indonesia, FinnWatch, Helsinki.* Retrived from https://www.somo.nl/wp-content/uploads/2009/06/Legal-and-illegal-Blurred.pdf

Reigber, A., Scheiber, R., Jäger, M., Prats-Iraola, P., Hajnsek, I., Jagdhuber, T., ... Moreira, A. (2013). Very-high-resolution airborne synthetic aperture radar imaging: Signal processing and applications. *Proceedings of the IEEE, 101*, 759–783. doi:10.1109/JPROC.2012.2220511

Richards, J. A. (2009). *Remote sensing with imaging radar.* New York: Springer.

Sanga-Ngoie, K., Iizuka, K., & Kobayashi, S. (2012). Estimating CO_2 sequestration by forests in oita prefecture, Japan, by combining LANDSAT ETM+ and ALOS satellite remote sensing data. *Remote Sensing, 4*, 3544–3570. doi:10.3390/rs4113544

Sharma, R. C., Tateishi, R., Hara, K., & Iizuka, K. (2016). Production of the Japan 30-m land cover map of 2013–2015 using a random forests-based feature optimization approach. *Remote Sensing, 8*, 429. doi:10.3390/rs8050429

Small, D. (2011). Flattening Gamma: Radiometric terrain correction for SAR imagery. *IEEE Transactions on Geoscience and Remote Sensing.* doi:10.1109/TGRS.2011.2120616

Tang, J., Wang, L., & Yao, Z. (2008). Analyses of urban landscape dynamics using multi-temporal satellite images: A comparison of two petroleum-oriented cities. *Landscape and Urban Planning, 87*, 269–278. doi:10.1016/j.landurbplan.2008.06.011

Villa, P. (2012). Mapping urban growth using soil and vegetation index and landsat data: The Milan (Italy) city area case study. *Landscape and Urban Planning, 107*, 245–254. doi:10.1016/j.landurbplan.2012.06.014

Weber, C. (2003). Interaction model application for urban planning. *Landscape and Urban Planning, 63*, 49–60. doi:10.1016/S0169-2046(02)00182-2

Wen, C., Zhang, Y., & Deng, K. (2009, September) Urban area classification in high resolution SAR based on texture features. International Conference on Geo-spatioal Solutions for Emergency Management (GSEM 2009) (pp. 281–285). Beijing, China.

Wu, Q., Li, H., Wang, R., Paulussen, J., He, Y., Wang, M., ... Wang, Z. (2006). Monitoring and predicting land use change in Beijing using remote sensing and GIS. *Landscape and Urban Planning, 78*, 322–333. doi:10.1016/j.landurbplan.2005.10.002

Zhou, G., Chen, W., Kelmelis, J. A., & Zhang, D. (2005). A comprehensive study on urban true orthorectification. *IEEE Transactions on Geoscience and Remote Sensing, 43*, 2138–2147. doi:10.1109/TGRS.2005.848417

5

Sedimentary facies and evolution of the lower Urho Formation in the 8th area of Karamay oilfield of Xinjiang, NW China

Hongwei Kuang[1]*, Guangchun Jin[2] and Zhenzhong Gao[3]

*Corresponding author: Hongwei Kuang, Institute of Geology, Chinese Academy of Geological Sciences, Beijing 100037, China

E-mail: kuanghw@126.com

Reviewing editor: Xiao-Jun Yang, China Unviersity of Mining and Technology, China

Abstract: On the basis of core descriptions, examinations of rock thin sections, and EMI logging, a fluvial-dominated fan delta depositional system consisting of the fan delta plain and front subfacies from the lower Urho Formation of the Upper Permian in the Karamay oilfield of Xinjiang, NW China, was recognized in this research. The typical characteristics of this kind of fan delta is the depositional processes that were dominated by the fluvial and tractive current structure that developed very well, while pro-fan delta and gravity deposition occurred very less. The key microfacies associations in the fan delta plain subfacies are braided channel, sheeted flow, mud flow, and sieve deposits, while the fan delta front subfacies commonly contains subaqueous channels, interdistributary channels, debris flow, grain flow, and subaqueous levee microfacies. A study of image logging, grain size, and compositions of rocks indicate that provenance directions are different from the early to the last depositional periods of the Urho Formation, i.e. it changed from the southwest initially to the northwest in the latest among the fifth period to the first period. In addition, with the corresponding provenance direction changing, the sedimentary facies also transform regularly. This paper illuminated the sedimentary facies, subfacies and microfacies and also discussed deeply the depositional

ABOUT THE AUTHOR

Hongwei Kuang is a professor working for the Institute of Geology, Chinese Academy of Geological Sciences (CAGS), Beijing, China. Her research area of interest is Mesozoic stratigraphy, sedimentology and dinosaur taphonomy, Precambrian sedimentary geology, and petroleum geology. Kuang obtained her PhD from the China University of Geoscience (Beijing) in 2003, and she also worked at the Yangtze University for 12 years. Currently, she is managing several research projects to study the stratigraphy and sedimentology of the Mesozoic and Precambrian.

PUBLIC INTEREST STATEMENT

When a river carrying a large amount of deposits influxes into the sea or a lake, those deposits would be unloaded and gradually develop into a fan-like alluvial plain on the mouth of the river, which is named delta. The aquatic part of a delta is the delta plain, whereas the continuous extension into the underwater is called the delta front. It is known as the prodelta, which mostly consists of muds, when the bottom of the delta reaches into the shallow lake or half deep lake area. The sedimentary area of a delta is usually a good oil and gas zone in the geological history because prodelta muds, always containing a large amount of organic matter, are good source rocks; delta plain and delta front sand bodies are favorable reservoir; oil and gas migrates upward to the inclination along the slope direction and can form good reservoirs. Therefore, the area of the delta deposited is the favorable zone for oil and gas exploration.

distribution and evolution of the lower Urho Formation. A sedimentary model of the fluvial-dominated fan delta was presented in this paper.

Subjects: Earth Sciences; Earth Systems Science; Geology - Earth Sciences; Mining, Mineral & Petroleum Engineering

Keywords: fluvial-dominated fan delta; sedimentary facies; provenance analysis; upper Permian; lower Urho Formation; Karamay oilfield, NW China

1. Introduction

The lower Urho Formation of the Upper Permian has been yielding high commercial oil and gas flow since 1960s. Sedimentary facies and its spatial distribution are the key control factors for oil and gas reservoir distribution and directly dominate the results of exploration and development. However, as a typical kind of conglomerate reservoir, the sedimentary facies research of the lower Urho Formation of the Lower Permian at the 8th zone of Karamay oilfield has a long history, but has been controversial. In the 1980s, although Larerson (1989) from SSI Petroleum Company of the United States defined the sedimentary facies of the lower Urho Formation as a wet fan deposit, a facies association from mountain alluvial fan, alluvial fan to fan delta was proposed by Xu (2001), while the Research Institute of the Petroleum Exploration and Development of CNPC (RIPED) and the Branch Company of the Xinjiang Oilfield (2004) concluded a depositional model of a braided river. In recent years, with great progress being made in the oil and gas exploration and development, sedimentary facies research of the lower Urho Formation has been highlighted. Multiple theories were presented. Most of the researchers thought that the conglomerate reservoirs are fan deltas (He, Qiu, Luo, & Wu, 2014; Pang, 2015; Yang, Wei, Abulimidi, Chen, & Bing, 2016). Pang (2015) further divided the fan delta into traction current sand-conglomerate and gravity current sand-conglomerate in the lower Urho Formation. Comparing the physical characteristics among different reservoirs, Zhang et al. (2015) found that the reservoir property of the lower Urho Formation (P_2w) is the best in the Kebai area of the northwestern Junggar Basin. Wu et al. (2013) presented a sub-lacustrine fan having a Bouma sequence, and the grain size distribution revealed a typical turbidity deposit. Other researchers thought it was an alluvial fan (Hu & Gong, 2016), a composite fan (Lu, Shi, Ge, & Zhang, 2012), and a subaqueous fan (Li et al., 2015), respectively. Obviously, there are still some ambiguous views on the identification of sedimentary facies in the lower Urho Formation. It is necessary to re-research the sedimentary characteristics of the lower Urho Formation of the Lower Permian for pinpointing the direction of oil and gas exploration.

Our research, which is based on observations and descriptions of more than 1,000-m-long cores and examinations of rock thin sections from 12 wells and on the analysis of well logging of more than 800 wells and especially electrical micro-imaging (EMI) logs of 15 wells, suggests that the lower Urho Formation of the Upper Permian would be a fluvial-dominated fan delta depositional system and corresponding facies association in the 8th area of the Karamay oilfield. Two subfacies and nine microfacies were recognized further and studied in detail.

2. Geological setting

The south fault of Baijiantan is located in the central south of the Karamay–Urho fault zone, which is in the northwestern margin of Junggar Basin of Karamay oilfield. The 8th zone is located in the lower plate of the south Baijiantan fault (Figure 1). The Lower Permian, includes the Fengcheng Formation and Jiamuhe Formation, and the Upper Permian consists of the upper Urho Formation and the lower Urho Formation, which is overlain by the Mesozoic Xiazijie Formation. The Permian lower Urho Formation is overlapped layer by layer in the 8th zone of the Karamay oilfield, Junggar Basin. It suffered from intense denudation because of the late strong thrusting and uplifting (Wu et al., 2015), so that the top and bottom of the lower Urho Formation were both bounded by a regional unconformity in the 8th zone. This study focused on the sedimentary facies and provenance of the lower Urho Formation, in which a number of oil reservoirs were explored. A series of structures bearing oil and gasses in the NE–SW trend were tectonically presented (Figure 1). The lower Urho Formation of

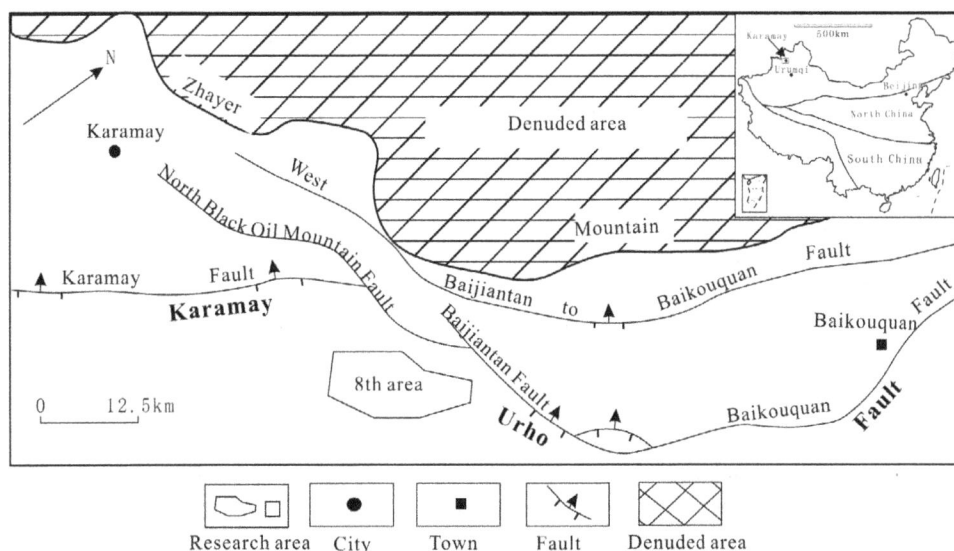

Figure 1. Map showing the location of the study area and schematic tectonic geology.

Table 1. Classification of lithological stratigraphy and sequence stratigraphy of the lower Urho Formation

Stratigraphy					Sequence stratigraphy			
Triassic	Series	Formation	Member	Layer	Parasequence	Parasequence set	System tract	Sequence
Permian	Upper Permian	Urho Formation	Member No. 1	1–1	6	Progradation	Highstand system tract (HST)	Sequence II
				1–2	5			
				1–3	4			
				1–4	3			
				1–5	2			
				1–6	1			
			Member No. 2	2–1	5	Retrogradation	Transgressive system tract (TST)	
				2–2	4			
				2–3	3			
				2–4	2			
				2–5	1			
			Member No. 3	3–1	4	Progradation	Highstand system tract (HST)	Sequence I
				3–2	3			
				3–3	2			
				3–4	1			
			Member No. 4	4–1	2	Retrogradation–Aggradation	Transgressive system tract (TST)	
				4–2	1			
			Member No. 5	5–1	5	Aggradation–Progradation	Lowstand system tract (LST)	
				5–2	4			
				5–3	3			
				5–4	2			
				5–5	1			
	L. Per.	Jiamuhe Formation						

the Upper Permian is made up mainly of coarse sediments of conglomerate and subdivided into five lithological members and 22 layers on the basis of studies of spectral logging, combination logging (Li, Wang, Zhou, Tan, & Li, 2004), seismic data, depositional evolution, and sequence stratigraphy; two sequences and five system tracts or parasequences were recognized (Table 1) (Lei et al., 2005).

3. Sedimentary facies

Two subfacies and nine microfacies were recognized, respectively, in the lower Urho Formation of the Upper Permian in the 8th zone of Karamay oilfield, Junggar Basin. Four microfacies in the fan delta plain subfacies are braided channel, sheeted flow, mud flow, and sieve deposits, while the fan delta front subfacies commonly contains subaqueous channels, interdistributary channels, debris flow, grain flow, and subaqueous levee microfacies. Detailed descriptions of these microfacies are shown in the text of this paper.

3.1. Braided channel

Braided channel deposits, occupying 80% in a total thickness of the fan delta plain, is the main microfacies in the fan delta plain and consists of mainly conglomerates in various grain size (Figure 2(A)) and occasionally pebble sandstones (Li, 1997). Parallel bedding, large-scale

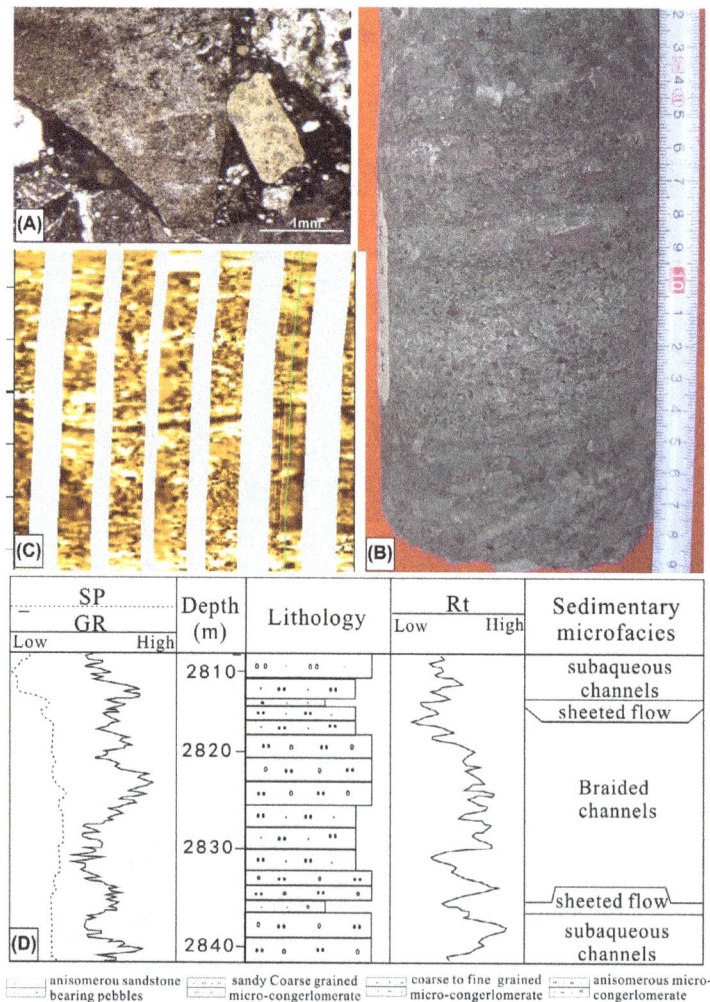

Figure 2. Sedimentary characteristics of braided channel showing in thin section, cores and well logging. (A) Micro-grained conglomerate, which is mainly acidic lavas, a little is ferrite rock; subangular to subround, good pores; siliceous cement, 2,588.96 m depth of the Jianwu well 3; (B) color micro-grained conglomerate with parallel bedding, braided channel, 2,592.44 m depth of Jianwu well 3; (C) braid channel deposit, thinning upward, 2,503.6–2,508.4 m depth of well T85601; (D) microfacies of braided channels and sheeted flow of Jianwu Well 3.

cross-bedding, and anti-dune cross-bedding in a fining-upward sequence and eroding surface on the bottom are common (Figure 2(B)). Generally, it appears a feature upward from light to dark in color and big to small in white points in the EMI logging and remarkable bedding or lamination and fining-upward in grain size, although cross-bedding and parallel bedding are common (Figure 2(C)). In addition, a gamma curve with great amplitude shows a teeth-like case or campaniform. Resistivity displays teeth-like campaniform or case-form (Figure 2(D)).

3.2. Sheeted flow

It is defined as finer grained sediments deposited by sheet flow or overbank flooding at the distal fan channels, which consists of reddish shale sandstone or siltstone (Figure 3(A) and (B)), and cracks are common because of the exposed environment (Figure 3(B)). Gamma curve of sheet flow deposits is always jigsaw with the number of teeth crests in small amplitude, which becomes smaller upward (Figure 2(D)). This microfacies is also identified as the dark areas in the alternation ribbon of bright and dark in the EMI image logging (Figure 3(C)).

3.3. Mud flow

Mud flow is a type of massive mixture sediments in varied grain size, i.e. clay, sand, and gravels, poorly sorted and huge floating conglomerates, rare bedding, and rich in mud (Figure 4(A)). This microfacies is generally thin in thickness and poor developed. Resistivity curve of this microfacies is jagged zigzag, the peak value is higher, and gamma curve appears jigsaw in moderate amplitude; both resistivity and gamma curves display gradual patterns on the top and bottom (Figure 4(C)), and the EMI logging represents white dotted floating images among gray mid-lower resistivity matrix (Figure 4(B)).

3.4. Sieve deposit

Sieve sediments, appearing occasionally, is mainly moderate–coarse conglomerate with well voids (Figure 4(D)).

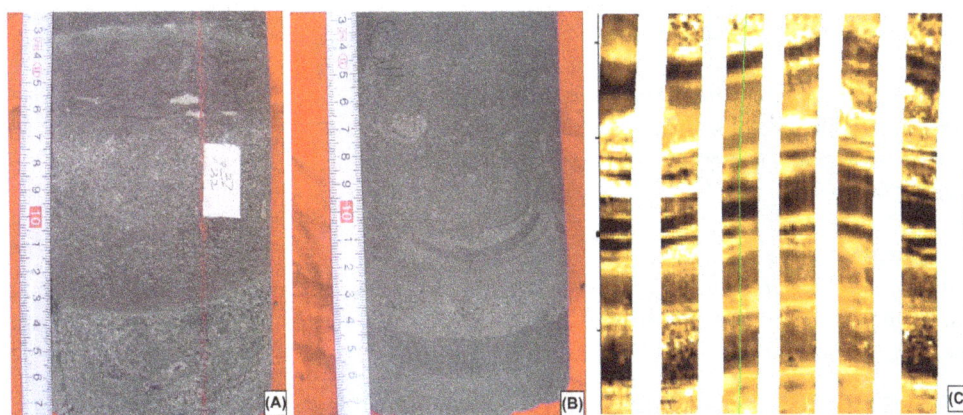

Figure 3. Sedimentary characteristics of sheeted flow showing in thin sections, cores and well logging. (A) Middle and upper part: pink reddish mudstone, clayey sandstone with gray green siltstone ribbons, sheet flow; deposit of braided channel on the bottom, mainly micro-grained conglomerate, 2,564.67 m depth of well T85722; (B) sandstone and mudstone, mud cracks in the middle part, sheet flow deposits, 2,828.79 m depth of well 3 of Jianwu; (C) sheeted flow deposit, 2,619.55–2,620.38 m depth of well T85689.

Figure 4. Sedimentary characteristics of mud flow and sieve deposit showing in thin sections, cores and well logging. (A) Colorized sandy conglomerate, mud flow deposits, 2,594.92 m depth of Jianwu well 3; (B) mud flow deposit, dotted and massive, white in color and with high resistivity, interpreted as conglomerate, and brown in color and with moderate resistivity matrix, 2,657.02–2,658.60 m depth of well T85689; (C) microfacies of braided channels and mud flow of Jianwu Well 3; (D) sieve deposit, 2,718.1 m depth of Jianwu well 3.

3.5. Subaqueous distributary channel

Deposits of subaqueous distributary channels, characterized by gray, finer sandstone with well-developed cross-bedding compared with braided channels, are dominantly in the fan delta front and consist of gray-green fine or coarse micro-conglomerates with normal rhythm. Coarse gravels appear generally on the bottom of the rhythm, whereas coarse sandstones with gravels on the top occasionally erode the surface (Figure 5(A)–(E)).

Its resistivity curves are lower zigzag and moderate amplitude, and gamma curves appear zigzag, case-like, or campaniform with high amplitudes (Figure 5(G)). The EMI log of this microfacies exhibits an obvious change in lightness from light at the lower part to dark at the upper portion (i.e. from coarse to fine sediments) (Figure 5(F) and (H)).

3.6. Subaqueous levees

Subaqueous levees are characterized by the interbedded thin deposits of sandstone and mudstone with limited extensions (Figure 6(A)). The gamma curve is sharp or straight with little amplitude and dark ribbons in the images of EMI log (Figure 6(C)).

3.7. Subaqueous interdistributary channel

It refers to the gray or dark-gray sandy mudstone or sandstones interbedded with conglomerates of subaqueous channels, with horizontal or parallel bedding, cross-bedding occasionally. The gamma curve is straight, and interbedded light and dark ribbons are presented on the well logging images (Figure 6(D)).

3.8. Debris flow

Except for the gray color, this microfacies is similar with reddish mud flow microfacies in grain size, texture, and sediment compositions (Figure 6(B) and (E)).

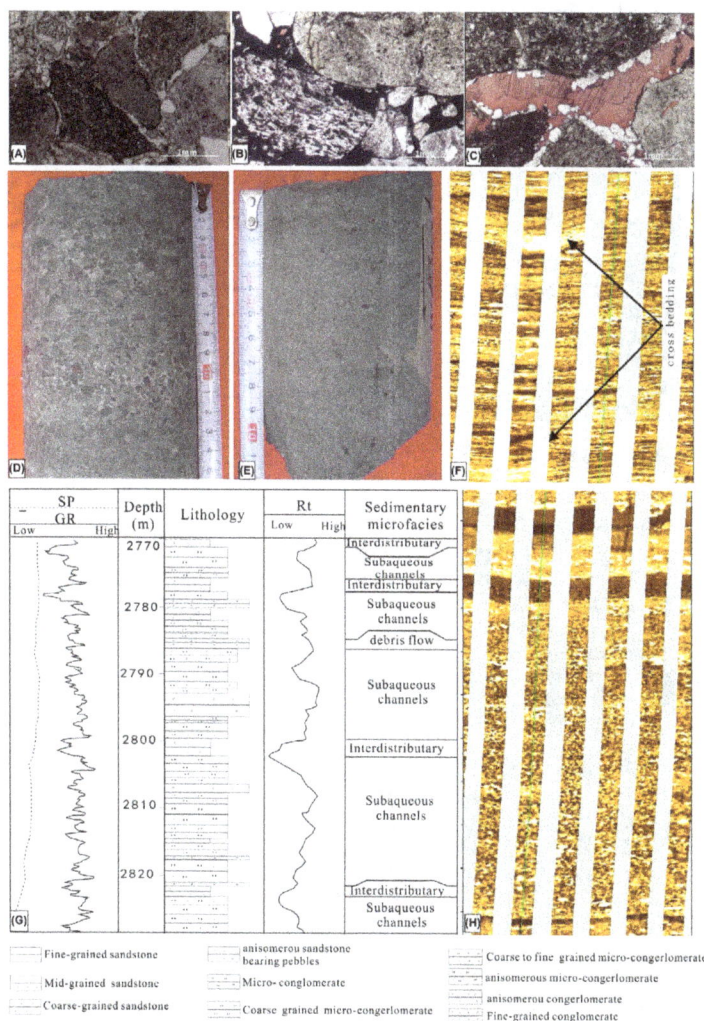

Figure 5. Sedimentary characteristics of subaqueous distributary channel showing in thin sections, cores and well logging. (A) Micro-grained conglomerate, which is mainly andesite, acidic lavas, partly chloritization; cement is tuff filled in the pores among gravels, 2,556 m depth of well T85722; (B) coarse to fine grained micro-conglomerate, which is mainly rhyolite and andesite, 2,628.1 m depth of well T85722; (C) coarse grained micro-conglomerate, good pores, two-phase cements, the early part is granular analcites, and the late portion is bulky calcite crystalline, 3,037.15 m depth of well 8,650; scale bar = 1 mm; (D) rhythmic deposits of coarse- and fine-grained micro-conglomerate, subaqueous channel deposit, 2,641.47 m depth of well T85722; (E) parallel bedding and cross-bedding, subaqueous channel deposit, 2,788.55 m depth of Jianwu well 3; (F) cross-bedding, subaqueous channel deposit, 2,680.97–2,684.90 m depth of Well T85024; (G) microfacies of subaqueous interdistributary channel and debris flow of well 85689A; (H) subaqueous channel deposit, thinning upward, 2,853.7–2,856.4 m depth of well T85024.

3.9. Grain flow

Grain flow sediments are characterized by a depositional succession of coarsening-upward sequences (Li & Guo, 2000). It is composed of coarse sandstones with pebbles, fine conglomerates, and coarse conglomerates from bottom to top and rare sedimentary structures. A layer of grain flow deposition is not too thick, i.e. sand grain flow deposits sere only several centimeters in thickness, although one layer of grain flow deposition mainly composed by gravels can be several centimeters to meter scale. Gamma curves of grain flow deposits present infundibuliform with serration. In the image curves, it displays color grading from dark to light from lower to upper or light points becoming bigger upward (Figure 6(F)).

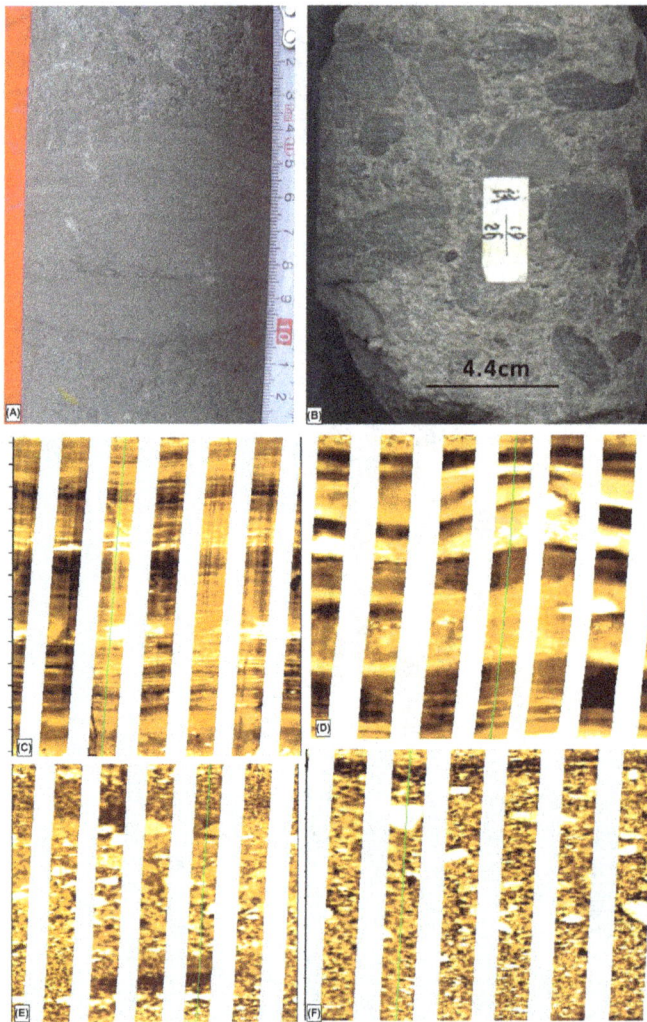

Figure 6. Sedimentary characteristics of subaqueous levees, subaqueous interdistributary channel, debris flow, and grain flow showing in thin sections, cores and well logging. (A) Subaqueous levee deposit, 2,881.64 m depth of Jianwu well 3; (B) debris flow deposit, 3,023.50 m depth of well 85095; (C) subaqueous levee deposit, 3,050.10–3,052.15 m depth of well T85462; (D) subaqueous interdistributary channel deposit, 2,641.89–2,644.30 m depth of well T86277; (E) debris flow deposit, 2,897.50–2,898.70 m depth of well T85024; (F) grain flow deposit, coarsening-upward sequence, 2,699.78–2,701.60 m depth of well T86166.

4. Discussion: Sedimentary model of fluvial-dominated fan delta

Previous studies have shown that the reservoir characteristics are better than fan delta or alluvial fan, while the reservoir distribution has some fan features (Zhang et al., 2015) in the lower Urho Formation. We also found that the lower Urho Formation has a series of sedimentary characteristics of fan delta, but it has a key difference from alluvial fan and subaqueous fan, i.e. gravity flow; characterization of fan sedimentary characteristics occurred rarely in the lower Urho Formation.

Table 2 lists the similarities and differences of the main sedimentary characteristics among the common alluvial fan, fan delta, fluvial-dominated fan delta, and braided delta. Coarse-grained depositional system includes alluvial fan, fan delta, braided river delta, etc. Currently, the progress of research on alluvial fan deposition mainly includes the deepened research on the main factors controlling the development and evolution of such deposition and the application of diversified means to the establishment of its sedimentary models (Zhu et al., 2016).

Table 2. Comparison with sedimentary characteristics among four coarse-grained depositional systems

Characteristics		Depositional model			
		Alluvial fan	Fan delta under arid climate conditions	Delta under wet climate conditions	Braided river delta
Type of channel of a river		Braided channel	Braided channel	Braided channel	Braided channel
Stability of channel		Without a fixed channel	Extremely Unstable	Extremely Unstable	Unstable
Variability of flux		Paroxysmal	Extremely largest	Larger	Large
Slope of landform		Biggest	Bigger	Gentle	Big or gentle
Mechanism of transportation		Gravity flow	Gravity flow mainly	Tractive current mainly	Tractive current mainly
Sedimentary structure		Massive bedding	Massive bedding and graded bedding	Cross-bedding and fining-upward succession	Lateral accretion cross-bedding and massive bedding
Texture of sandstone	Grain size	Mainly disordered conglomerate	Mainly disordered conglomerate and sandy conglomerate	Mainly disordered conglomerate and sandy conglomerate	Conglomeratic sandstone and sandstone with conglomerate
	Sorting	Worst	Worse	Middle to bad	Middle to bad
	Matrix	Most	More	A few	A few
Ratio of sandstone and mudstone		High	Higher	Higher	Highest
Distributary mouth bar and sheet sandstone in front of the delta		No	Extremely rare	Rare	Few
The distance from resource area		Nearest	Nearer	Nearer	Near

The main factors controlling the development and evolution of alluvial fan mainly include piedmont tectonic activities, climate, material source, mountain pass landforms, base level lift, etc., and piedmont tectonic activities and climate are the main factors controlling the development of alluvial fan. The tectonic uplift amplitude controls the pattern and scale of the alluvial fan. Climate conditions influence the weathering extent of the mother rock and control the volume of clastic materials and hydrodynamic conditions in that drainage area, thus controlling the sedimentary characteristics of the alluvial fan. Geographic relief controls the pattern of the alluvial fan and its sedimentary characteristics; the steeper topographic slope shall render larger alluvial fan thickness, and such fan is dominated by conglomerate and sandstone; relatively gentle topographic slope shall render larger alluvial fan area, and such fan is dominated by conglomerate, sandstone, and siltstone. Base level lift influences the development pattern of the alluvial fan; base level ascending shall render thick alluvial fan deposition with a small area, whereas base level descending shall push such fan forward toward the basin, and therefore, the area of alluvial fan deposition is large.

Usually, a fan delta deposit is a kind of coarse clastic sedimentary system from the continental surface to the subaqueous formed by the alluvial fan into the lake (Galloway & Hobday, 1989; Xue & Galloway, 1991). It usually develops a series of microfacies, such as mud flow, braided channel, debris flow, subaqueous distributary channel, distributary mouth bar, etc. in the front delta; there is often some coarse-grained bore in the pre-delta mud. It is the important sedimentary characteristics for fan delta system that gravity flow deposits are dominant in it, in which the grain size is always coarse and mainly sandy conglomerates, few distributary channels distributed in the gravity flow deposits taking the form of small lenses in most cases, which is the product of the temporary channel for the fan on the flood. Sediments experienced only a little modification and a few distributary mouth bars. So this kind of fan delta resulted from a paroxysmal catastrophic incident bearing abundant material resources from a nearer area (Zhu et al., 2016).

Braided river delta is a common coarse-grained sedimentary type in continental basins. Its sedimentary models and main controlling factors are different from those of the fan delta (Xue & Galloway, 1991). Oil/gas exploration practices and scientific researches in recent years show that depositional systems of braided river deltas exist in Ordos, Tarim, and Bohai Bay basins in China (Dong, Yang, Chen, Wang, & Cao, 2014; Zhu, Deng, et al., 2013; Zhu, Pan, et al., 2013). In the margin of large-scale depression lake basin and faulted lake basin, the deltaic sedimentary area is closed to the source area, which is favorable for the sand/gravel-rich braided river with bedload to prograde into the Lake Basin and form a delta. Shallow-water sedimentary structures like erosion surface and large cross-bedding as well as intermittent depositional positive rhythm have been developed on such delta, the underwater distributary channels are well ramified, the sand body is thick and extends to a long distance; under the common control of tectonic activities, sedimentary source, seasonal flooding, lake level fluctuation and other factors, a coarse-grained composite sand body with the superimposition of multi-stage distributary channels is easily formed (Dong et al., 2014; Zhu, Pan, et al., 2013).

We learn that based on the number of microfacies recognized and described above, the lower Urho Formation of the Upper Permian in the 8th zone of the Karamay oilfield mainly consist of subaqueous channel, sieve and sheet flow deposits that cover a vast area, and rare mud flow deposits. A fluvial-dominated fan delta depositional system, consisting of the fan delta plain and front subfacies from the lower Urho Formation of the Upper Permian in the Karamay oilfield of Xinjiang, NW China, was recognized. The typical characteristics of this kind of fan delta are the depositional processes that were dominated by the fluvial and tractive current structure that developed very well, while pro-fan delta and gravity deposition occurred to a lesser degree. Pebbly braided rivers formed braided river plains for the fan delta plain in wet regions with perennial river (Hasiotis, Charalampakis, Stefatos, Papatheodorou, & Ferentinos, 2006; Yu, 2002) and landform with a genital slope, with much more distributary channels with a shallow cutting and fast migration (Reading, 1978); there developed dispersed parallel cross-bedding in the conglomerates bar, while it developed diversity cross-beddings in the channels' pebbles and sandstones. The river channel extends to the subaqueous. The mud flow (and debris flow) deposits are rare due to perennial surface runoff. Because mud flow (and debris flow) deposits often developed in the drought area where there is a lack of rain, a lot of hillside wastes from weathering products accumulating will mix with the flood and then form high-density, high-viscosity flow—mud flow (or debris flow) when heavy rains come in the summer. Therefore, debris flow deposits in this area is less, and the river deposition, sieve deposition, filter flow deposits formed by low-viscosity flow distributed widely. Typical depositional succession of these deposits in the study area represents a normal fining-upward rhythm, i.e. conglomerate, sandstone, and mudstone from bottom to top. Beddings formed in high-energy regime, for example, cross-bedding of retrograde sand wave, are common. The braided channel that extended into the lake formed the subaqueous distributary channel with a variety of tractive current structures, which dominated the front delta.

Tectonic conditions and the size of the catchment basin have a strong influence on the distribution of the surface runoff, and the size of the basin area is proportional to the scope of the distribution of the fan delta (Yu, 2002). Meanwhile, around a depositional area of 40 km^2 of the extended lower Urho Formation, there are no sediments of predelta subfacies and rare deposits of distal front delta such as channel mouth bar and distal bar (Reading, 1978), where the front delta deposits is dominant and the distal front delta is out of the study area. All of these indicate that the depositional range of the fan delta in study area is very large. Therefore, this study suggests that the lower Urho Formation of the Upper Permian in the 8th zone of the Karamay oilfield is a depositional system of fluvial-dominated fan delta. It also suggests that the sedimentary water system is widely distributed, and terrain slope is not very big. In this context, even in drought areas, it can also develop fluvial-dominated fan delta.

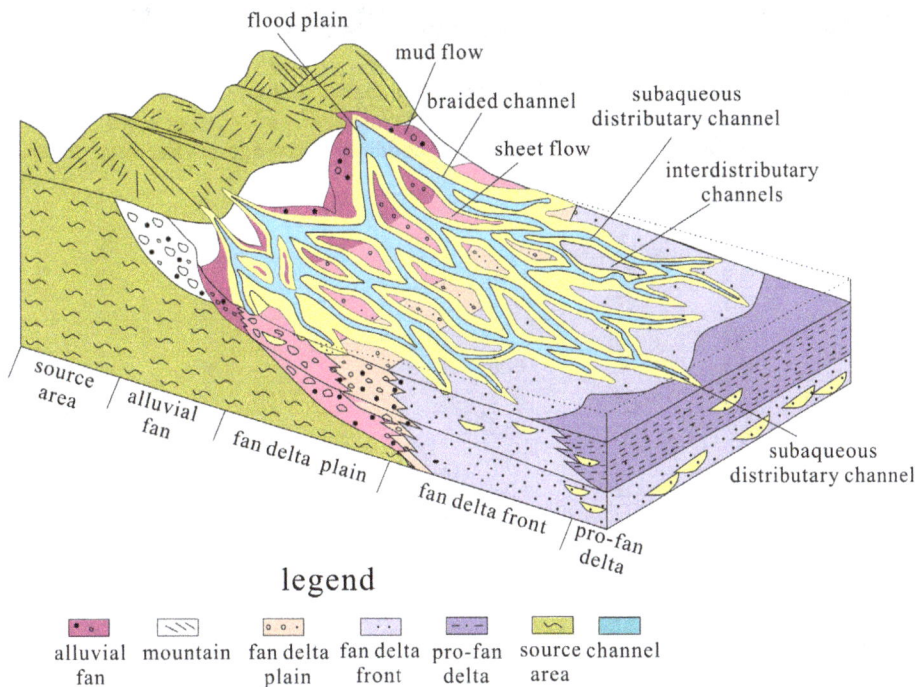

Figure 7. Sedimentary model of the fluvial-dominated fan delta (modified from He et al., 2014).

The term of the fluvial-dominated fan delta here can also be defined as a wet fan delta; it differs from the dry fan delta or fan delta (Dang, Yin, & Zhao, 2004) in sediments and facies association. The former is characterized by the dominant deposits of channel microfacies, and the latter is mainly gravity flow deposits. In other words, gravity flow (including mainly mud flow in the plain delta and debris flow in the front delta) developed very well in the dry fan delta environment (Chen, Sun, & Jia, 2006; Meng et al., 2006), are rare or a very small-scale, while braided channel and subaqueous distributary channel microfacies are perfectly developed in the fluvial-dominated fan delta. Furthermore, the covering area of sediments from the fluvial-dominated fan delta may be much larger than the dry fan delta, and the cumulative thickness of sediments is also great. Thus, the fluvial-dominated fan delta is different from the common fan delta. Figure 7 is the sedimentary model of the fluvial-dominated fan delta.

5. Provenance analysis and sedimentary evolution

5.1. Provenance analysis

In this study, based on the EMI log of 15 wells, plenty of cross-bedding structures were recognized, and then the trend and dip of 443 cross-beddings and stratigraphy were measured and revised by Wulff net calibration. The total measured data from cross-bedding structures of the lower Urho Formation are listed in Table 3. The measured data number of rose maps of cross-bedding dip and average dip distribution of cross-bedding of the lower Urho Formation were worked out (Figure 8).

Figure 8(A) shows that the dips of cross-beddings from wells of T86120 and T85024 (southwest), and T85015 (south) and T85722 (southeast) present one dominant trend, while those of wells of T85006, T85072, T85713, 85607, etc. display two centralized trends, i.e. northeast and south or southeast. This fact indicates that at least two provenances exist in the periods when the lower Urho Formation was deposited.

Table 3. Statistic table of the average cross-bedding dip in the lower Urho Formation

Well	The 1st member		The 2nd member		The 3rd member		The 4th member		The 5th member	
	Azimuth of average trend (°)	Number of cross-bedding	Azimuth of average trend (°)	Number of cross-bedding	Azimuth of average trend (°)	Number of cross-bedding	Azimuth of average trend (°)	Number of cross-bedding	Azimuth of average trend (°)	Number of cross-bedding
T85722			108.0	3	135.3	6				
T85601			100.3	3	68.0	5	91.8	4		
T85689			111.0	3	126.5	9	66.2	5	56.2	2
T86277	171.2	3	137.5	10	92.4	8	91.3	3		
T85607	106.5	4	115.2	10	113.3	6	78.9	8		
T86245	114.3	7	126.3	8	82.0	9				
85090A	140.8	2	107.6	13	90.2	7	75.5	2		
T85713	125.6	20	92.8	13	86.5	11	48.6	5	102.5	3
T85027	157.7	3	117.4	10	108.4	7	65.5	2	56.5	2
T85006	184.7	3	139.9	16	134.6	5	71.8	5	56.8	6
T85462	164.3	21	163.8	20	116.1	8	51.0	8	36.2	3
T85015	150.9	7	165.2	6	162.0	1				
T85024	167.7	25	176.2	11	160.0	4				
T86120	200.1	18	172.7	15	169.4	7				
T86166	155.7	18	123.4	17	159.3	3				
Total amount	153.3	131	130.5	158	120.3	96	71.2	42	61.6	16

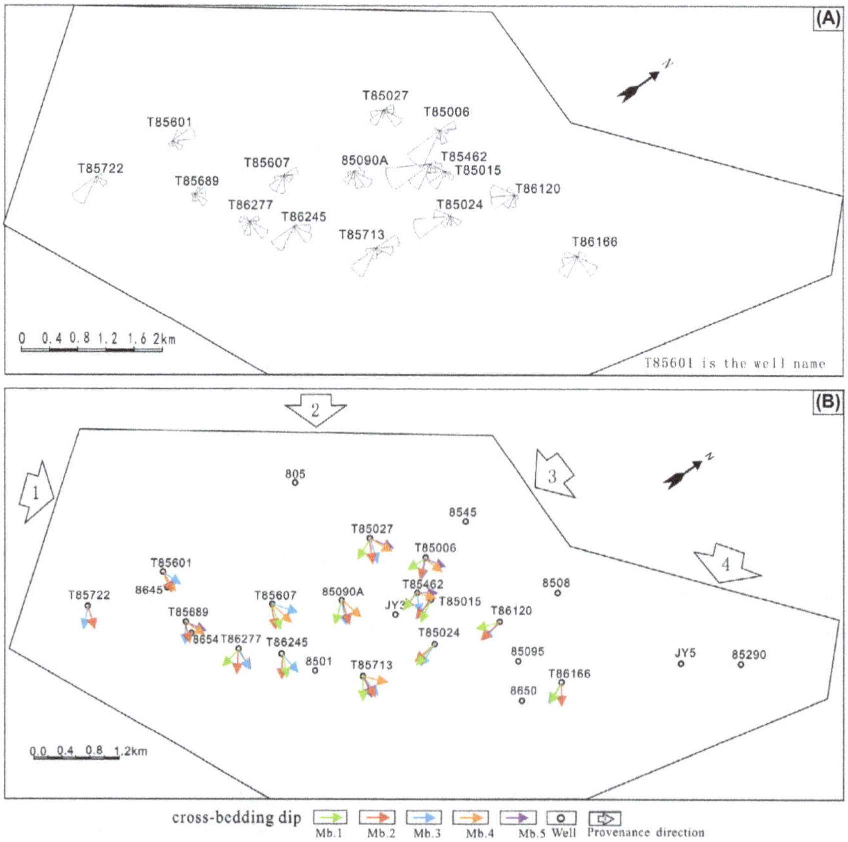

Figure 8. Rose map of cross-bedding dip and the current direction of the lower Urho Formation in the study area.

For the 5th member of the lower Urho Formation, except well T85713 in which the average dip trend is east, the other four wells' average dip trend of cross-beddings is northeast, which supports a southwest provenance. The provenance of the 4th member succeeded the 5th; seven wells' average dip trend of cross-beddings is mainly northeast and a southwest provenance except for wells of T85601 and T86277's eastern average dip trend (Figure 8(B)). Although average dip trend from the 3rd member changed to southeast, some wells represent northeast average dip trend, which suggests that the southwest provenance in the depositional periods of the 5th and 4th members was replaced gradually by a northwest provenance (Figure 8(B)). As for the depositional period of the 2nd member, northeast provenance was completely replaced by a southeast one, and a poor south provenance was added (Figure 8(B)).

The above statement indicates that three changeable provenances had prevailed in the period of the late Permian. They were, respectively, located at the southwestern and northern area of wells 805, 8545, and JY5 and northwestern region around well 805 (Figure 8(C)). These three provenances have played different roles in different periods during which the lower Urho Formation was deposited.

5.2. Sedimentary evolution

Combining the stratigraphy distribution (Lei et al., 2005), sedimentary facies (Xing, Zhu, Kuang, Aa, & Liu, 2006) and palaeocurrent research, sedimentary evolution of the lower Urho Formation experienced two big cycles and formed two sequences. Members 5 to 3 comprise sequence I; members 1 and 2 constitute sequenceII. Member 5 is mainly distributed in the lower part of the paleotopography, which gradually overlapped upward to the top of the monoclinic slope (Figure 9(A)). The depositional area of the 4th member was most widely distributed by fan delta front subfacies; moreover, the microfacies types were most complete than other members (Figure 9(B)). After the 3rd member deposition, the southwestern part of the study area rose up and suffered from erosion, and the local unconformity formed between member 2 and member 3. Corresponding to lake level fluctuation and changing of provenance directions, the sedimentary facies also changed. From period 5 to 4, deposited range increased gradually, and the proportion of the plain delta decreased, while the proportion of the front delta increased and reached the maximum at the end of the fourth member deposition. From period 4 to 3, the depositional range of the front delta decreased, accompanying the plain delta continuously extending from the southwestern edge to the northwestern edge. After the 3rd member deposition, which was affected by rising and suffering from erosion in the southwest of the study area, the plain deposit remained less in the southwestern edge (Figure 9).

The 1st and 2nd members remained incomplete, especially missing much more on the uppermost part. At the beginning of the 2nd member deposition, water quickly overlapped to the southwest direction, which makes the range of sedimentary records in member 2 larger than those in the period of member 3 deposition. Member 1 remained less because of late denudation. During the 2nd member deposition, the plain delta had quite a wide range, but later, the plain area decreased, and the front of the delta deposits increased with the transgression toward the southwest. With the lake level down during the 1st member deposition, the range of the front delta reduced, and the plain deposits increased. But now, member 1 maintains a narrower range due to late erosion (Figure 9(D) and (E)).

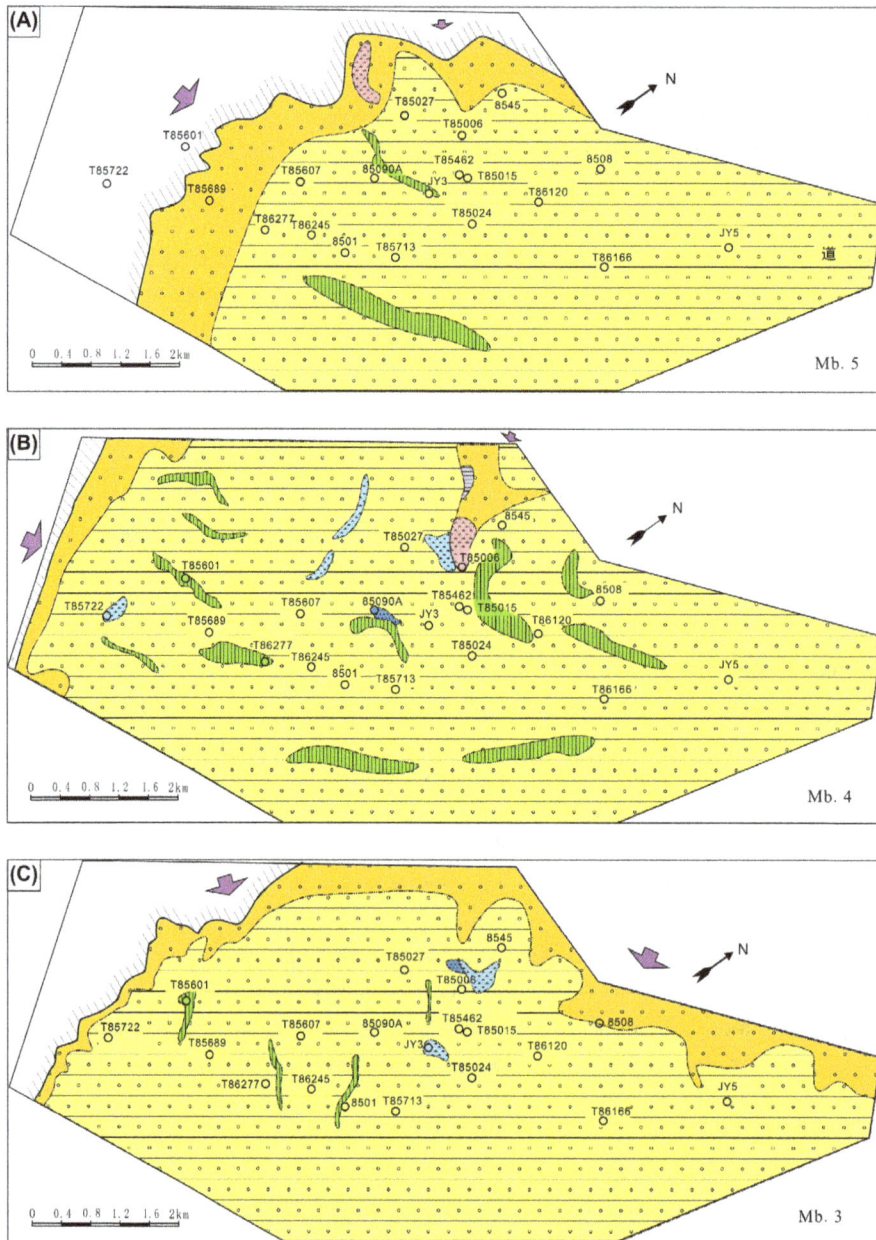

Figure 9. Sedimentary evolution of the lower Urho Formation in the 8th area of Karamay oilfield.

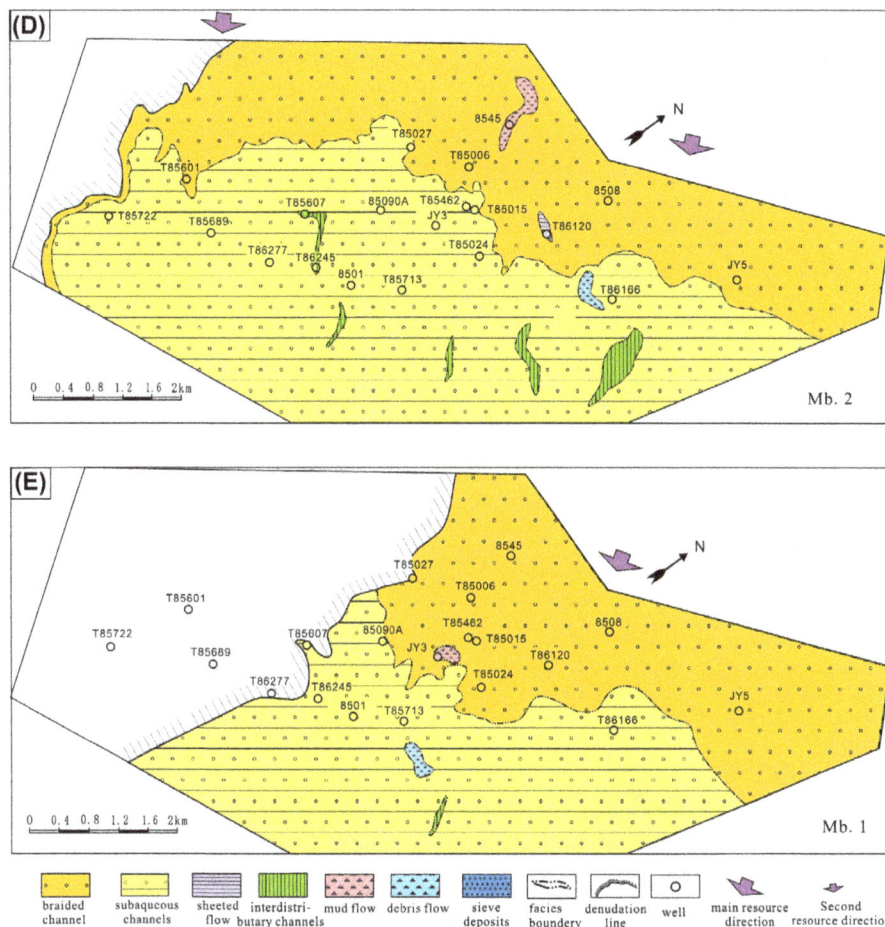

Figure 9. (Continued)

6. Conclusions

The lower Urho Formation of the Upper Permian is considered as sediments of a fluvial-dominated fan delta facies, which mainly consists of fan delta plain and fan delta front subfacies. This delta is controlled by river and main channel sediments and rare gravity flow deposits. Nine key microfacies of fluvial-dominated fan delta were recognized, i.e. braided channel, sheeted flow, mud flow and sieve deposits in fan delta front subfacies and subaqueous channel, inter-distributary, debris flow, grain flow, and subaqueous levee in fan delta plain subfacies. The distribution of cross-beddings dip, based on EMI log, indicates that four provenances prevailed in the period of Late Permian: south-western, northwestern, northern and northeastern, which implies a paleo-landform with high in the west and lower in the east that had existed in the Late Permian in the northern Junggar Basin.

Funding

This research was supported by the National Natural Science Foundation of China Projects [grant numbers 41272021, 41372109]; China Geological Survey Project [grant number 12120115068901].

Author details

Hongwei Kuang[1]
E-mail: kuanghw@126.com
Guangchun Jin[2]
E-mail: jgcjky@163.com
Zhenzhong Gao[3]
E-mail: Gaozz@163.com

[1] Institute of Geology, Chinese Academy of Geological Sciences, Beijing 100037, China.

[2] Oil & Gas Survey of China Geological Survey, Beijing 100029, China.

[3] School of Earth Science of Yangtze University, Wuhan 434023, Hubei, China.

References

Chen, C., Sun, Y., & Jia, A. (2006). Development and application of geological knowledge database for fan-delta front in the dense spacing area. *Acta Petrolei Sinica, 27*, 53–57.

Dang, Y., Yin, C., & Zhao, D. (2004). Sedimentary facies of the Paleogene and Neogene in Western Qaidam Basin. *Journal of Paleogeography, 6*, 297–306 (in Chinese with English abstract).

Dong, Y., Yang, S., Chen, L., Wang, Q., & Cao, Z. (2014). Braided river delta deposition and deep reservoir characteristics in

Bohai Bay Basin: A case study of Paleogene Sha 1 Member in the south area of Nanpu Sag. *Petroleum Exploration and Development, 41*, 385–393.

Galloway, W. E., & Hobday, D. K. (1989). *Terrigenous clastic sedimentary system, application in the oil, coal and Uranium exploration* (pp. 22–44) (X. Gu, J. Gu, & X. Gao, Trans.). Beijing: Petroleum Industry Press.

Hasiotis, T., Charalampakis, M., Stefatos, A., Papatheodorou, G., & Ferentinos, G. (2006). Fan delta development and processes offshore a seasonal river in a seismically active region, NW Gulf of Corinth. *Geo-Marine Letters, 26*, 199–211. https://doi.org/10.1007/s00367-006-0020-8

He, C., Qiu, Z., Luo, G., & Wu, R. (2014). Sedimentary facies analysis of the Urho Formation in 5–8 District in the Northwestern Junggar Basin. *Xinjiang Oil & Gas, 10*(3), 1–9.

Hu, Z., & Gong, F. (2016). Study on reservoir types of Upper Wuerhe Formation in the east of 53 district of Karamay Oilfield. *Petrochemical Industry Technology*, (3), 168–169+14.

Larerson, W. N. (1989). *Report on geology and reservoir engineering of the lower Urho Formation in the 8th area of Kamaray oilfield, Xinjiang*. The 2nd oil production of Xinjiang Oilfield. Unpublished manuscript.

Lei, C., Zhang, B., Peng, J., Yu, C., Wang, C., & An, Z. (2005). Sequence stratigraphy of the lower Urho in the 8 District of Karamay oilfield. *Journal of Oil and Gas Technology, 27*, 142–145+1.

Li, G., Wang, X., Zhou, S., Tan, H., & Li, X. (2004). Application of the EMI image logging technique to the Jurassic reservoir of west Sichuan. *Fault Block Oil & Gas Field, 11*, 79–80 (in Chinese with English abstract).

Li, Q. (1997). *Series of oilfield development in China—Conglomerate oilfield development* (pp. 20–50). Beijing: Petroleum Industrial Publishing House (in Chinese).

Li, X., Xiao, C., Yuan, S., Xu, H., Dong, X., & Hou, W. (2015). Percolated mechanism research of a lower permeable conglomerate reservoir—A case study from the lower Urho Reservoir in the 8 District of Karamay Oilfield. *Xinjiang Oil & Gas, 11*, 37–41+2.

Li, Z., & Guo, H. (2000). Storm deposits in the Sinian Jiayuan formation of Xuzhou area (in Chinese with English abstract). *Journal of Paleogeography, 4*, 19–27.

Lu, X., Shi, J., Ge, B., & Zhang, S. (2012). Characteristics of glutenite reservoir of Permian Upper Urho Formation in Zhong Guai—Wuba area in the in the northwestern margin of Junggar Basin. *Northwest Oil & Gas Exploration, 24*, 54–59.

Meng, Y., Gao, J., Niu, J., Sun, H., Yin, X., Xiao, L., … Wang, Y. (2006). Controls of the fan delta sedimentary microfacies on the diagenesis in the south of western Liaohe Depression, Bohai Bay Basin. *Petroleum Exploration and Development, 33*, 36–39.

Pang, D. (2015). Sedimentary genesis of sand–conglomerate reservoir and its control effect on reservoir properties: A case study of the lower Urho Formation in Ma 2 well block of Mahu Depression. *Northwest Oil & Gas Exploration, 27*, 149–154.

Reading, G. J. (1978). *Sedimentary environment and facies* (pp.

20–34) (M. Zhou, C. Chen, J Zhang, Trans.). Beijing: Science Press.

Research Institute of Petroleum Exploration and Development of CNPC (RIPED) and Branch Company of Xinjiang Oilfield. (2004). *Reservoir regulative Project of the Lower Urho Formation in the 8th area of Kamaray oilfield, Xinjing*. The 2nd oil production of Xinjiang Oilfield. Unpublished manuscript.

Wu, T., Wu, C., Qi, Y., Yao, A., Zhang, S., Xu, Y., & Shi, J. (2015). Quantitative resumption method of stratum denudation thickness and its application in Junggar Basin: A case study on the Permian lower Urho Formation in Block 8 of Karamay oilfield. *Journal of Paleogeography, 17*, 81–90.

Wu, T., Zhao, C., Wu, C., Su, Y., Zhou, Z., Zhang, S., & Shi, J. (2013). Sedimentary characteristics and geophysical responses of sublacustrine fan in the Junggar Basin: A case study on the lower Urho Formation of the Permian in the Wuba block. *Karamay oilfield. Oil & Gas Geology, 34*, 85–94.

Xu, H. M. (2001). *Reservoir description of the lower Urho Formation in the 8th area of Kamaray oilfield, Xinjiang*. The 2nd oil production of Xinjiang Oilfield. Unpublished manuscript.

Xue, L., & Galloway, W. E. (1991). The classification of fan-delta, braided delta and delta systems. *Acta Geologica Sinica, 41*, 141–154.

Xing, F., Zhu, S., Kuang, H., Aa, Z., & Liu, Y. (2006). Application of EMI image logging to study of sedimentary facies—An example of Lower Wuerhe conglomerate reservoir of Upper Permian in Karamay Oilfield. *Xinjiang Petroleum Geology, 27*, 607–610.

Yang, F., Wei, Y., Abulimidi, Y., Chen, G., & Bing, B. (2016). Optimization of favorable reservoir–caprock assemblages of Middle-Upper Permian in Mahu Sag, Junggar Basin. *Xinjiang Petroleum Geology, 37*, 131–137+3.

Yu, X. (2002). *Detrital oil and gas reservoir sedimentology* (pp. 125–144). Beijing: Petroleum Industrial Publishing House (in Chinese).

Zhang, S. C., Huang, Z., Lu, X., Zou, N., Zhang, S. Y., Liu, S., & Shi, J. (2015). Main controlling factors of Permian sandy conglomerate reservoir in the northwestern Junggar Basin. *Journal of Lanzhou University (Natural Sciences), 51*, 20–30.

Zhu, X., Deng, X., Liu, Z., Sun, B., Liao, J., & Hui, X. (2013). Sedimentary characteristics and model of shallow braided delta in large-scale lacustrine: An example from Triassic Yanchang Formation in Ordos Basin. *Earth Science Frontiers, 20*, 19–28.

Zhu, X., Pan, R., Zhao, D., Liu, F., Wu, D., Li, Y., & Wang, R. (2013). Formation and development of shallow-water deltas in lacustrine basin and typical case analyses. *Journal of China University of Petroleum, 37*, 7–14.

Zhu, X., Zhong, D., Yuan, X., Zhang, H., Zhu, S., Sun, H., … Xian, B. (2016). Development of sedimentary geology of petroliferous basins in China. *Petroleum Exploration and Development, 43*, 890–901. https://doi.org/10.1016/S1876-3804(16)30107-0

6

Numerical modeling of virus transport through unsaturated porous media

Kandala Rajsekhar[1], Pramod Kumar Sharma[1,2]* and Sanjay Kumar Shukla[2]

*Corresponding author: Pramod Kumar Sharma, Department of Civil Engineering, Indian Institute of Technology, Roorkee, Roorkee 247667, India; Discipline of Civil and Environmental Engineering, School of Engineering, Edith Cowan University, Perth, WA 6027, Australia
E-mail: drpksharma07@gmail.com

Reviewing editor: Craig O'Neill, Macquarie University, Australia

Abstract: This paper describes the movement of virus in one-dimensional unsaturated porous media. The governing virus transport equations consider the inactivation in liquid phase, liquid–solid interface, air–liquid interface, and sorption in both liquid–solid and air–liquid interfaces. Finite-volume method has been used for solving the advection and dispersion processes of the virus transport equation. The effects of transport parameters on virus concentration profiles have been investigated for virus present in liquid phase, adsorbed liquid–solid and liquid–air phases. The results show that the movement of viruses in three phases is affected by soil moisture, inactivation rate, pore velocity, and mass transfer coefficients. It is found that the magnitude of virus sorption is higher at the air–liquid interface as compared to the liquid–solid interface. A higher value of mass transfer coefficient leads to an increase in the virus concentration in both liquid–solid and air–liquid interfaces.

Subjects: Hydraulic Engineering; Pollution; Water Engineering

Keywords: virus transport; numerical method; unsaturated media; virus concentration profiles

1. Introduction

In the past few decades, it has been noticed that the level of groundwater contamination has increased due to increase in industrial and agricultural activities. The presence of viruses in drinking water causes human diseases and it originates from septic tanks, sewage sludge, sanitary landfills, and agricultural practices. The experimental studies reported in the past indicate that the virus survives for a certain period of time in unsaturated porous media before reaching into the subsurface water (Chu, Jin, Baumann, & Yates, 2003; Schaub & Sorber, 1977; Yates & Ouyang, 1992; Yates, Yates, Wagner, & Gerba, 1987). Hence, it is essential to understand the transport process of viruses through the unsaturated porous media to prevent further contamination of subsurface aquifer system.

ABOUT THE AUTHORS

Kandala Rajsekhar was formerly an MTech student and Pramod Kumar Sharma is an associate professor in Department of Civil Engineering at Indian Institute of Technology Roorkee, India. Sharma worked as a postdoctoral fellow from May 2015 to October 2015, at Edith Cown University, Perth, Australia and his research interest is modeling contaminant transport through fractured porous media. Sanjay Kumar Shukla is an associate professor at Edith Cowan University, Perth, Australia and his research interest is Geosynthetics and Ground improvement.

PUBLIC INTEREST STATEMENT

In this study, finite-volume method has been used to get the numerical solution of virus transport equations considering inactivation in liquid phase, liquid–solid interface, air–liquid interface, and sorption in both liquid–solid and air–liquid interfaces. The present model has been used to investigate the behavior of virus concentration in liquid phase, adsorbed in liquid–solid and air–liquid interfaces with depth. Present model can be used to simulate the experimental data of viruses in the field and also to predict the movement of viruses in the subsurface media.

Unsaturated porous media consist of liquid, solid, and air phases and for water-wet solid surfaces, both liquid–solid and air–liquid interfaces exist (Freeze & Cherry, 1979). It is also known that the unsaturated zone plays an important role in the transport of fluids and contamination from the surface to the groundwater. Virus sorption within unsaturated porous media is affected by the presence of two interfaces. Viruses are sorbed onto liquid–solid interfaces via physical adsorption, chemical sorption, or ion exchange (Vilker & Burge, 1980). Vilker (1981) has suggested that non-equilibrium kinetic sorption is appropriate for describing virus attachment to the liquid–solid interfaces and for viruses with size similar to the size of solutes. This sorption process represents the rate of approach to equilibrium between adsorbed and liquid-phase virus concentration considering virus transport to the outer layer of a solid particle by mass transfer. Powelson, Simpson, and Gerba (1990) have suggested that the virus sorption is greater at air–liquid than liquid–solid interfaces. Virus sorption takes place at the liquid–solid interfaces due to electrostatic double-layer interactions and van der Waals forces (Teutsch, Herbold-Paschke, Tougianidou, Hahn, & Botzenhart, 1991). Tim and Mostaghimi (1991) developed a numerical model for water flow and virus transport in variably saturated porous media assuming that the virus sorption is an equilibrium process. Park, Blanford, and Huyakorn (1992) developed a semi-analytical model for both steady state and transient vertical virus transport in the unsaturated media and along the flow lines in the saturated zone considering equilibrium sorption and first-order inactivation. Yates and Ouyang (1992) developed a one-dimensional numerical model for flow of water, viruses, and heat in unsaturated porous media considering moisture-independent sorption, filtration, and temperature-dependent inactivation. Adsorption and inactivation are two different processes of virus removal, and viruses can get detached from the soil media because of the ions present in the groundwater (Bales, Li, Maguire, Yahya, & Gerba, 1993). Virus sorption at air–liquid interfaces is controlled by virus particle surface hydrophobicity, solution ionic strength, and particle charge (Wan & Wilson, 1994). Poletika, Jury, and Yates (1995) have showed that the viruses at an air–liquid interface may be desorbed under high interfacial shear stresses induced by fast interstitial fluid flow.

Virus inactivation is generally considered as a first-order irreversible sink mechanism (Sim & Chrysikopoulos, 1996). Recently, the mathematical models have been developed for virus transport in saturated porous media considering different inactivation rates for viruses in different phases (Sim & Chrysikopoulos, 1996, 1998, 1999). Schijven and Šimůnek (2002) have shown that the factors like size, attachment characteristics, and rate of inactivation affect the movement of viruses in porous media. Due to advection and dispersion, viruses get spread in the soil media which reduces its concentrations. The removal of viruses from the groundwater occurs due to the processes of adsorption and inactivation (Chattopadhyay, Chattopadhyay, Lyon, & Wilson, 2002; Chu, Jin, Flury, & Yates, 2001). Various factors like temperature, moisture content, pH, hydraulic conditions affect the transport of viruses below the ground surface. However, the temperature significantly affects the transport of viruses and a relationship between increase in temperature and inactivation rate of virus for different viruses has been given by Gerba and Rose (2003), and Gerba and Smith (2005). Torkzaban, Hassanizadeh, Schijven, de Bruin, and de Roda Husman (2006) studied the transport of bacteria through saturated and unsaturated porous media and their results demonstrate that the attachment to the air–water interfaces is reversible. Anders and Chrysikopoulos (2009) conducted soil column experiments under both saturated and unsaturated conditions. Their results indicate that even for unfavorable attachment conditions within a sand column, saturation levels can affect the virus transport through porous media. Further, many researchers have used the numerical method to investigate movement of viruses through saturated and unsaturated porous media and also estimated virus transport parameters considering liquid and solid phases (Joshi, Ojha, Sharma, & Surampalli, 2013; Ratha, Prasad, & Ojha, 2009; Sharma & Srivastava, 2011). Syngouna and Chrysikopoulos (2015) studied the effect of colloids and water saturation level on the attenuation and transport of colloids and viruses in unsaturated porous media. Thus, from the past research works, it is found that the effects of transport parameters on viruses in three phases, i.e. liquid phase, adsorbed liquid–solid and liquid–air interfaces have not been studied in detail. Hence, the present study describes the finite-volume method approach to solve the governing transport equation and investigates the behavior of movement of viruses in three phases through unsaturated porous media.

In this study, the numerical finite-volume method has been used to solve the governing equations for virus transport in one-dimensional, unsaturated porous media considering virus sorption on liquid-solid and air-liquid interfaces. An attempt has been made to investigate the effects of moisture content, inactivation rate constant, pore velocity, and mass transfer coefficients on the variation of virus transport along with vertical depth.

2. Governing equations

2.1. Virus transport
The governing differential equation for virus transport in one-dimensional, homogeneous, unsaturated porous media can be written as (Sim & Chrysikopoulos, 2000):

$$\frac{\partial}{\partial t}[\theta C] + \rho \frac{\partial C_s}{\partial t} + \frac{\partial}{\partial t}[\theta C_a] = \frac{\partial}{\partial z}\left[D_z \theta \frac{\partial C}{\partial z}\right] - \frac{\partial}{\partial z}[qC] - \lambda \theta C - \lambda_s \rho C_s - \lambda_a \theta C_a \qquad (1)$$

where C represents the virus concentrations in liquid phase (ML^{-3}); C_s represents the adsorbed at the liquid-solid interface (MM^{-1}); C_a represents the adsorbed at the air-liquid interface (ML^{-3}); θ represents the volumetric moisture content (L^{3}L^{-3}); q represents the specific discharge (LT^{-1}); λ, λ_s, and λ_a are the inactivation rate constants of viruses in the liquid phase, adsorbed in the liquid-solid interface and adsorbed in the air-liquid interfaces (T^{-1}), respectively; and ρ is the bulk density of the soil media (ML^{-3}). The hydrodynamic dispersion coefficient D_z can be expressed as (L^{2}T^{-1}) (Nielsen, Van Genuchten, & Biggar, 1986):

$$D_z = \alpha_z \frac{q}{\theta} + D_0 \qquad (2)$$

where α_z is the dispersivity (L), and D_0 is the molecular diffusion coefficient (L^{2}T^{-1}).

2.2. Virus sorption at interfaces
The expressions for the viruses adsorbed at solid-liquid and air-liquid interfaces can be written as (Sim & Chrysikopoulos, 1999):

$$\rho \frac{\partial C_s}{\partial t} = k\theta(C - C_g) - \lambda_s \rho C_s \qquad (3)$$

$$\frac{\partial}{\partial t}(\theta C_a) = k_a \theta C - \lambda_a \theta C_a \qquad (4)$$

where k represents the mass transfer rate of liquid to liquid-solid (T^{-1}); k_a liquid to air-liquid interface (T^{-1}); and C_g is the virus concentration in liquid phase (ML^{-3}), which is in close contact with soil solids. It was assumed a linear equilibrium relationship by Sim and Chrysikopoulos (1996) and it can be expressed as:

$$C_s = K_d C_g \qquad (5)$$

$$k = \kappa \cdot a_T \qquad (6)$$

where K_d is the distribution coefficient (L^{3}M^{-1}); κ is the mass transfer coefficient for liquid to liquid-solid interface (LT^{-1}); and a_T is the specific area of liquid-solid interface (L^{2}L^{-3}). It can be defined as the ratio of total surface area of soil particles to the bulk volume of the porous medium and it is expressed as (Fogler, 1992):

$$a_T = \frac{3(1 - \theta_{sat})}{r_p} \qquad (7)$$

where r_p is the average radius of soil particles (L); and θ_s is the water content of a saturated porous medium. The mass transfer rate coefficient (k_a) for liquid to air-liquid interface (T^{-1}) is given as:

$$k_a = \kappa_a a_{Ta} \tag{8}$$

where κ_a is the mass transfer coefficient of liquid to air–liquid interface (LT^{-1}); a_{Ta} is the specific area of air–liquid interface (L^{2}L^{-3}); and the expression can be given as (Cary, 1994):

$$a_{Ta}(\theta) = \frac{2\theta_{sat}^b}{r_0}\left(\zeta\theta_{res}\frac{\theta_{sat}^{-b} - \theta^{-b}}{-b} + \frac{\theta_{sat}^{1-b} - \theta^{1-b}}{1-b}\right) \tag{9}$$

The expression for r_0, i.e. the effective pore radius can be expressed as:

$$r_0 = \frac{2\sigma}{\rho_w g h_0} \tag{10}$$

where ζ and b are the soil constants; σ is the surface tension of water (MT^{-2}); ρ_w is the density of water (ML^{-3}); g is the gravitational constant (LT^{-2}); and h_0 is the air-entry value (L), which is defined as the pore water head where air begins to enter water-saturated pores. It can be observed from Equation (9) that a_{Ta} is a function of moisture content and its value decreases as the available moisture content increases and becomes zero when the available moisture content is equal to the saturated moisture content, i.e. $\theta = \theta_{sat}$.

2.3. Initial and boundary conditions for virus transport
Initially at time $t = 0$, it is assumed that there is a negligible concentration of viruses present in all the three phases of soil media.

$$C(0,z) = C_s(0,z) = C_a(0,z) = 0 \tag{11a}$$

$$C(t,0) = C_0 \tag{11b}$$

$$\frac{\partial C(t,L)}{\partial z} = 0 \tag{11c}$$

where C_0 is the source concentration of virus on the ground surface.

3. Numerical model
Numerical method admits any arbitrary boundary condition and it can be used in cases of dealing with complex problem for which analytical solution cannot be obtained. The numerical method can also be used for real-field problem such as complex geometry, partial variation of hydraulic conductivity and non-linear problem. However, analytical method has limitations in case of non-linear problem; in such cases, a numerical method works very well.

The finite-volume method has been used to get the solution of governing equations for virus transport. Using Equations (3) and (4), the simplified form of Equation (1) can be expressed as:

$$\frac{\partial C}{\partial t} = D_z\frac{\partial^2 C}{\partial z^2} - v\frac{\partial C}{\partial z} - (\lambda + k + k_a)C + \frac{k}{K_d}C_s \tag{12}$$

The advective term of Equation (12) can be given as (Putti, Yeh, & Mulder, 1990):

$$\frac{\partial C}{\partial t} = -v\frac{\partial C}{\partial z} \tag{13}$$

The dispersion transport term can be expressed as:

$$\frac{\partial C}{\partial t} = D_z\frac{\partial^2 C}{\partial z^2} \tag{14}$$

The operator split approach reduces the reaction terms in Equation (12) to a coupled ordinary differential equation which can be expressed as:

$$\frac{\partial C}{\partial t} = -(\lambda + k + k_a)C + \frac{k}{K_d}C_s \tag{15}$$

3.1. Procedure of numerical solution

The finite-volume method is used for solving the advective transport and is based on monotone upwind schemes for conservation laws (MUSCL) (Van Leer, 1977). Equation (12) has been broken into three parts, based on the suggestion by Putti et al. (1990), in which the explicit numerical scheme is used for advective transport and an implicit numerical scheme is used for dispersive transport, and explicit method for reaction part. The advantage of this method is to handle either advection dominated or dispersion dominated for solute transport through porous media accurately. This method is globally high-order accurate and non-oscillatory, and the detailed procedure for solution of the advective part has been given by Ratha et al. (2009), and Sharma, Joshi, Srivastava, and Ojha (2014).

3.2. Dispersive transport equation

The resulting output concentrations of the advective transport are used as the initial condition for dispersive transport. A conventional fully implicit finite-difference scheme, which is unconditionally stable, is used to obtain the final concentration at the end of time step. The implicit finite-difference formulation of dispersive transport equation can be expressed as:

$$\frac{C_j^{n+1} - C_j^n}{\Delta t} = \frac{D_z}{(\Delta z)^2}\left(C_{j+1}^{n+1} - 2C_j^{n+1} + C_{j-1}^{n+1}\right) \tag{16}$$

where Δz and Δt are spatial grid size and time step, respectively.

Remaining Equations (3), (4), and (15) are solved using the explicit numerical method and the formulation is given below:

$$\rho\left(\frac{C_{sj}^{n+1} - C_{sj}^n}{\Delta t}\right) = k\theta\, C_j^n - \left(\lambda_s \rho + \frac{k\theta}{k_d}\right)C_{sj}^n \tag{17}$$

$$\frac{C_{aj}^{n+1} - C_{aj}^n}{\Delta t} = k_a C_j^n - \lambda_a C_{aj}^n \tag{18}$$

$$\frac{C_j^{n+1} - C_j^n}{\Delta t} = (\lambda + k + k_a)C_j^n + \left(\frac{k}{k_d}\right)C_{sj}^n \tag{19}$$

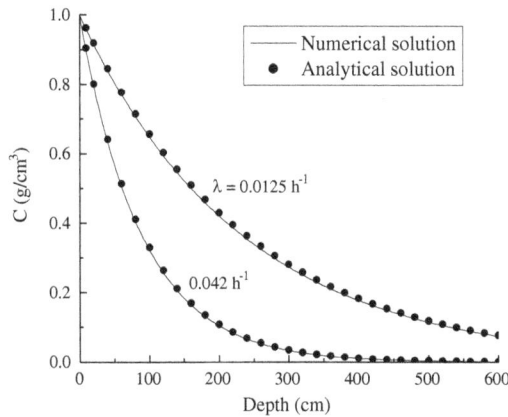

Figure 1. Comparison of numerical results with analytical solution for spatial virus concentration with two different values of inactivation rate constant (t = 240 h, v = 4 cm h^{-1}, D = 15 cm^2 h^{-1}, and k = 0.005 h^{-1}).

Table 1. Model parameters used for simulation of virus distribution with depth

Parameters	Values
C_0	1 g cm^{-3}
b	2
D_0	1.542E-5 cm^2 h^{-1}
g	980 cm s^{-2}
h_0	2 cm
K_d	20 cm^3 g^{-1}
r_p	0.1 cm
α_z	0.5 cm
ζ	160
θ_{sat}	0.45
κ	0.006 cm h^{-1}
ρ	1.5 g cm^{-3}
σ	74.2 × 10^{-3} N m^{-1}

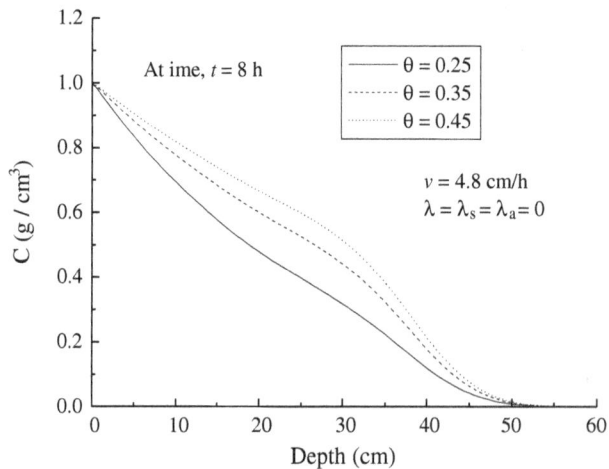

Figure 2a. Variation of liquid-phase virus concentration with depth for different values of moisture content.

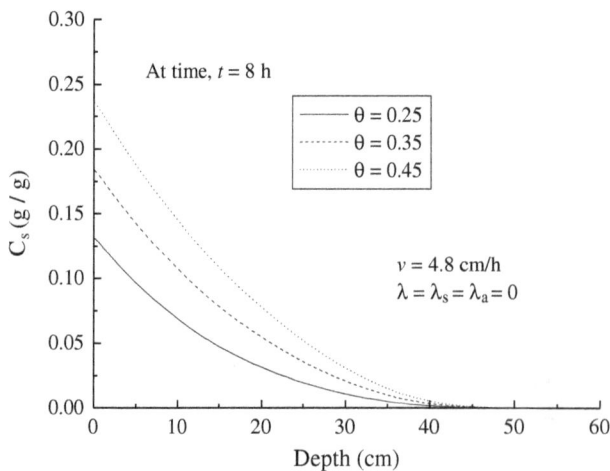

Figure 2b. Variation of adsorbed liquid–solid interface virus concentration with depth for different values of moisture content.

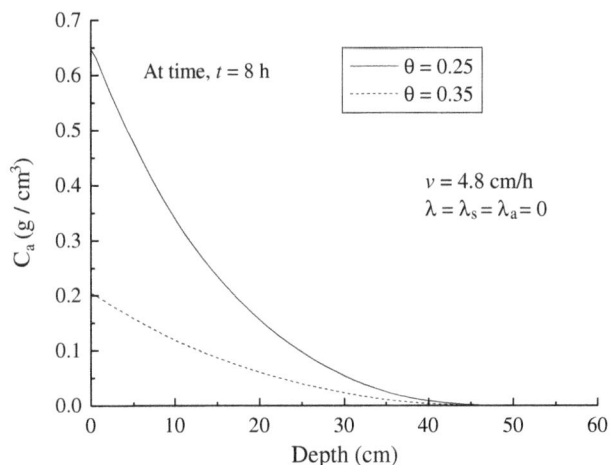

Figure 2c. Variation of adsorbed liquid–air interface virus concentration with depth for different values of moisture content.

4. Verification of model

The present numerical model is verified using the analytical solution given by Sim and Chrysikopoulos (1995). The numerical and analytical results of spatial virus concentration in liquid phase have been shown in Figure 1 with two different values of inactivation rate constants. The grid size $\Delta z = 0.2$ cm and time step $\Delta t = 0.025$ h have been used in this simulation. It can be seen that the numerical results match very well with the analytical solution. To reduce the numerical error, the values of both the Peclet number and Courant number are kept less than one.

5. Results and discussion

The numerical model has been used to investigate the behavior of virus concentration profiles through unsaturated porous media. The input parameters used for simulation are shown in Table 1 (Sim & Chrysikopoulos, 2000), and a continuous source of virus is injected at the ground surface. Figures 2a–2c show the model results for spatial variation of concentration of viruses in liquid phase, liquid–solid interface and air–liquid interfaces for different values of soil moisture contents. The value of constant pore velocity and negligible inactivation rate coefficients are used during simulation. The results of virus concentration have been predicted at transport time of 8 h. The behavior of virus concentration is not uniform along the depth. The magnitude of virus concentration (in three phases) increases with an increase in the value of moisture content for intermediate depth of soil and the values remain same at end depth. As the moisture content of porous media reduces from its full saturation to partial saturation, there is a significant decrease in the liquid-phase virus concentration and significant increase in the virus concentration sorbed at air–liquid interface. This increase is due to the increase in specific air–liquid interface.

Figures 3a–3c show the variation of virus concentration along with depth for different values of inactivation rate constants. In this study, the inactivation rate coefficients λ, λ_a, and λ_s all are assumed to be constant. However, in reality, its value depends on temperature and time (Sim & Chrysikopoulos, 1996). It is assumed that $\lambda_a = \lambda$ and $\lambda_s = \lambda/2$ as suggested by Yates and Ouyang (1992), and Thompson, Flury, Yates, and Jury (1998). The simulation is carried out with constant pore velocity $v = 4$ cm h^{-1}, uniform water content $\theta = 0.35$, and saturated water content $\theta_{sat} = 0.45$. As expected, with an increase in the value of inactivation rate coefficients, the concentration of viruses in liquid phase, liquid–solid interface and air–liquid interface decreases. The behavior of virus concentration is non-uniform along the depth.

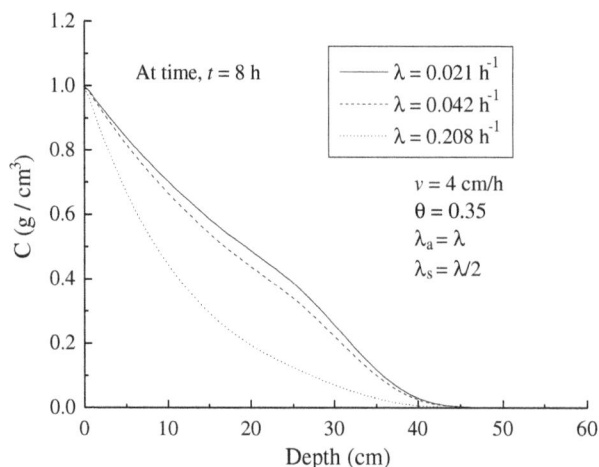

Figure 3a. Variation of liquid-phase virus concentration with depth for different values of inactivation rate constant.

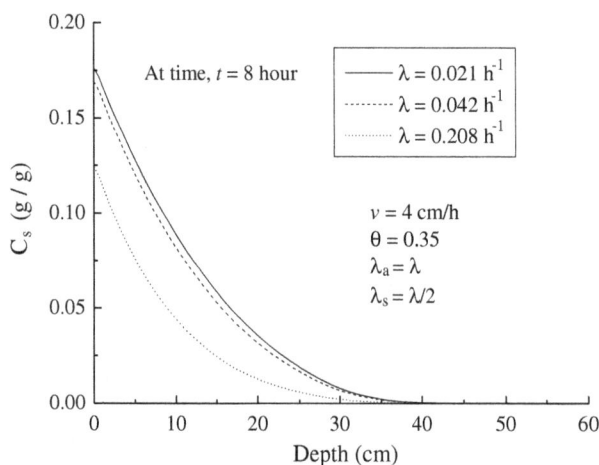

Figure 3b. Variation of adsorbed liquid–solid interface virus concentration with depth for different values of inactivation rate constant.

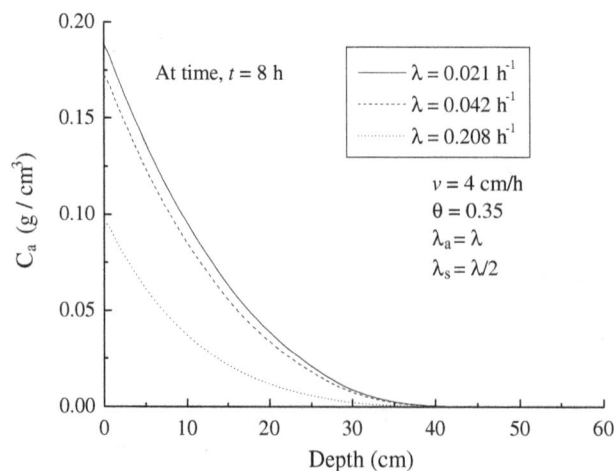

Figure 3c. Variation of adsorbed liquid–air interface virus concentration with depth for different values of inactivation rate constant.

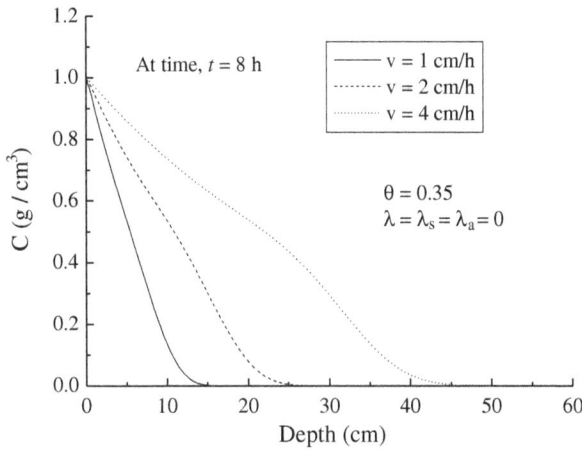

Figure 4a. Variation of liquid-phase virus concentration with depth for different values of pore velocity.

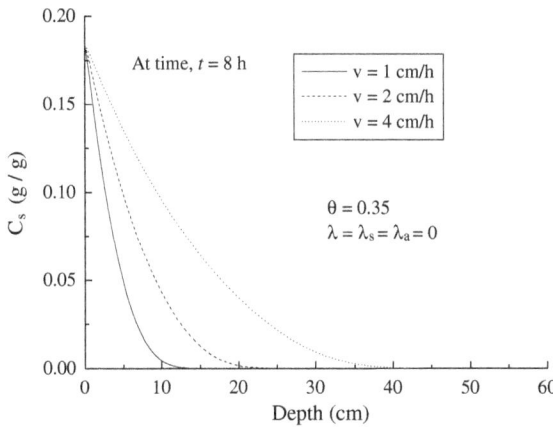

Figure 4b. Variation of adsorbed liquid–solid interface virus concentration with depth for different values of pore velocity.

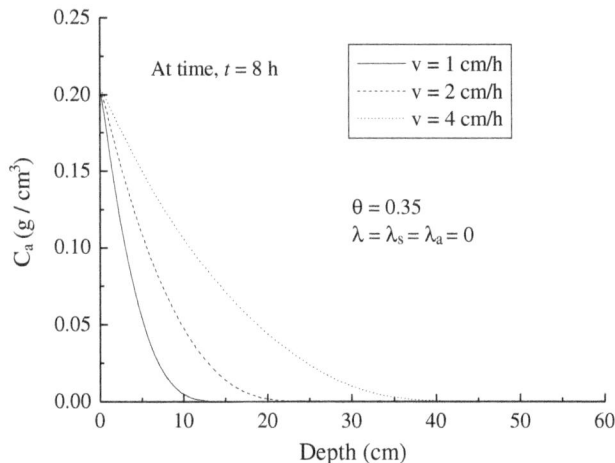

Figure 4c. Variation of adsorbed liquid–air interface virus concentration with depth for different values of pore velocity.

Figures 4a–4c show the effect of pore water velocity on the variation of concentration of viruses in liquid phase, liquid–solid interface and air–liquid interface, respectively. The simulations are carried out for three different pore water velocities of 1, 2, and 4 cm h^{-1} with negligible inactivation coefficients (i.e. $\lambda = \lambda_s = \lambda_a = 0$) and soil moisture content of $\theta = 0.35$. As expected, with an increase in the value of pore water velocity, there is a considerable increase in the value of virus concentration with

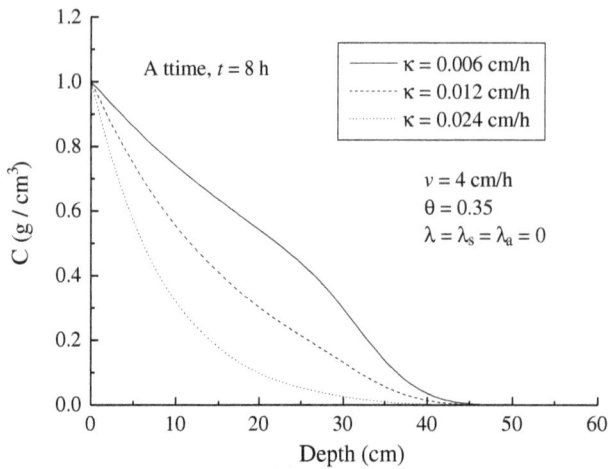

Figure 5a. Variation of liquid-phase virus concentration with depth for different values of mass transfer coefficients.

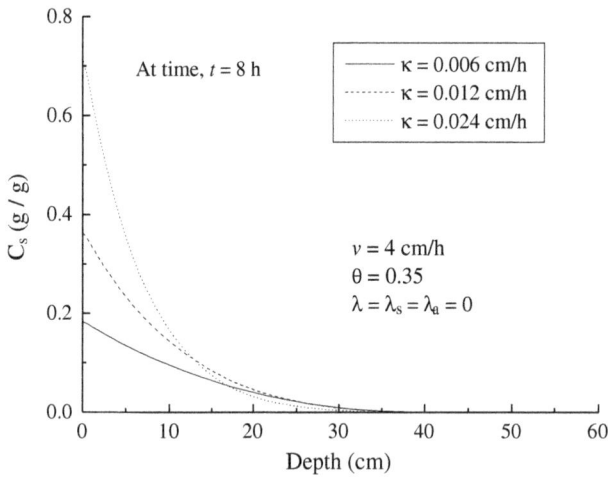

Figure 5b. Variation of adsorbed liquid–solid-phase virus concentration with depth for different values of mass transfer coefficients.

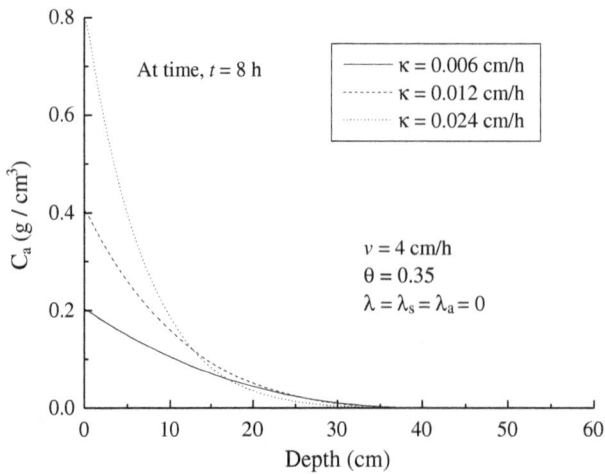

Figure 5c. Variation of adsorbed liquid–air-phase virus concentration with depth for different values of mass transfer coefficients.

depth. The behavior of virus concentration is non-uniform for intermediate depth of soil. It means that the variation of virus concentration with depth is different for virus present in three phases.

Figures 5a–5c show the effect of mass transfer coefficient on the variation of concentration of viruses in the liquid phase, liquid–solid interface and air–liquid interface. The simulations are carried out with constant pore velocity of 4 cm h^{-1}, considering negligible inactivation constants. It is assumed that the mass transfer coefficient of liquid to air–liquid interface is five times the mass transfer coefficient of liquid to liquid–solid interface as considered by Sim and Chrysikopoulos (2000). Three different values of liquid to liquid–solid mass transfer coefficients, i.e. $\kappa = 0.006$, 0.012, and 0.024 cm h^{-1}. It is seen that on increasing the value of mass transfer coefficient (κ), the concentration of viruses in liquid phase decreases and concentration of viruses in both liquid–solid interface and air–liquid interface increases. The magnitude of virus concentration in air–liquid interface (C_a) is higher in comparison to the concentration of viruses in liquid–solid interface (C_s) for the same depth. This has occurred due to the assumption of $k_a = 5\ k$.

6. Summary and conclusions

In this study, a finite-volume method has been used to develop a numerical model to analyze the virus transport in one-dimensional unsaturated porous media. The model accounts for the virus sorption on the liquid–solid and air–liquid interfaces as well as inactivation of viruses suspended in the liquid phase and virus attached to both interfaces. The effects of transport parameters on virus concentration profiles were investigated and the findings are listed below:

(1) In the field, the moisture content of porous media reduces from its full saturation to partial saturation; there is a significant decrease in concentration of viruses in liquid phase and liquid–solid interface, and an increase in concentration of viruses in air–liquid interface. This increase is due to the increase in liquid to air–liquid mass transfer rate, as it depends on the specific air–liquid interface area, which increases with a decrease in the value of moisture content.

(2) As expected, with an increase in the value of inactivation rate coefficients, the magnitude of virus concentration in liquid phase and both adsorbed interfaces reduce. A higher value of pore water velocity leads to increase in the movement of viruses in depth. It is also seen that the variation of virus concentration along the depth is non-uniform.

(3) A higher value of mass transfer coefficient decreases the magnitude of virus concentration in liquid phase, while it increases virus concentration in both liquid–solid and air–liquid interfaces. Also, the magnitude of viruses in air–liquid interface is higher in comparison to concentration of virus in liquid–solid interface at the same depth. Finally, this model may be used for simulation of experimental data of viruses in the field.

Funding
The authors received no direct funding for this research.

Author details
Kandala Rajsekhar[1]
E-mail: raj.sekhar005@gmail.com
ORCID ID: http://orcid.org/0000-0003-4283-6642
Pramod Kumar Sharma[1,2]
E-mail: drpksharma07@gmail.com
Sanjay Kumar Shukla[2]
E-mail: s.shukla@ecu.edu.au
[1] Department of Civil Engineering, Indian Institute of Technology, Roorkee, Roorkee 247667, India.
[2] Discipline of Civil and Environmental Engineering, School of Engineering, Edith Cowan University, Perth 6027, Australia.

References
Anders, R., & Chrysikopoulos, C. V. (2009). Transport of viruses through saturated and unsaturated columns packed with sand. *Transport in Porous Media, 76*, 121–138. http://dx.doi.org/10.1007/s11242-008-9239-3

Bales, R. C., Li, S., Maguire, K. M., Yahya, M. T., & Gerba, C. P. (1993). MS-2 and poliovirus transport in porous media: Hydrophobic effects and chemical perturbations. *Water Resources Research, 29*, 957–963. http://dx.doi.org/10.1029/92WR02986

Cary, J. W. (1994). Estimating the surface area of fluid phase interfaces in porous media. *Journal of Contaminant Hydrology, 15*, 243–248. http://dx.doi.org/10.1016/0169-7722(94)90029-9

Chattopadhyay, D., Chattopadhyay, S., Lyon, W. G., & Wilson, J. (2002). Effect of surfactants on the survival and sorption

of viruses. *Environmental Science & Technology, 36,* 4017–4024.
http://dx.doi.org/10.1021/es0114097

Chu, Y., Jin, Y., Baumann, T., & Yates, M. V. (2003). Effect of soil properties on saturated and unsaturated virus transport through columns. *Journal of Environment Quality, 32,* 2017–2025.
http://dx.doi.org/10.2134/jeq2003.2017

Chu, Y., Jin, Y., Flury, M., & Yates, M. V. (2001). Mechanisms of virus removal during transport in unsaturated porous media. *Water Resources Research, 37,* 253–263.
http://dx.doi.org/10.1029/2000WR900308

Fogler, H. S. (1992). *Elements of chemical reaction engineering* (2nd ed.). Englewood Cliffs, NJ: Prentice-Hall.

Freeze, R. A., & Cherry, J. A. (1979). *Groundwater.* Englewood Cliffs, NJ: Prentice-Hall.

Gerba, C. P., & Rose, J. B. (2003). International guidelines for water recycling: Microbiological considerations. *Water Science & Technology: Water Supply, 3,* 311–316.

Gerba, C. P., & Smith, J. E. (2005). Sources of pathogenic microorganisms and their fate during land application of wastes. *Journal of Environmental Quality, 34,* 42–48.

Joshi, N., Ojha, C. S. P., Sharma, P. K., & Surampalli, R. Y. (2013). Parameter identification of virus transport in porous media using equilibrium and non-equilibrium models. *Journal of Environmental Chemical Engineering, 1,* 1099–1107.
http://dx.doi.org/10.1016/j.jece.2013.08.023

Nielsen, D., Van Genuchten, M., & Biggar, J. (1986). Water flow and solute transport processes in the unsaturated zone. *Water Resources Research, 22,* 89S–108S.
http://dx.doi.org/10.1029/WR022i09Sp0089S

Park, N., Blanford, T., & Huyakorn, P. (1992). *VIRALT: A modular semi-analytical and numerical model for simulating viral transport in ground water.* HydroGeologic Inc. for USEPA/ International Ground Water Modelling Centre/Colorado School of Mines.

Poletika, N. N., Jury, W. A., & Yates, M. V. (1995). Transport of Bromide, Simazine, and MS-2 Coliphage in a Lysimeter containing undisturbed, unsaturated soil. *Water Resources Research, 31,* 801–810.
http://dx.doi.org/10.1029/94WR02821

Powelson, D., Simpson, J., & Gerba, C. (1990). Virus transport and survival in saturated and unsaturated flow through soil columns. *Journal of Environment Quality, 19,* 396–401.
http://dx.doi.org/10.2134/jeq1990.0047242500190003 0008x

Putti, M., Yeh, W., & Mulder, W. A. (1990). A triangular finite volume approach with high eesolution upwind terms for the solution of ground water transport equations. *Water Resources Research, 26,* 2865–2880.

Ratha, D. N., Prasad, K. H., & Ojha, C. S. (2009). Analysis of virus transport in groundwater and identification of transport parameters. *Practice Periodical of Hazardous, Toxic, and Radioactive Waste Management.* doi:10.1061/_ASCE_1090-025X_2009_13:2_98_

Schaub, S. A., & Sorber, C. A. (1977). Virus and bacteria removal from wastewater by rapid infiltration through soil. *Applied and Environmental Microbiology, 33,* 609–619.

Schijven, J., & Šimůnek, J. (2002). Kinetic modeling of virus transport at the field scale. *Journal of Contaminant Hydrology, 55,* 113–135.
http://dx.doi.org/10.1016/S0169-7722(01)00188-7

Sharma, P., & Srivastava, R. (2011). Numerical analysis of virus transport through heterogeneous porous media. *Journal of Hydro-environment Research, 5,* 93–99.
http://dx.doi.org/10.1016/j.jher.2011.01.001

Sharma, P. K., Joshi, N., Srivastava, R., & Ojha, C. S. P. (2014).

Reactive transport in fractured permeable porous media. *Journal of Hydrologic Engineering.* doi:10.1061/(ASCE) HE.1943-5584.0001096

Sim, Y., & Chrysikopoulos, C. V. (1995). Analytical models for one-dimensional virus transport in saturated porous media. *Water Resources Research, 31,* 1429–1437.
http://dx.doi.org/10.1029/95WR00199

Sim, Y., & Chrysikopoulos, C. V. (1996). One-dimensional virus transport in porous media with time-dependent inactivation rate coefficients. *Water Resources Research, 32,* 2607–2611.
http://dx.doi.org/10.1029/96WR01496

Sim, Y., & Chrysikopoulos, C. V. (1998). Three-dimensional analytical models for virus transport in saturated porous media. *Transport in Porous Media, 30,* 87–112.
http://dx.doi.org/10.1023/A:1006596412177

Sim, Y., & Chrysikopoulos, C. (1999). Analytical models for virus adsorption and inactivation in unsaturated porous media. *Colloids and Surfaces A: Physicochemical and Engineering Aspects, 155,* 189–197.
http://dx.doi.org/10.1016/S0927-7757(99)00073-4

Sim, Y., & Chrysikopoulos, C. (2000). Virus transport in unsaturated porous media. *Water Resources Research, 36,* 173–179.
http://dx.doi.org/10.1029/1999WR900302

Syngouna, V. I., & Chrysikopoulos, C. V. (2015). Experimental investigation of virus and clay particles cotransport in partially saturated columns packed with glass beads. *Journal of Colloid and Interface Science, 440,* 140–150.
http://dx.doi.org/10.1016/j.jcis.2014.10.066

Teutsch, G., Herbold-Paschke, K., Tougianidou, D., Hahn, T., & Botzenhart, K. (1991). Transport of microorganisms in the underground—Processes, experiments and simulation models. *Water Science & Technology, 24,* 309–314.

Thompson, S. S., Flury, M., Yates, M. V., & Jury, W. A. (1998). Role of air-water interface in bacteriophage sorption experiments. *Applied and Environmental Microbiology, 64,* 304–309.

Tim, U. S., & Mostaghimi, S. (1991). Model for predicting virus movement through soils. *Ground Water, 29,* 251–259.
http://dx.doi.org/10.1111/gwat.1991.29.issue-2

Torkzaban, S., Hassanizadeh, S. M., Schijven, J. F., de Bruin, H. A. M., & de Roda Husman, A. M. (2006). Virus transport in saturated and unsaturated sand columns. *Vadose Zone Journal, 5,* 877–885.
http://dx.doi.org/10.2136/vzj2005.0086

Van Leer, B. (1977). Towards the ultimate conservative difference scheme. IV. A new approach to numerical convection. *Journal of Computational Physics, 23,* 276–299.
http://dx.doi.org/10.1016/0021-9991(77)90095-X

Vilker, V. L. (1981). Simulating virus movement in soils. In I. K. Iskandar (Eds.), *Modeling waste renovation: Land treatment* (pp. 223–253). New York, NY: John Wiley.

Vilker, V. L., & Burge, W. D. (1980). Adsorption mass transfer model for virus transport in soils. *Water Research, 14,* 783–790.
http://dx.doi.org/10.1016/0043-1354(80)90256-0

Wan, J., & Wilson, J. L. (1994). Colloid transport in unsaturated porous media. *Water Resources Research, 30,* 857–864.
http://dx.doi.org/10.1029/93WR03017

Yates, M., & Ouyang, Y. (1992). VIRTUS, a model of virus transport in unsaturated soils. *Applied and Environmental Microbiology, 58,* 1609–1616.

Yates, M., Yates, S., Wagner, J., & Gerba, C. (1987). Modeling virus survival and transport in the subsurface. *Journal of Contaminant Hydrology, 1,* 329–345.
http://dx.doi.org/10.1016/0169-7722(87)90012-X

7

Effect of shear rate on the residual shear strength of pre-sheared clays

Farzad Habibbeygi[1]* and Hamid Nikraz[1]

*Corresponding author: Farzad Habibbeygi, Faculty of Science and Engineering, Department of Civil Engineering, Curtin University, Perth, Australia

E-mail: farzad.habibbeygi@postgrad. curtin.edu.au

Reviewing editor: Silvana Irene Torri, Universidad de Buenos Aires, Argentina

Abstract: This paper presents the result of an experimental study on the shear behaviour of a pre-sheared clayey soil. The effect of shear rate on the residual strength of pre-sheared clays was investigated in a ring shear apparatus. A pre-sheared surface was initially developed at a slow rate of 0.1 mm/min. Fast shear rates were then applied to the pre-sheared specimens to investigate the impact of the shear rates on the results. The laboratory results show that there is an immediate tendency for the residual strength to increase with increasing shear rate. Following this, the fast residual strength continues to increase with further displacement, reaching the peak value of fast residual strength (i.e. positive rate effect). Finally, the fast residual strength drops with increasing displacement to a value less than slow residual strength (i.e. negative rate effect). Overall, the relationship between the residual strength and the fast shear rates indicates a positive and negative rate effect based on the shear rates and the shear displacement.

Subjects: Earth Sciences; Earth Systems Science; Applied & Economic Geology; Geology - Earth Sciences; Civil, Environmental and Geotechnical Engineering

Keywords: clay; laboratory test; shear rate; residual strength; expansive

1. Introduction

A residual shear strength of clayey soil under a prescribed normal stress is defined as the minimum shear strength of the soil under relatively large shear displacement (ASTM D647) (American Society for Testing and Materials [ASTM], 2015). The displacement required to develop the residual strength is greater than the displacements corresponding to the peak shear strength (maximum shear strength) and the critical state shear strength (fully softened) in over consolidated soils. Accordingly, the residual shear strength is not related to the first slide of slopes but related to the existence of slip

ABOUT THE AUTHOR

Farzad Habibbeygi is a PhD student at Department of Civil Engineering, Faculty of Science and Engineering, Curtin University and a member of Engineers of Australia. His research area of interest is the mechanical and compression behaviour of expansive clays, numerical modelling and artificial neural network. He has also published several research papers in international journals in his area of interest.

PUBLIC INTEREST STATEMENT

This paper presents the result of an experimental study on the shear behaviour of a pre-sheared clayey soil. The residual shear strength is the centre of focus of many researchers for slope stability analyses of old landfills where a slip surface exists. A residual shear strength of clayey soil is defined as the minimum shear strength of the soil along an existing shear surface. The dependency of residual shear strength and shear rate was investigated in this paper using a ring shear apparatus. Based on the experimental results, there is an increase in residual shear strength in comparison with the slow residual shear strength when the shear displacement rates increase.

surface such as old landslides (Skempton, 1985). The orientation of clay particles parallel to the shear direction results in post peak drop of undrained shear strength and the shear strength reaches to the critical state value at the end of the first step (smaller shear displacement). When the particle reorientation completes at larger displacement, the shear strength reaches the residual value and remains constant with the increase in displacement (Skempton, 1985).

Moreover, the residual shear strength of the clayey soil is the shear strength along a pre-developed shear zone. In fact, this pre-existing movement can be developed by various factors, such as an old landslide, tectonic forces, a change in groundwater, earthquake forces, blasting and dynamic/seismic loadings.

The residual shear strength of cohesive soils has been the centre of focus for many researchers all over the world for the past few decades to predict the behaviour of landslides (Gratchev & Sassa, 2015; Li, Wen, Aydin, & Ju, 2013; Tika & Hutchinson, 1999; Tika, Vaughan, & Lemos, 1996; Vulliet & Hutter, 1988). It is anticipated that the soil skeleton, clay type, particle size distribution, displacement rate and pore water have considerable effect on the residual shear strength of cohesive soils (Gratchev & Sassa, 2015; Mesri & Olson, 1970).

Tika et al. (1996) performed a series of experimental tests on natural clays to investigate the effect of fast shearing on their shear behaviour. Their findings indicate that the shear strength of the soils studied increases when the shear rates are high. However, the shear strength of the fast rate test drops to a residual of shear strength, which may be greater, equal or even less than the residual shear strength at a slow rate. Tika et al. (1996) also proposed three modes of failure (i.e. sliding, transitional and turbulent failure) to explain the dissimilar tendency observed for the shear rate effect.

Tika and Hutchinson (1999) carried out several ring shear tests on two natural samples collected from a slip surface failure in Italy. Their laboratory results show that there is a negative correlation between the residual stress and the shear displacement rate for the samples studied. In fact, the residual shear strength at a fast rate of displacement was assessed and found to be up to 60% less than that performed at a slow rate.

Li et al. (2013) carried out 27 large ring shear tests on the specimens collected from the slip surfaces of three existing landslides. Three different shear rates were employed in their investigation to study the effect of fast rate on the results (i.e. 0.1, 1 and 10 mm/s). According to Li et al. (2013), the Atterberg limits and the particle size distribution, as well as the gravel and the fine fraction are additional factors that affect the residual shear strength. It was also found that the residual shear strength for the natural soils studied has a negative correlation with the plasticity index and the surface smoothness. This means that the residual strength decreases with an increase in either of these parameters. On the other hand, the particle symmetry defined by the elongation of the soil particles has a positive effect on the residual strength of the soil studied.

Gratchev and Sassa (2015) studied the shear strength of three different clays at different rates of shear displacement. It was concluded that the displacement rate had a considerable impact on all three clays studied. Based on their findings, the shear strength increases slightly when the shear rate decreases. It was also discovered that the increase of the shear strength depends on the confining pressure. Additionally, the increase in the shear strength is significant for confining pressures greater than 100 kPa.

Scaringi and Di Maio (2016) performed a series of controlled, direct shear tests on several mixtures of bentonite, kaolin and sand. In their study, the shear rate varied between 0.0001 and 100 mm/min. It was concluded that there is a positive correlation between the shear displacement rate and the residual strength of clayey soils when the clay fraction is higher than 50% and the displacement rate is greater than 1 mm/min. However, the displacement rate has no effect on the results for rates ranging from 0.0001 to 1 mm/min. Ring shear tests with various rates were also performed on a mixture of the soils and sodium chloride solution to study the effect of salt concentration on the results. The results revealed that the salt concentration has a positive rate effect. In other words, the rate impact increases with an increasing salt concentration. It was also established that this increase can be explained well by the decrease in the void ratio and the increase in the solid particle contacts due to the solution concentration.

Most studies on the shear-rate effect show that the excess pore-water pressure developed at the shearing stage affects the undrained shear strength of clayey soils (Casacrande & Wilson, 1951; Gratchev & Sassa, 2015; Mesri & Olson, 1970; Richardson & Whitman, 1963). Casacrande and Wilson (1951) proposed that the pore-water pressure generated in the sample at the low shear rate is higher than at higher shear rate. Accordingly, a change in the generated pore-water pressure during the shearing stage results in the dependency of undrained shear strength (Casacrande & Wilson, 1951; Richardson & Whitman, 1963).

However, many studies have been carried out on the residual shear strength of cohesive soils, and some inconsistent or even opposite results have been reported in the literature (Gratchev & Sassa, 2015; Scaringi & Di Maio, 2016), which indicates that there is still a lack of experimental data on this topic. Correspondingly, more laboratory tests are required to ascertain the shear behaviour of clayey soils at large shear displacements. In this study, the residual shear strength of an expansive clay was measured after pre-shearing at various shear rates and normal stresses to investigate the effect of these factors on the shear behaviour of the soil studied.

2. Materials and test procedure

A clayey soil collected from the south of Perth, Western Australia was used in this study. This clay referred herein as "black clay", is a highly expansive clay containing 68% clay, 20% sand and 12% silt. The clay had a high value liquid limit (w_L) of 82, and a plasticity index (I_p) of 47 (Habibbeygi, Nikraz, & Chegenizadeh, 2017; Habibbeygi, Nikraz, & Verheyde, 2017). Analyses of the mineralogy of the samples indicate that the predominant clay mineral is smectite. The results of one-dimensional consolidation tests illustrate a high potential of expansion and compressibility for the black clay (Habibbeygi et al., 2017). The in situ water content of the black clay was measured shortly after sample collection, and then assessed and found to be approximately 40% (Habibbeygi et al., 2017).

The ring shear device used in this study is shown in Figure 1. This device can provide predefined torsional shear on an annular specimen under a fully saturated condition to measure the residual shear strength of the specimen. The specimen is placed vertically between two porous stones. A counter balance (10:1 ratio lever) loading system is employed to apply the required pressure. A variable speed motor controlled by computer applies the rotation to the lower platen. A pair of load cells automatically measure the torque transmitted to the specimen. Both load cells and linear transducer are connected to a data acquisition system to record and process data. The shear strength of the specimen and the vertical load can be calculated for the predefined normal stress and shear rate using Equations (1) and (2):

$$P = 4\pi\left(R_2^2 - R_1^2\right) \cdot \sigma_n' \tag{1}$$

$$\tau = \frac{0.75L(F_1 + F_2)}{4\pi(R_2^3 - R_1^3)} \tag{2}$$

where τ is the shear stress (MPa), F_1, F_2 are the shear forces being measured by load cells (N), R_1, R_2 are the inner radius and outer radius of the specimens respectively (mm), L is the length of the torque arm (mm), P is the vertical load being applied to the specimens (N), and σ_n' is the normal stress being applied vertically to the specimens (MPa).

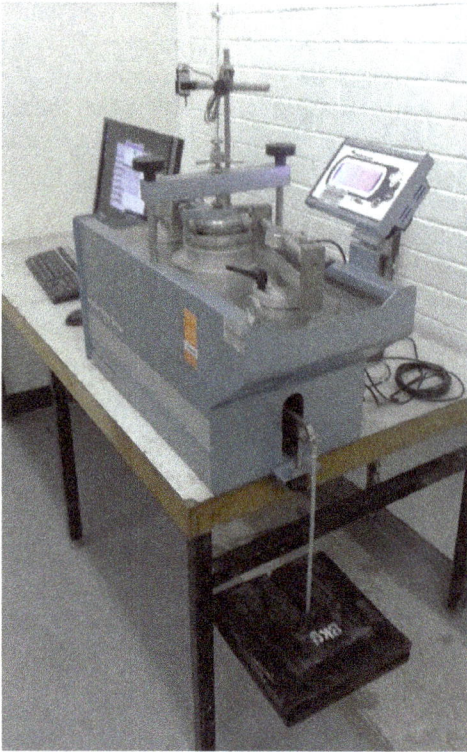

Figure 1. Ring shear apparatus used in this study.

First, the specimens were prepared following Head's procedure for preparing a remoulded sample (Head, 1986). Remoulded specimens were prepared by crushing an air-dried sample using a rubber pestle and mortar. The predefined amount of water was then added to the oven-dried soil to achieve a water content equal to the *in situ* water content. All samples were kneaded on a glass plate in order to be mixed uniformly and were stored in a three-layer air-tight bag for one day to mature prior to torsional shear tests. Finally, the specimens were placed into an annular mould using a spatula. The specimen filling was carried out in small horizontal layers to avoid any voids in the cell. The top of the specimen was flattened with the specimen container. They were then kept for 24 h in the container fully filled with water to be saturated before applying the normal stress increments (Figure 2). The container was kept full during the consolidation stage.

Secondly, the specimens were consolidated under the stress increments to achieve the final desired normal stresses of 50, 100, and 200 kPa. Prior to proceeding to the pre-shear step, the normal stress increment at each stage of consolidation and the completion of each increment was verified following the test method, ASTM D2435. The vertical load required to be applied to the annular specimen was calculated using Equation (1).

Following this, the specimens were pre-sheared to develop a shear surface. The specimens were sheared slowly at the rate of 0.1 mm/min for one day. A slow rate of shear displacement was selected to prevent any sample extrusion or to generate any pore water pressure during this step. To investigate the effect of shear rate on the shear behaviour of the pre-sheared specimens, various shear rates were applied to the specimens (i.e. 200, 100, 50, 10, 1 mm/min). The test procedure used in this study is schematically presented in Figure 3. According to this figure, the pre-sheared specimens were sheared subsequently at five faster shear rates. The shear rates begin with a high value of 200 mm/min and eventually decrease to a relatively slow rate of 1 mm/min. The shear strength of the specimens at each step was then calculated for the predefined normal stress and shear rate of that step using Equation (2).

Figure 2. Sample prepared before the consolidation stage.

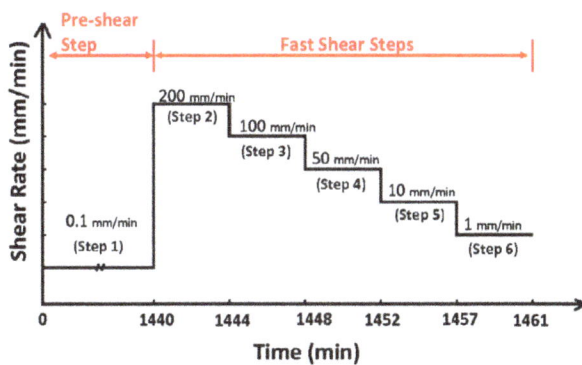

Figure 3. Shear rate and test procedure in this study.

3. Results and discussion

The ring shear tests under controlled shear displacement were performed on the black clay specimens consolidated at various normal stresses (i.e. 50, 100, 200 kPa). The specimens were sheared initially at a slow rate of 0.1 mm/min to develop a predefined shear surface. The shear rate was retained at a slow rate to prevent any sample extrusion, and to develop pore water pressure (step 1). In accordance with ASTM D6467, at least one revolution of the ring was employed to create a pre-sheared surface. At this stage, the shear strength neither increased nor decreased when the shear displacement increased. Therefore, it can be concluded that an existing sheared surface was present before conducting fast shearing stages. The experimental tests show that the residual shear strength of the clayey soil studied is 14.9, 45.3 and 51.1 kPa for the applied normal stresses of 50, 100 and 200 kPa, respectively. Figure 4 illustrates the results of the shear behaviour for the fast shear steps for different normal stresses (i.e. steps 2 to 6). As illustrated, the fast residual strength increases the slow residual strength at all fast shear rates after a relatively small shear displacement. For example, the increase in percentage relative to the slow residual strength is approximately between 5 and 10% for various shear rates at the normal stress of 200 kPa. This increase in comparison with the slow residual strength occurs at the maximum shear rate (i.e. 200 mm/min) and step 4 (i.e. 50 mm/min). The increase to the slow residual strength decreases with a decreasing shear rate, from 50 mm/min to the minimum shear rate of 1 mm/min. However, the fast residual strengths are greater than the slow residual strengths under all normal stresses and at the shear rates.

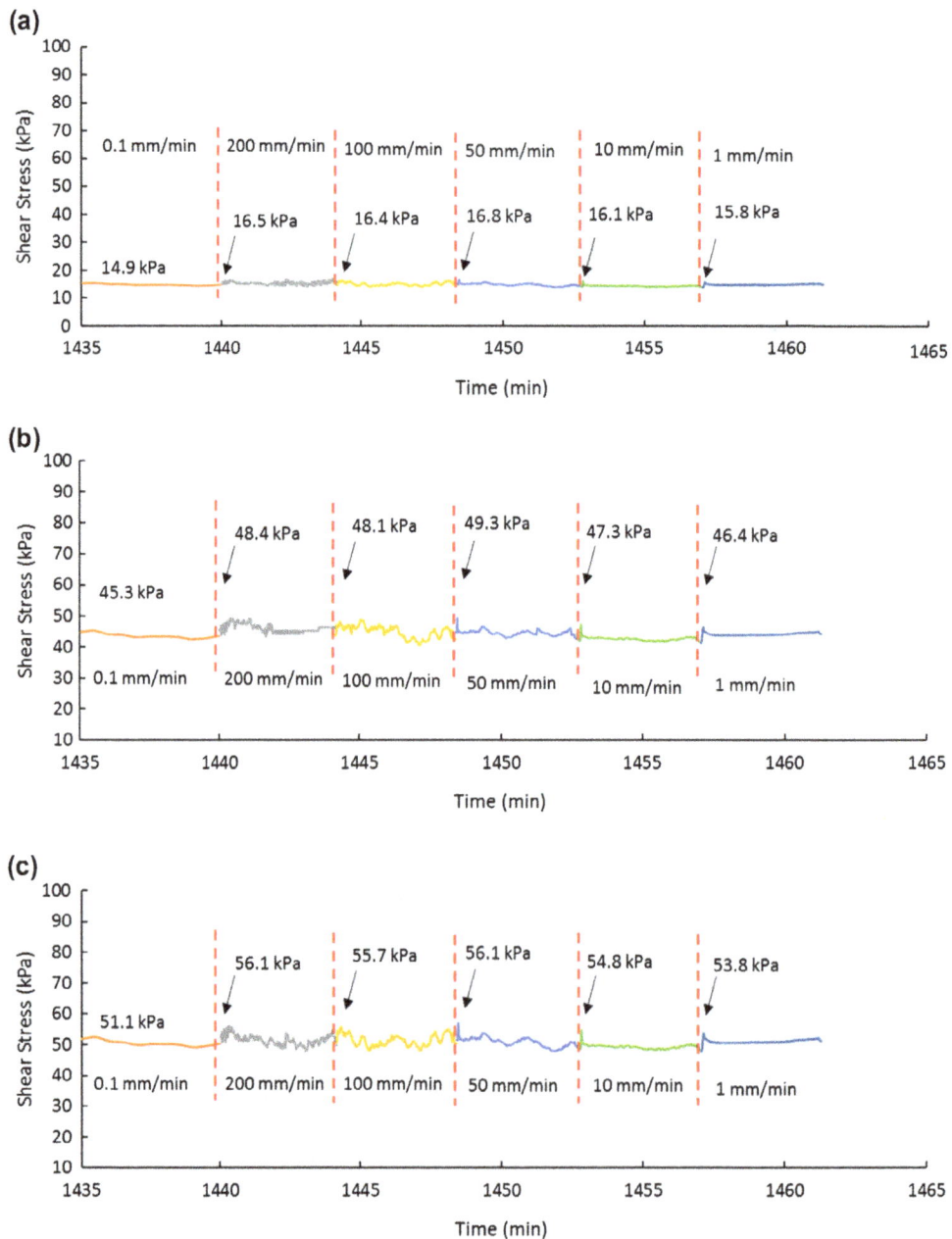

Figure 4. Residual shear strength vs. time at various shear rates. (a) σ'_n = 50 kPa, **(b)** σ'_n = 100 kPa **and (c)** σ'_n =200 kPa.

The fast residual strengths of the clay studied vs. time were replotted in Figure 5 to illustrate the effect of the applied normal stress on the results. As expected, the residual strength increases with increasing applied normal strength. In fact, under the normal stress of 200 kPa, the slow residual strength of the soil studied is equal to 51.1 kPa, which is 342 and 112% of those under the normal stresses of 50 and 100 kPa, respectively. Similarly, the same trend is nearly identical for the peak residual strength at faster rates. The fast residual strength under 100 kPa is nearly 10% higher than that under 50 kPa. Furthermore, the fast residual strength increases to three times the relative strength when the normal stress increases to 200 kPa.

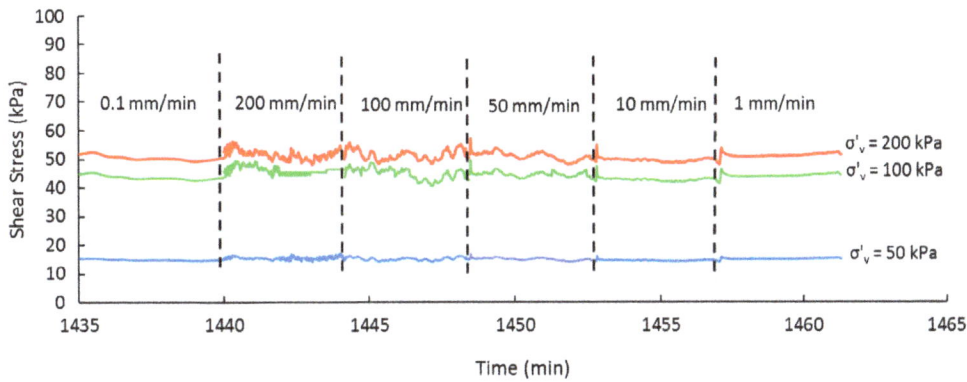

Figure 5. A comparison of residual shear strength at different normal stresses.

Figure 6 presents the relationship between the fast residual strengths and time for various shear rates under the normal stress of 200 kPa. As illustrated, the fast residual strength increases immediately to a value greater than the slow residual stress (i.e. 54.1 kPa in comparison with 51.1 kPa) when the shear rate increases. The fast residual strength at the rate of 200 mm/min then continues to increase with time and reaches the peak fast residual strength (56.1 kPa). Tatsuoka, Di Benedetto, Enomoto, Kawabe, and Kongkitkul (2008) called this positive impact of fast shear rates on the residual strength of the clayey soils, "Isotach behaviour". This positive effect is likely followed by a negative effect with a further displacement. In fact, the fast residual strength decreases from 51.1 to 47.9 kPa (i.e. 6% drop). Such an increase and decrease in the residual strength was also observed by other researchers for clayey soils (Tika et al., 1996). Likewise, such similar behaviour is also observed in this study for other rates of shear displacement. However, for the last two steps (steps 5 and 6), there is a sharp increase in the fast residual strength when the rate changes, which is followed by a drop in the fast residual strength. The positive and negative effect of shear rate is virtually the same for other normal stresses. However, only the results of the normal stress of 200 kPa are presented herein for the sake of brevity and simplicity.

Overall, when a pre-sheared surface developed at residual shear strength by slow shearing was subjected to more rapid shear displacement rate, the Isotach behaviour was observed. The observed behaviour includes an increase in shear strength (positive effect) at a negligibly small displacement followed by a drop reaches the minimum value (negative effect). The studies on the mechanism of the negative effect suggest that the main reason for the observed loss of strength is the penetration of water along the shear surface. Fast shearing results in an increase in the void ratio and thus the water content. The water is then free to penetrate along the shear surface and reduces the shear strength (Tika et al., 1996). On the other hand, the increase in the shear strength at small displacement may be related to volume and structural changes within the shear surface during the fast shearing stage. Fast shearing can cause disordering the structure of the shear surface and result in an increase in the strength (Figure 6).

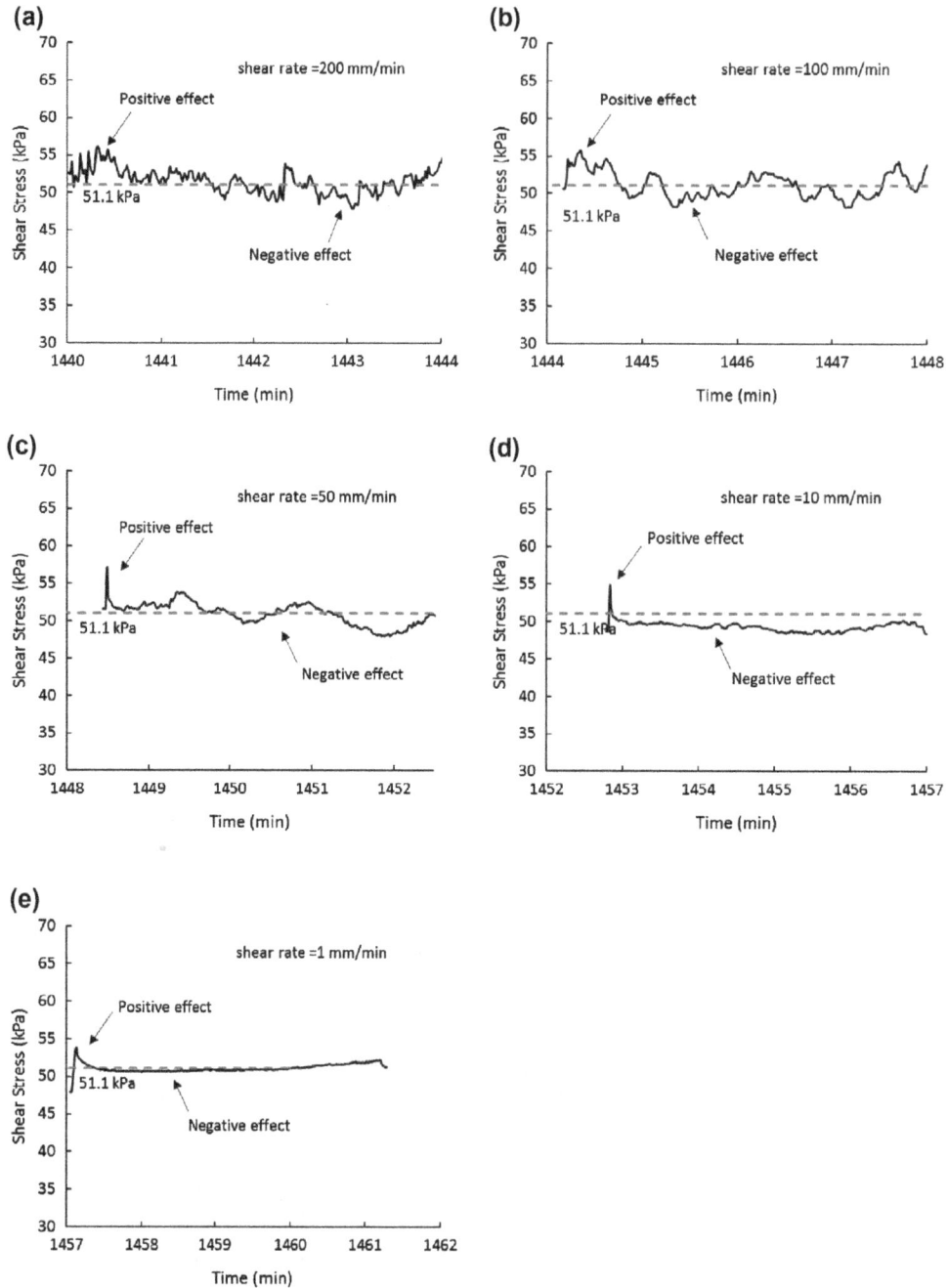

Figure 6. Residual shear behaviour under σ'_n = 200 kPa at the various shear rates of (a) 200 mm/min, (b) 100 mm/min, (c) 50 mm/min, (d) 10 mm/min and (e) 1 mm/min.

4. Conclusion

The shear behaviour of a pre-sheared clay was investigated in this study to examine the effect of shear rate on residual shear strength. Based on the results of the ring shear tests, the following has been concluded:

• There is an immediate increase of residual strength in comparison with the slow residual strength when the shear rates increase. In fact, there is a positive rate effect on the residual strength at a small displacement.

• The fast residual strength of pre-sheared clays increases with an increasing applied normal stress.

- The peak residual strengths of the pre-sheared clay in the fast shear steps are dependent on the rates of shear displacement. The fast residual strengths increase the slow residual strength when the rates increase. Moreover, the maximum peak residual strength of the fast steps occurs at the displacement rates of 50 and 200 mm/min.

- There is a drop in the fast residual strength on top of the increase in the displacement for all various shear rates and the applied normal stresses.

- There is a positive and negative effect of displacement rate on the residual strength for the rates greater than 50 kPa, which requires more research on the shear behaviour of pre-sheared clayey soils at fast shearing conditions.

Funding
This work was supported by the Australian Government Research Training Program.

Author details
Farzad Habibbeygi[1]
E-mail: farzad.habibbeygi@postgrad.curtin.edu.au
ORCID ID: http://orcid.org/0000-0002-5231-3397
Hamid Nikraz[1]
E-mail: H.Nikraz@curtin.edu.au
[1] Faculty of Science and Engineering, Department of Civil Engineering, Curtin University, Perth, Australia.

References
American Society for Testing and Materials. (2015). *Specifications*. West Conshohocken, PA: Author.
Casacrande, A., & Wilson, S. (1951). Effect of rate of loading on the strength of clays and shales at constant water content. *Géotechnique, 2*(3), 251–263. https://doi.org/10.1680/geot.1951.2.3.251
Gratchev, I. B., & Sassa, K. (2015). Shear strength of clay at different shear rates. *Journal of Geotechnical and Geoenvironmental Engineering, 141*(5), 06015002. doi:10.1061/(asce)gt.1943-5606.0001297
Habibbeygi, F., Nikraz, H., & Chegenizadeh, A. (2017). Intrinsic compression characteristics of an expansive clay from Western Australia. *International Journal of GEOMATE, 12*(29), 140–147. doi:10.21660/2017.29.20455
Habibbeygi, F., Nikraz, H., & Verheyde, F. (2017). Determination of the compression index of reconstituted clays using intrinsic concept and normalized void ratio. *International Journal of Geomate, 13*(39), 54–60. doi:10.21660/2017.39.98271

Head, K. H. (1986). *Manual of soil laboratory testing* (Vol. 2). London: Pentech Press.
Li, Y. R., Wen, B. P., Aydin, A., & Ju, N. P. (2013). Ring shear tests on slip zone soils of three giant landslides in the Three Gorges Project area. *Engineering Geology, 154*, 106–115. doi:10.1016/j.enggeo.2012.12.015
Mesri, G., & Olson, R. (1970). Shear strength of montmorillonite. *Géotechnique, 20*(3), 261–270. https://doi.org/10.1680/geot.1970.20.3.261
Richardson, A. M., & Whitman, R. V. (1963). Effect of strain-rate upon undrained shear resistance of a saturated remoulded fat clay. *Géotechnique, 13*(4), 310–324. https://doi.org/10.1680/geot.1963.13.4.310
Scaringi, G., & Di Maio, C. (2016). Influence of displacement rate on residual shear strength of clays. *Procedia Earth and Planetary Science, 16*, 137–145. doi: https://doi.org/10.1016/j.proeps.2016.10.015
Skempton, A. W. (1985). Residual strength of clays in landslides, folded strata and the laboratory. *Géotechnique, 35*(1), 3–18. https://doi.org/10.1680/geot.1985.35.1.3
Tatsuoka, F., Di Benedetto, H., Enomoto, T., Kawabe, S., & Kongkitkul, W. (2008). Various viscosity types of geomaterials in shear and their mathematical expression. *Soils and Foundations, 48*(1), 41–60. https://doi.org/10.3208/sandf.48.41
Tika, T. E., & Hutchinson, J. (1999). Ring shear tests on soil from the Vaiont landslide slip surface. *Géotechnique, 49*(1), 59–74. https://doi.org/10.1680/geot.1999.49.1.59
Tika, T. E., Vaughan, P., & Lemos, L. (1996). Fast shearing of pre-existing shear zones in soil. *Géotechnique, 46*(2), 197–233. https://doi.org/10.1680/geot.1996.46.2.197
Vulliet, L., & Hutter, K. (1988). Viscous-type sliding laws for landslides. *Canadian Geotechnical Journal, 25*(3), 467–477. https://doi.org/10.1139/t88-052

8

La Pintada landslide—A complex double-staged extreme event, Guerrero, Mexico

María Teresa Ramírez-Herrera[1]* and Krzysztof Gaidzik[1‡]

*Corresponding author: Maria Teresa Ramirez-Herrera, Laboratorio Universitario de Geofísica Ambiental & Instituto de Geografía, Universidad Nacional Autónoma de México, Ciudad Universitaria, Coyoacán, 04510 Ciudad de México, México
E-mails: tramirez@igg.unam.mx, maria_teresa_ramirez@yahoo.com
Reviewing editor:Jia-wen Zhou, Sichuan University, China

Abstract: Extreme storms commonly trigger landslides in regions of humid, warm tropical climate causing loss of life and economic devastation. The tropical mountainous areas of Guerrero in southwest Mexico are frequently hit by extreme hurricanes and cyclones and thus prone to landslides. On 16 September 2013, a huge landslide resulted in 71 fatalities and destroyed a large part of La Pintada Village. We applied remote sensing techniques using the LIDAR DEM and high-resolution images of the La Pintada area, a post-landslide field survey, geotechnical laboratory tests of colluvium material from the landslide, and a slope stability analysis. We also interviewed eyewitnesses accounts of the event. Our results suggest that the 2013 La Pintada landslide was a complex and two-stage event. An intense four-day-long rainfall event related to the landfall of Hurricane Manuel resulted in the oversaturation of soil, which was the main factor that caused the landslide. The effect of rainfall was amplified by the lack of high and dense vegetation on the 250-m-high slope. The lack of vegetation and slope-under-cutting likely contributed to the decreased slope stability. We suggest that increased intensity of extreme storms has contributed to increased landslides in this area. Furthermore, in tropical climate areas, where significant population lives in mostly developing countries, the combination of these phenomena makes them highly vulnerable to extreme storms and landslide hazards.

ABOUT THE AUTHORS

María-Teresa Ramírez-Herrera and Krzysztof Gaidzik's group research key research activities focus on the study of active tectonics, paleoseismology and tsunami deposits, and geohazards with emphasis on landslides. The reported research relates to a wider project on the Mexican forearc active deformation and geohazards as well to landslide automatic susceptibility mapping.

PUBLIC INTEREST STATEMENT

Landslides are one of the most important natural hazards causing loss of life and economic devastation. They are commonly triggered by extreme hurricanes and cyclones in tropical climate areas, such as Guerrero, southwest Mexico. Increasing intensity of extreme storms related to climate change increments landslide hazard in these areas and favorizes the occurrence of deadly landslides. This research was undertaken to provide a better understanding of a landslide that occurred on 16 September 2013, and resulted in 71 fatalities and destroyed a large part of La Pintada Village. The results of a detailed analysis of high-resolution digital elevation models, a post-landslide field survey, laboratory tests of soil material, and eyewitness interviews suggest that the studied landslide was a complex event triggered by an intense four-day-long rainfall related to the landfall of Hurricane Manuel. The lack of vegetation and slope undercutting likely contributed to the decreased slope stability.

Subjects: Earth Sciences; Natural Hazards & Risk; Earth Systems Science; Geomorphology

Keywords: landslide; hurricane; extreme events; LIDAR; hazard

1. Introduction

Natural hazards can lead to disasters that result in human death, economic loss, and environmental impacts (e.g. Scawthorn, Schneider, & Schauer, 2006). Extreme precipitation events related to climate change have been increasing over the last decades (e.g. Allan & Soden, 2008; Intergovernmental Panel on Climate Change [IPCC], 2012, 2013; Knutson et al., 2010). The frequency, intensity, spatial coverage, and duration of extreme events have increased, corresponding with a greater number of natural hazards and disasters (UNISDR, 2012). Increasing extreme precipitation events can influence the occurrence of landslides (IPCC, 2012). Studies of the impacts of climate change on mass movements in different regions around the world have suggested that the number, temporal occurrence and size of mass movement phenomena will likely be altered by changes in precipitation and temperature (e.g. Crozier, 2010; Fischer, Amann, Moore, & Huggel, 2010; Ravanel & Deline, 2011; Stoffel, Tiranti, & Huggel, 2014). However, the societal impact of mass movements is not equally distributed around the world; over the last few decades, most fatalities (more than 95%) related to extreme events have been recorded in developing countries (UNEP, 2012).

A landslide occurred in La Pintada Village in Guerrero, southwest Mexico, on 16 September 2013 due to the unprecedented rainfall (ca. 500 mm in 4 days) produced by Hurricane Manuel (Pasch & Zelinsky, 2014). This landslide crossed the entire floodplain of the Coyuca River (Figure 1), nearly buried an entire village, and resulted in 71 fatalities, leading to its ranking as one of the deadliest landslides to occur in Mexico and as a significant landslide worldwide (Haque et al., 2016; Petley, 2012). The La Pintada landslide occurred on the S-SSE facing slope, which was ca. 250 m-high with a slope of 25°. The same slope had failed previously, and the most recent failure occurred in 1974 (García Herrera, 2014). Here, we present the physical characteristics and processes that influenced the failure of this landslide. We focus on the analysis of a landslide at La Pintada as a complex and two-stage event because of its mobility, shape and deposits. We aim to provide a better understanding of this event and its implications for landslide hazard evaluation.

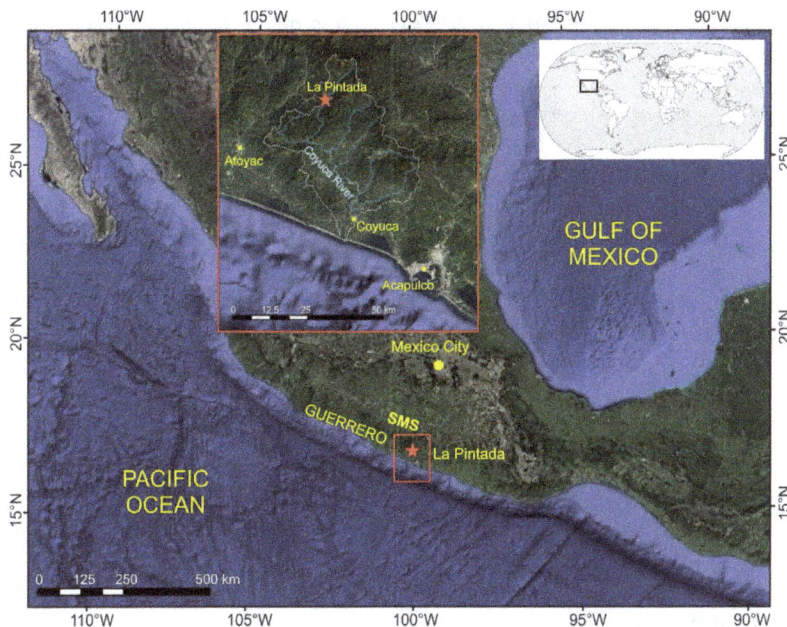

Figure 1. Location of the La Pintada landslide (red star) in southern Mexico on the right bank of the Coyuca River.

Notes: SMS: Sierra Madre del Sur. Data © LDEO-Columbia, NSF, NOAA, Image Landsat, Data © SIO, NOAA, U.S. Navy, NGA, GEBCO.

2. Study area

La Pintada is located ~60 km NW of Acapulco Guerrero State in southern Mexico (Figure 1) in the NW part of the Coyuca drainage basin (1,300.8 km^2) and on the western bank of the Coyuca River (locally named the Pintada River). La Pintada Village is located at 1,073 m amsl in the Sierra Madre del Sur Mountains ~40 km from the coast (Figure 1). The elevation of the mountains surrounding La Pintada Village varies from ~1,150 to 1,200 m amsl on the eastern side of the river to almost 1,300 m amsl on the western side of the river above the village. The local base level, i.e. the elevation of the Coyuca River, is at ca. 1,050–1,060 m amsl near La Pintada Village, and the local relief reaches 240–250 m above the village on the S-SSE facing slope that produced the landslide. The hills on the other side of the river are much lower, reaching only 100–150 m above the level of the Coyuca River.

2.1. Lithology and soils

The lithology of the Coyuca drainage basin is homogenous. The La Pintada area sits on igneous rocks of mainly granite (Figure 2(b)). The following three main units can be distinguished in this area with increasing depth: red clay with low permeability (10–12 m thick), strongly weathered and fractured altered granite, i.e. regolith, with high permeability (ca. 10 m thick), and fresh, impermeable granite (García Herrera, 2014). Soils in the Coyuca River Basin are generally poorly developed, usually with little or no profile development, and shallow, gravelly or sandy with a very thin layer of organic matter. This is especially true for the mountainous part of the basin, where young soils with no profile development are predominant (IUSS, 2015). The area of La Pintada is characterized by cambisols, i.e. moderately developed soils with some initial horizon differentiation (Driessen, Deckers, Spaargaren, & Nachtergaele, 2000). In the field, we observed reddish, very oxidized and clayey soil with a deeply weathered regolith, which is common of tropical environments.

2.2. Tectonic and seismic setting

The Coyuca drainage basin is in the Mexican subduction forearc where the oceanic lithosphere of the Cocos plate subducts beneath the North American continental plate at a convergence rate of 6.4–6.7 cm/year (DeMets, Gordon, & Argus, 2010) (Figure 2(a)). The seismicity pattern of the Guerrero section is distinctive. This section of the Mexican subduction forearc, known as the Guerrero seismic gap, has experienced no significant thrust earthquakes since 1911 (Anderson, Singh, Espindola, & Yamamoto, 1989; Kostoglodov & Ponce, 1994). Prior to 1911, four large earthquakes of Mw > 7 occurred in this section: 14 January 1900, 20 January 1900, 15 April 1907, and 16 December 1911

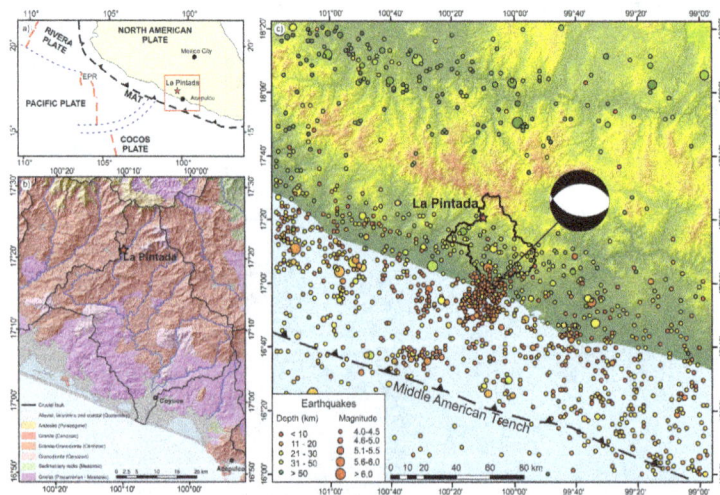

Figure 2. (a) Tectonic setting of the study area (red star). (b) Geology of the Coyuca drainage basin; (c) Seismicity map showing epicenters of earthquakes of magnitude Mw > 4.0 from time period 1998–2015, data from SSN (https://www.ssn.unam.mx). Location and focal mechanism solution for the Mw 5.8 Coyuca earthquake of 8 October 2001.

Notes: MAT: Middle American Trench and EPR: East Pacific Rise.

(Ramírez-Herrera, Corona, & Suárez, 2016). Thus, this section has the potential for a large subduction earthquake with a magnitude of Mw 8.1–8.4 (Suárez, Monfret, Wittlinger, & David, 1990). The small magnitude seismicity in this area is characterized by shallow, interplate thrust events near the trench (mainly offshore) and crustal thrusts and normal events situated ~80–105 km from the trench (Pacheco & Singh, 2010) near the area of La Pintada (Figure 2(c)). Among the largest events to occur here is the Coyuca earthquake, Mw = 5.8, 8 October 2001, with an epicenter located near the town of Coyuca (Pacheco, Iglesias, & Singh, 2002; Pacheco & Singh, 2010) and approximately 40 km from La Pintada (Figure 2(c)).

2.3. Climate and precipitation

The climate of the study area is sub-humid, tropical, warm to very warm (Instituto de Geografía de la UNAM, 2007), with average annual temperatures above 20°C. On Figure 3 is shown the average annual precipitation, except for a station near Coyuca, is more than 1,000 mm/yr (Figure 3). The temporal distribution of rainfall is uneven, with most precipitation occurring during five months of the rainy/wet season, i.e. from June to October. The maximum monthly rates of precipitations during a year, which reached nearly 400 mm at the Rio Santiago station, were recorded in the month of September (Figure 3). The annual number of days with rain varies from ca. 50 near the Pacific Ocean to more than 70 in the central and western portions of the Coyuca River Basin. Notably, no meteorological stations are located upstream in the mountainous area of the Coyuca River Basin near La Pintada. We expect that the precipitation rates are higher uphill than downhill.

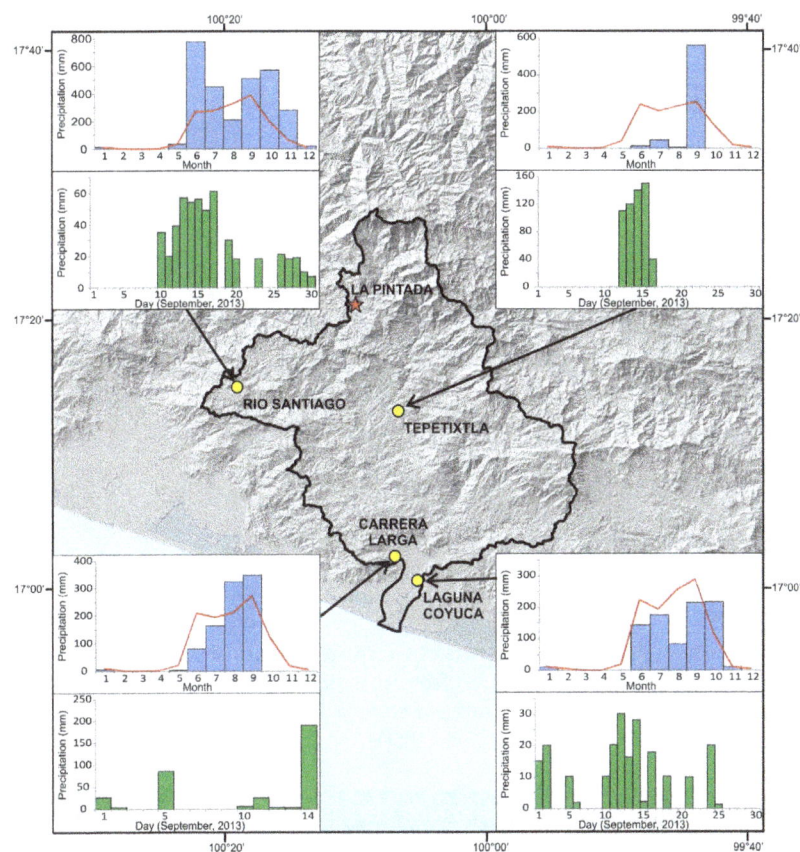

Figure 3. Precipitation: Rio Santiago, Tepetixtla, Carrera Larga and Laguna Coyuca stations in the Coyuca drainage basin, data from SMN (https://smn.cna.gob.mx/index. php? option=com_content&view =article&id=182:guerrero&catid =14:normales-por-estacion).

Notes: Average monthly mean precipitation (red line) and monthly precipitation in 2013 (blue columns); daily precipitation in September 2013 (green columns). Note that for the Carrera Larga station, the daily data ends on 14 September 2013.

September has the highest average rainfall and the highest number of days with rain and is when hurricanes and/or tropical cyclones occur. In 2013, a category 1 hurricane (on the Saffir-Simpson Hurricane Wind Scale), known as Hurricane Manuel, made landfall as a tropical storm on the coast of Guerrero from 13 to 19 September (Pasch & Zelinsky, 2014). The total amount of rainfall that fell on the Guerrero coast during this event reached the monthly average precipitation. In the mountains in the La Pintada area, Tepetixtla station recorded rainfall during 5 days, from 12 to 16 September, that was nearly three times higher than the average rainfall that occurred throughout the month of September (Figure 3). The amount of precipitation related to this hurricane at the Carrera Larga station nearest La Pintada is unknown because no data are available after 14 September 2013.

The exceptionally large amount of rainfall produced by Hurricane Manuel strongly impacted the discharge of the rivers draining the Sierra Madre del Sur and flowing into the Pacific Ocean on the Guerrero coast, destroying two stations. Thus, we combined data from the La Sabana River (east of La Pintada) to show the impacts of the rainfall related with Hurricane Manuel. On Figure 4 is illustrated the discharge of La Sabana River from 14–19 September that reached up to 900 times the average discharge for the years 1955–2012 and reached more than 10 times the previous record from 1984 (Figure 4). The water level rose to > 7 m, i.e. approximately 4 m higher than the average maximum water level (Table 1). The impact of the hurricane is also reflected in the runoff volume, which was more than 15 times the average annual runoff volume for the month of September 2013 (Table 1).

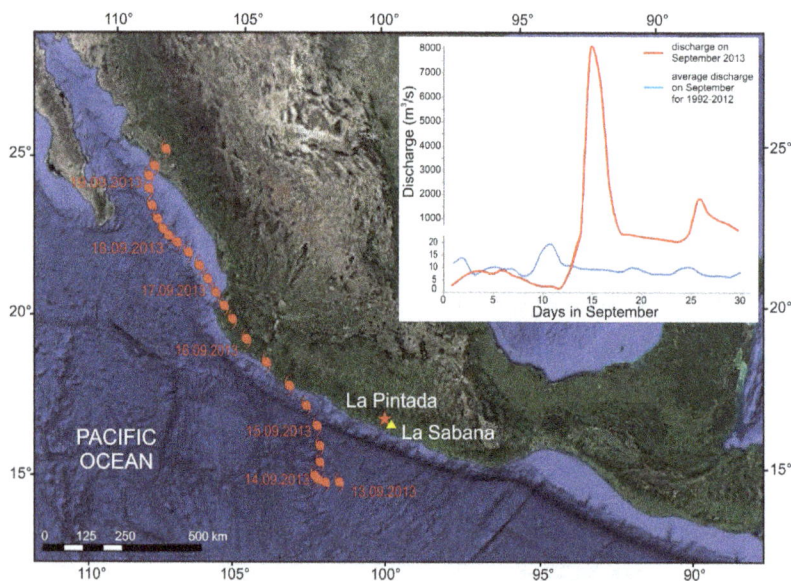

Figure 4. The best track position for Hurricane Manuel, 13–19 September 2013 (Pasch & Zelinsky, 2014).

Notes: Insert shows discharge from the La Sabana River (station km. 21,000; yellow triangle) for September 2013 (red line) and the average discharge for September 1992–2012 (blue line) (CONAGUA, https://www.conagua.gob. mx). Image from Google Earth. Based on Data © LDEO-Columbia, NSF, NOAA, Image Landsat, Data © SIO, NOAA, U.S. Navy, NGA, GEBCO.

Table 1. Discharge, water level and runoff volume for the La Sabana River (station km. 21,000; data from CONAGUA)		Max. discharge (m³/s)	Max. water level (m)	Annual runoff volume (m³)	Average annual discharge (m³/s)
Parameter **Time period**					
1955–2012	Maximum (date)	1,096.6 (14 September 1984)	5.7 (14 September 1984)	436,549.3 (1984)	13.8 (1984)
	Average	321	3.1	119,566.6	3.8
2013		13,941.2 (15 September 2013)	7.3 (15 September 2013)	2,571,553.1 (September—2,102,494.5)	82.2 (September—811.2)

3. Materials and methods

We applied remote sensing techniques using the LIDAR DEM and high-resolution images of the La Pintada area, a post-landslide field survey, geotechnical laboratory tests of colluvium material from the landslide, and a slope stability analysis.

The aim of the post-landslide field survey was to recognize the landscape and landforms, analyse the landslide features and material, obtain samples for geotechnical laboratory tests and gather residents accounts of the event. In the field, we took 356 photographs of the landslide from different perspectives and angles to produce a high-resolution DEM using the Structure from Motion (SfM) procedure (e.g. James & Robson, 2012; Johnson et al., 2014; Snavely, Seitz, & Szeliski, 2008) implemented by AgiSoft Photoscan software. We collected six samples from three different locations along the landslide: 3 samples at different depths (from the surface up to a depth of 43 cm) from above the headscarp, 2 samples (from the surface to a depth of 55–60 cm) from the middle point of the landslide, and 1 sample from the toe slope (Figure 5(b)). We were unable to obtain colluvium from the landslide because it had been removed before we reached the site; thus, we sampled the material on the slope nearest the slide. At sites where we took samples for further analysis, we also obtained compaction measurements using the Lang Penetrometer (with a scale ranging from 1 to 20).

In the field, we studied a second, smaller, landslide near the town of El Paraiso and west of the landslide in La Pintada shown on Figure 5(a) and (c). We took 72 photographs from various perspectives and angles to produce a DEM of the landslide using the SfM method (Figure 5(c)). We collected a sample of the material moved by this slide for laboratory geotechnical tests.

During fieldwork, we also gathered La Pintada residents' accounts. Among others, we talked with Mr. Galdino Alvarez Ocampo, who was the local governor (Comisario Municipal) of La Pintada at the time of fieldwork (April 2015).

3.1. Morphometric analysis and LIDAR derived DEM

To calculate the main morphometric parameters (e.g. steepness of slope) of the Coyuca River Basin, we used a 15-m-resolution DEM provided by INEGI. However, the resolution of this model was not sufficient to study the landslide itself. Therefore, we acquired LIDAR data for the study area of 15.7 km^2 in the upstream section of the Coyuca River Basin (Figure 5(b)). We used an airborne laser scanner (RIEGL Q-780) with a laser pulse repetition rate of 400 Hz and a field of view (FOV) of 60° (+30°/−30°) and gathered data with a density of 8 points/m^2 and 35 cm of horizontal and vertical a priori precision. A CESSNA TU206H aircraft flying at an elevation of 700 m above the ground was used to scan pulsed laser beams across the study area. The LIDAR data were collected on 19 March 2015. RIEGL software (e.g. RiAcquire, RiProcess, RiAnalyze and RiWord) was used to control the scanner and for data acquisition, visualization, projection, georeferentiation, and exportation. After adjusting the point cloud to the ground, the data were converted into "LAS" format. Later, the data were classified into ground and default points (e.g. vegetation, infrastructure, buildings, etc.) using TerraScan and TerraModeler. The automatic classification was later verified and improved using manual classification. To estimate the length and width of the landslide, the steepness of the slope that produced the landslide and the morphology of the La Pintada area, we produced a 1-m-resolution LIDAR-derived digital terrain model (DTM) from point cloud data classified and filtered in the previous stage (Figure 5(b)). We also obtained orthophotomaps using a Digital Airborne Scanner with one nadir RGB-sensor 1-DAS-1 from GeoSystem and digital elevation models derived from SfM using terrestrial photography (e.g. Figure 5(b) and (c)).

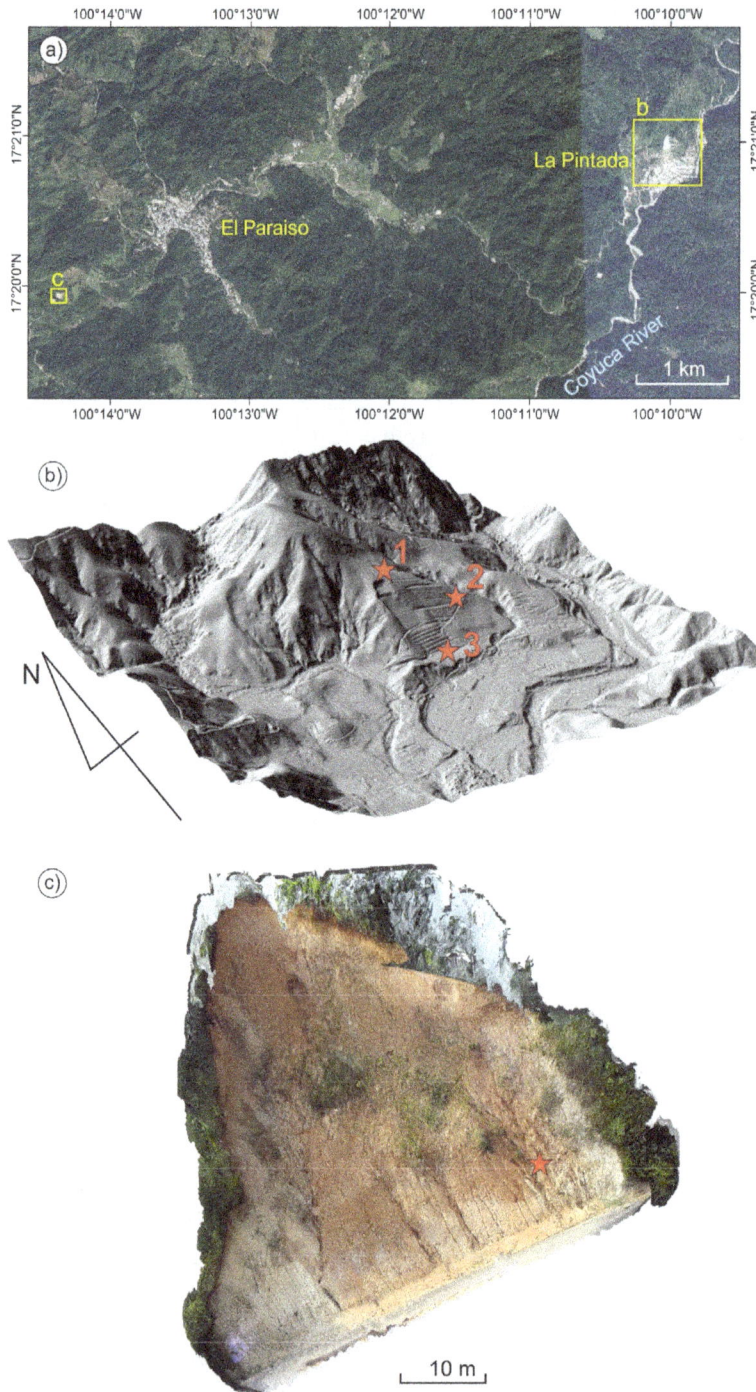

Figure 5. (a) Location of the studied landslides in La Pintada (b) and El Paraiso (c) [Image © 2016 DigitalGlobe]. (b) LIDAR DTM showing the main La Pintada landslide, shaded relief LIDAR (1 m-resolution), red stars and numbers from 1 to 3—location of samples for geotechnical analysis; and, (c) Digital model of the landslide in El Paraiso constructed from

To calculate the morphometric parameters and visualize DEMs, we used ArcGIS and GlobalMapper software. Calculations based on our LIDAR-derived DEM are an estimate and show minimum values because the LIDAR postdates the event and was obtained after the slide material had been removed and the slope was covered with concrete. However, we reconstructed the morphology and other slope morphometric values based on air-photos (1: 20,000) and satellite images (DigitalGlobe) taken before the event. To calculate the depth of the slide, we subtracted the current topography (LIDAR) from the theoretical pre-slide surface (obtained by interpolating points on the edges of the niche).

3.2. Geotechnical data analysis

Geotechnical tests (shear strength, unconsolidated—undrained shear strength) on seven samples of slope material adjacent to the landslide collected in the field (6 from the landslide in La Pintada and one from landslide in El Paraiso) were conducted at the Laboratory of Engineering Solution for Electronic Measuring Devices (IDEM), Synergy Engineering Equipment and Supplies Company. Geotechnical tests resulted in the determination of the following parameters: water content (moisture), angle of friction, cohesion, dry and saturated sample weights, percentage of fine soils, percentage of saturation, Atterberg limits (liquid limit, plastic limit, plasticity index), and the soil classification according with Unified Soil Classification System, USCS.

3.3. Slope stability analysis and safety factor (F)

We conducted a slope stability analysis using the geotechnical test (mentioned above) results and the values of the morphometric parameters derived from the LIDAR DEM. To determine the slope conditions and the susceptibility of the slope to failure, we calculated the safety factor (F), which is usually defined as the ratio of forces resisting movement (strength) to the forces driving movement (load; e.g. Verruijt, 2012). Lower safety factors correspond with more unstable slopes, with a lower limit of 1. Thus, F values greater than 1 indicate a stable slope, i.e. forces resisting movement prevail over forces driving movement, and values less than 1 suggest a slope that is prone to failure, i.e. forces driving movement overpower forces resisting movement. We calculated the minimum safety factor using the following different methods: Bishop's method implemented by SLOPE software, the Fellenius and Spencer method, Janbu's method, and the Morgenstern-Price method, implemented by GEO5 2016 Slope stability software (Bishop, 1955; Fellenius, 1927; Janbu, 1954, 1973; Morgenstern & Price, 1965, 1967; Spencer, 1967; Verruijt, 2012).

4. Results

4.1. La Pintada landslide—historical development

The landslide that occurred at La Pintada on 16 September 2013 was not the first slide event to occur in this area. The occurrence of a landslide in this area during the rainy season in 1974 coincides well with the occurrence of Hurricane Dolores (HURDAT2, 2016), which is regarded as one of the strongest hurricanes to strike the coast of Guerrero in the twentieth century. We also corroborated the presence of at least three ancient slides on the same slope using satellite images. In the southern part of the village and outside the extent of the 2013 event, we observed large angular granite blocks of up to several cubic meters (up to ca. 60 m³) in size that probably resulted from older landslides.

4.1.1. Pre-failure slope characteristics

Next to the La Pintada landslide site, the slope gradient is approximately ≥25°; thus, the pre-failure gradient of the slope was probably like this gradient. We reconstructed the pre-failure surface using LIDAR and the elevations of points at the approximate positions of the headscarp of the slide. This surface shows similar slope steepness.

Dense mountain forests with pine trees and evergreen oaks are predominant near the coffee-growing village of La Pintada. However, the landslide slope is devoid of vegetation, as shown on the INEGI orthophotomap from 1995 shown on Figure 6(a) and the satellite images from 2010 to 2011 shown on Figure 6(c) and (d), i.e. preceding the event. The headscarp was initiated at about the boundary between the lower part of the slope that was free of vegetation and the densely forested upper slope.

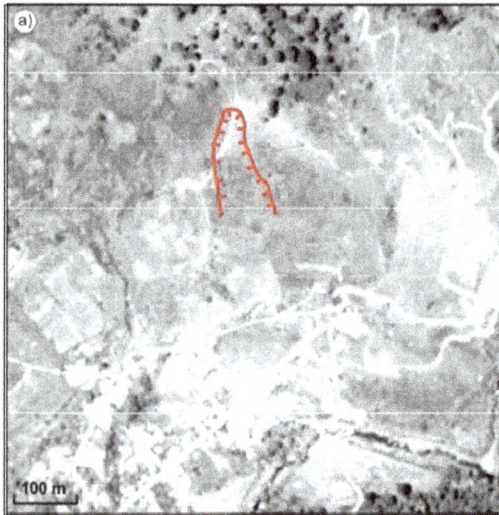

1995

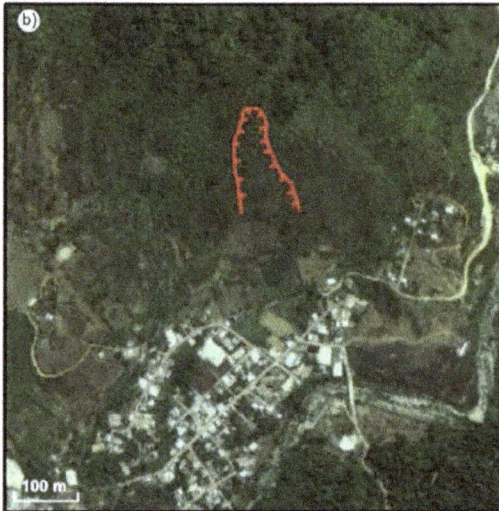

6/5/2010

Figure 6. Orthophotomap (a) and satellite images (b and c; Image © 2016 DigitalGlobe) of the La Pintada area from before (a, b and c). (a) Orthophotomap E14C36E, scale 1: 20,000, 2 m resolution (Source: INEGI, Ortofoto Digital E14C36E, El Paraiso 1999), with the contours of the La Pintada landslide, (b) and (c) images showing the slope before the event almost completely devoid of vegetation.

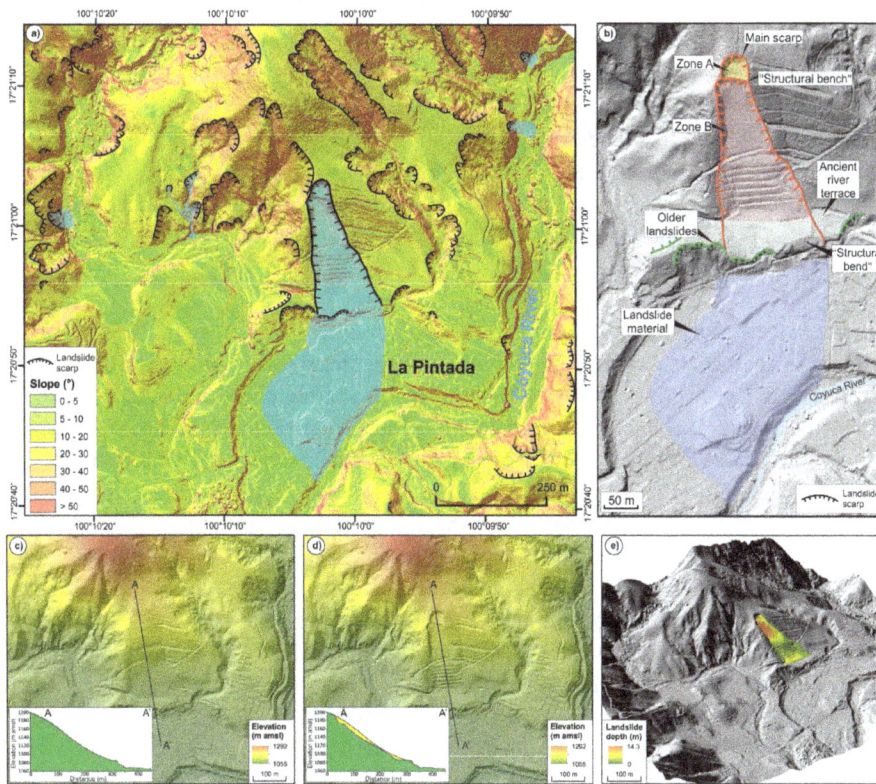

Figure 7. (a) Landslide inventory imposed on the slope steepness near La Pintada Village (LIDAR, 1 m-resolution), blue shades: landslides triggered by Hurricane Manuel. (b) Main characteristics of the La Pintada landslide. (c) Model of pre-failure and theoretical morphology of the slope above La Pintada, insert – the pre-failure slope profile. (d) Shaded relief of the slope after failure (1 m-resolution LIDAR), insert – the slope profile before failure (yellow) and after the landslide (green); and (e) Depth of the landslide.

4.2. Landslide characteristics

The landslide in La Pintada occurred on a moderately to steeply dipping (≥25°) S—SSE facing slope just north of La Pintada Village (Figure 7). The main morphometric parameters are presented in Table 2. Based on photographs taken after the event (Figures 7(b) and 8(b) and (d)), a clear bend and break occurs in the lower part of the slope that is emphasized by the presence of two ponds. The upper break marks a bench that extends beyond the landslide and likely corresponds to the outer scarp of an ancient river terrace that also has traces of ancient landslides.

Figure 7(b) and Table 2 show the La Pintada landslide presents two main zones divided by a "structural bench" and bend in the slope an upper zone (A) in the form of a symmetrical niche facing south and a lower zone (B) that continues up to the toe of the slope.

Table 2. Morphometric parameters of the La Pintada landslide							
	Height (m amsl)	Length (m)	Width (m)	Depth [m]	Area (m²)	Volume (m³)	Orientation of slope
Entire landslide	1,190–1,085	300	Up to ca. 100	Up to ~15	~23,375	75,000	Facing S-SSE
Zone A (upper)	1,190–1,170	40	45	~10	~1,795	4,500	Facing S
Zone B (lower)	1,170–1,085	260	Up to ca. 100	~15 in the upper part to 2–3 m downhill	~21,580	70,500	Facing SSE

According to our calculations based on LIDAR-derived DEM, the volume of remobilized material is approximately 75,000 m³. The colluvium material (i.e. rocks, mud, soils, and trees) transported by the landslide was deposited at the bottom of the slope and the ancient Coyuca River flood plain, burying the central part of La Pintada Village, with an area of approximately 50,000 m², reaching the Coyuca River and ~500 m from the headscarp of the landslide (Figure 7(a), (b) and 8). The thickness of the material reached up to 10 m, with an average thickness of approximately 5 m. Considering the average density of the samples collected in the field from the landslide material (clay) of ca. 2,645 kg/m³ and the calculated volume of the material, the weight of the colluvium material that buried La Pintada reached up to ca. 200,000 tons.

4.3. Geotechnical slope characteristics

The results from the geotechnical tests of six samples obtained from three sites on the slope adjacent to the landslide at La Pintada and from one sample obtained from a smaller landslide near the town of El Paraiso (Figure 5(b) and (c)) are summarized in Table 3. According with these results, the material on the slope above La Pintada consists of silt with low plasticity and silty sand (Table 3). The material taken from El Paraiso slope can be described as clay with high plasticity, i.e. fat clay. Samples taken from the surface or near the surface show an angle of internal friction of more than 30° or nearly 30° (samples 1a, 1b, 2a and 3, and sample from landslide in El Paraiso; Table 3); whereas, samples at depths of 55–60 cm showed smaller angles of up to 13° (Table 3, sample 2b). Differences between surface and deeper samples are also reflected in the values of the neutral stress coefficient, with deeper samples corresponding with larger values. The estimated values of cohesion range widely from ~4 kN/m² for the surface sample from the middle part of the landslide (sample 2a) to nearly 25 kN/m² for the deepest sample, i.e. sample 2b (Table 3). The weight of samples varied insignificantly, i.e. from less than 15 kN/m³ to almost 16 kN/m³ for dry samples and from

Figure 8. La Pintada Landslide. (a–d) La Pintada landslide of 13 September 2013 from different perspectives, all show the landslide, partially buried La Pintada Village, and Coyuca River; (e and f) Examples of more than a thousand smaller landslides related to the rainstorm of 13 September 2013 Manuel hurricane on the Guerrero Mountains.

~25 kN/m³ (samples 1a and 1c) to almost 27 kN/m³ (sample 3) for saturated samples. The samples had variable fine grain contents (<0.074 mm). The slope adjacent to the location of the La Pintada landslide contains fine particles ranging from 20% (sample 2a) to 65% (samples 1b and 2b) among the studied samples. For the landslide in El Paraiso, the content of small grains is higher reaching ca. 75% (Table 3).

Using the results of geotechnical tests conducted on seven samples taken in the field (six from La Pintada landslide and one from the landslide in El Paraiso) and measurements taken in the field and using the LIDAR DEM, we conducted a slope stability analysis to determine the safety factor (F). The results of this analysis are presented in Table 4, where we summarized the values of the minimum safety factor calculated using different methods (i.e. Bishop, 1955, Fellenius, 1927; Janbu, 1954, 1973; Morgenstern & Price, 1965, 1967; Spencer, 1967; Verruijt, 2012). The slope adjacent to the location of the La Pintada landslide shows minimum and maximum safety factors of 0.95 and 1.36, respectively (Table 4). These results differ because different methods use different formulas and require different input data.

Table 3. Results of field studies and laboratory tests conducted on seven samples taken from the landslide in La Pintada and El Paraiso (for location see Figure 5)

Parameters	Samples						
	La Pintada						El Paraiso
	1a	1b	1c	2a	2b	3	1
Latitude	17°21'04"	17°21'04"	17°21'04"	17°20'59"	17°20'59"	17°20'56"	17°20'08"
Longitude	100°10'02"	100°10'02"	100°10'02"	100°09'57"	100°09'57"	100°10'01"	100°14'33"
Depth (cm from the surface)	0–2	20–26	37–43	0–5	55–60	0	0
Results of field studies							
Penetrometer values	15.5	18	4.5	17.5	18.5	11.5	
	15	12.5	8.5	14.5	16.5	11.5	
	16	14.5	6.5	16	16.5	12	
Average value	15.5	15.0	6.5	16.0	17.2	11.7	
Results of laboratory tests							
Angle of friction (φ)	32.9	28	23	32	13	34	34
Neutral stress coefficient (K_0)[a]	0.46	0.53	0.61	0.47	0.78	0.44	0.44
Cohesion (kN/m²)	0	0	21.58	3.92	24.52	13.73	13.73
Weight of dry sample (kN/m³)	15.79	15.79	15.79	15.79	14.51	15.89	15.89
Weight of saturated sample in (kN/m)	24.91	26.18	25.01	25.99	26.09	26.87	25.79
Water content (moisture) of natural sample (ω%)	10.70	11.00	8.60	5.80	11.60	2.80	7.20
Water content (moisture) of saturated sample (ω%)	49.85	52.70	52.35	53.81	52.01	55.26	36.00
Percentage of fine soils (<0.074 mm)	51.30	64.60	46.70	21.20	63.90	58.30	74.60
Percentage of saturation (%Sr)	91.78	90.95	93.47	92.13	90.52	91.36	91.39
Liquid limit (%Ll)			73.60				53.20
Plastic limit (%Pl)			48.50				23.70
Plasticity index (%Pi)			25.10				29.50
Classification according to USCS	ML	ML	SM	SM	ML	ML	CH
Description	Silt, low plasticity	Silt, low plasticity	Silty sand	Silty sand	Silt, low plasticity	Silt, low plasticity	Clay of high plasticity, fat clay

Notes: Values from laboratory tests were used in the slope stability analysis. USCS: Unified soil classification system.

[a] $K_0 = 1 - \sin\varphi$ (Verruijt, 2012).

Table 4. Values of safety factor (F) calculated using different methods

Method	SLOPE software	GEO5 2016[a]	GEO5 2016[b]
Bishop analysis[c]	1.11	1.36	1.04
Fellenius[d]		1.29	0.95
Spencer[e]		1.33	0.96
Janbu[f]		1.36	0.99
Morgenstern-Price[g]		1.35	0.99

[a]Using data from geotechnical tests.

[b]Using average values based on soil classification according to USCS (see Table 3).

Sources: [c]Bishop (1955), [d]Fellenius (1927), [e]Spencer (1967), [f]Janbu (1954, 1973), [g]Morgenstern and Price (1965, 1967).

4.4. Seismic signals produced by La Pintada landslide

We analyzed the seismograms from the nearest seismological station, i.e. El Cayaco (CAIG), which is located ca. 40 km from La Pintada, to determine if the landslide caused any visible ground motion effects that were reflected on seismic record (Servicio Sismológico Nacional [SSN], 2017). We found that the landslide produced seismic signals at 3:23.50 pm on 16 September 2013 (Figure 9). A difference of nearly 7 s between the time of arrival of the P and S waves suggests that the landslide was located some 50 km from the El Cayaco station, which approximately corresponds with the location

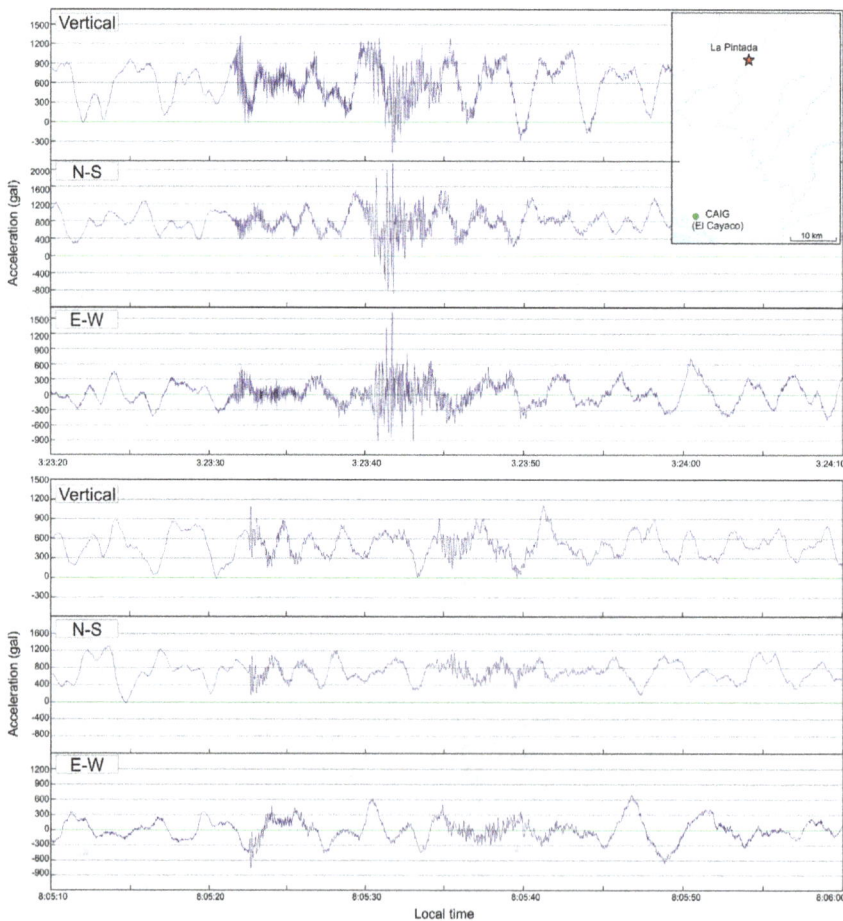

Figure 9. Shaking related to the La Pintada landslide registered on seismograms at the CAIG (El Cayaco) station on 16 September 2013, data from SSN (https://www.ssn.unam.mx).

Notes: Upper graphs refer to the first stage of the landslide at 3:23 pm local time, whereas the lower graph refers to the second stage (8:05 pm).

of the La Pintada landslide. The timing of the seismic signals nearly corresponded with the time range indicated by most of the eyewitnesses, around 3:30 pm.

At 8:05 pm, additional smaller seismic signals were recorded by the seismogram at approximately the same time and on the same day according to people that experienced near the landslide. These two seismic signals recorded in one station, El Cayaco, 50 km away from La Pintada, suggest a surficial ground motion, equivalent to two small magnitude Mc 3.2 and 3.1 events (SSN, 2017).

4.5. Failure mechanisms of the 16 September 2013 La Pintada landslide

Based on our post-landslide survey observations and interviews to people that experienced near the landslide, we hypothesized that the 2013 La Pintada landslide was complex and occurred in two sequential stages of movement. The different styles of sliding are marked by the two zones described above; an upper zone (A) with a rotational plane that produced a deep-seated slide (rocks, trees, and mud), and a lower zone (B) with a semi-rotational to planar plane that produced a debris slide (Figure 7(b)–(d)). The two hypothesized stages of sliding are supported by testimonies of people that experienced near the landslide and the seismic signals, though small in amplitude, recorded in the seismograph at Cayaco station that show two sequential events. The first event was significantly larger than the second event, producing a thunderous noise likely related to slope failure, the impacts of huge blocks on the ground, and the flow of mud and trees over the houses and flood plain topography (Figure 8). This event occurred in ca. 17 s, which did not leave enough time for people to leave their houses. The second stage consisted of a smaller landslide event that occurred 5 h after the first sliding episode. This event consisted of a smaller debris slide. The seismic signal of the second event was smaller and lasted less than 10 s. These seismic signals of the first and second events, have been attributed to landslide related ground vibrations resulting from flow over smaller scale topographic features, such as the scarp at the La Pintada failure plane, frictional processes, and impacts of individual blocks, such as the ones observed at La Pintada (Figure 8) (e.g. Allstadt, 2013; Wartman et al., 2016).

5. Discussion

The main contributing factor to trigger the deep-seated La Pintada landslide was extreme and long-duration rainfall, i.e. the amount of accumulated rainfall. Although residents of La Pintada have witnessed evidence of the landslide creeping prior to the large-scale event, our data indicate that the main direct factor that initiated the large-scale landslide in La Pintada, on Monday afternoon, 16 September 2013, was an immense rainfall, 600–800 mm in 4 days, related with Hurricane Manuel that affected the western coast of Mexico. Detailed mapping based on satellite images revealed that more than 1,500 landslides were provoked by this hurricane in the central and eastern parts of Guerrero State (Aranza Rodríguez, 2014). However, the exact number of landslide events for the entire state could be much higher. Most landslide events provoked by the hurricane are represented by debris-flows and shallow soil slips, which are typical for high-intensity, short-duration rainfall events. Deep-seated landslides, such as the studied event at La Pintada, are less common. These events are generally triggered by low-intensity, long-duration rainfall events (Larsen & Simon, 1993). The already creeping La Pintada landslide and many other large deep-seated landslides throughout Guerrero State in southwest Mexico resulted from prolonged rainfall over five days (12–16 September 2013) during Hurricane Manuel that pushed a section of the slope over the failure point. Four-day-long rainfall events resulted in oversaturated soil, which significantly increased the weight of the slope material (Table 3). Small landslides and debris flows that were observed in Guerrero State after Hurricane Manuel, such as the studied landslide near El Paraiso, likely resulted from local severe storm cells that formed during the passage of Hurricane Manuel.

The impacts of Hurricane Manuel can be observed in the variations of discharge of the main rivers flowing into the Pacific Ocean. Data from the station nearest the La Sabana River were more than 10 times greater than from the previously recorded data from 1984 (Table 1), and more than 900 times greater than the average discharge recorded for 1955–2012 (Figure 4). Extreme meteorological events, such as Hurricane Manuel, play a major role in landslide activity in regions of humid, warm tropical climate, provoking thousands of landslides, from shallow soil slips and debris flows to cata-strophic, large deep-seated landslides (e.g. Buckham et al., 2001; Cannon et al., 2001; Harp, Hagaman, Held, & McKenna, 2002; Larsen & Simon, 1993; Larsen & Torres-Sanchez, 1992). For ex-ample, Hurricane Mitch triggered a large, deep-seated landslide (volume of 400,000 m^3) in El Reparto, Honduras (Harp et al., 2002), and 11,500 landslides in Guatemala that were triggered by the same meteorological event (Buckham et al., 2001).

The calculated safety factors, for the slope adjacent to the location of the La Pintada landslide, minimum and maximum, are 0.95 and 1.36, respectively (Table 4), indicating that the slope was meta-stable under normal conditions and that a powerful direct factor was needed to initiate the landslide processes. Rainfall related to Hurricane Manuel was the triggering factor. Furthermore, the material that failed had low safety factor values. Thus, the slope adjacent to the La Pintada landslide is meta-stable to unstable and is a high landslide hazard for the local community.

The La Pintada landslide originated on a slope that was devoid of vegetation (Figure 6). Almost absent vegetation on the steeply inclined slope above the La Pintada Village potentially contributed to more intense erosion and, consequently, slope failure. The lack of vegetation played a significant role in reducing slope stability, i.e. it was a preparatory and causal factor of the landslide and played a role in the susceptibility of the slope to movement over time without initiating it (e.g. Knapen et al., 2006). In addition, the landslide headscarp is located roughly on the border between high and dense forest areas uphill and the slope without vegetation downhill (Figure 6). As mentioned earlier, resi-dents in La Pintada also indicated that the cutting the vegetation contributed to the generation of the landslide and that the slope should have been forested to prevent the soil from moving down the slope. Thus, this absence of trees on the studied slope was an important factor that contributed to the occurrence of this devastating landslide. As proven in many different regions, deforestation could lead to increased slope instability and erosion, a lower safety factor due to root decay (e.g. Knapen et al., 2006), and result in increased rates of landslide activity (e.g. Kamp, Growley, Khattak, & Owen, 2008). Studies in various environments worldwide have demonstrated a clear correlation between the absence of vegetation (or the presence of only small amounts of vegetation) and the occurrence of landslides; e.g. New Zealand (Glade, 2003; Phillips & Watson, 1994), Andes region (Vanacker et al., 2003), Himalaya (Kamp et al., 2008), and African highlands (Davies, 1996; Knapen et al., 2006; Mugagga, Kakembo, & Buyinza, 2012; Nyssen, Moeyersons, Poesen, Deckers, & Mitiku, 2003). The continued deforestation of landslide-prone areas is a major factor that contributes to a constant increase in landslide activity in the twenty-first century (Schuster, 1996). Mexico and Central America are not exceptions. Increases in landslide activity due to deforestation and changes in land-use have been reported for Mexico and Central America (e.g. Caballero et al., 2006; Restrepo & Alvarez, 2006). The landslide in La Pintada followed a similar pattern, where the absence of veg-etation decreased the slope stability and rainfall related with Hurricane Manuel was the direct factor that triggered the landslide.

Two important questions are: How often do extreme meteorological events happen, and do all extreme meteorological events trigger devastating landslides? Exceptionally high rainfall produced by extreme meteorological events has occurred before in Guerrero. Hurricane Pauline (also known as Paulina) triggered landslides and other damage in Oaxaca and Guerrero from the 5th to 10th of October in 1997 (Lawrence, 1997; Matias Ramírez, 1998). Moreover, six other hurricanes and tropical storms occurred between 1921 and 1996 on the coast of Guerrero and produced fatalities and land-slides (Matias Ramírez, 1998). A more complete record suggests the occurrence of 35 hurricanes that had catastrophic impacts on the Guerrero coast between 1949 and 2015 (HURDAT2, 2016). Thus, considering that only the largest hurricanes are reported (HURDAT2, 2016; Matias Ramírez, 1998), the recurrence time for extreme meteorological events is approximately 3–10 years. Therefore, the main direct factor that contributes to landslide activity, when most often soil creeping has already spanning days to years prior to a major slip, in the study area is the occurrence of ex-treme meteorological events (tropical storms, cyclones and hurricanes). Numerous examples from Central America and other regions of humid, warm tropical climate show that extreme meteorological events, such as hurricanes and tropical cyclones, are major factors that contribute to the landslide activity in these areas (e.g. Buckham et al., 2001; Cannon et al., 2001; Harp et al., 2002; Larsen & Simon, 1993; Larsen & Torres-Sanchez, 1992). Mass movements triggered by Hurricane Manuel on September 2013, such as the catastrophic landslide at La Pintada that caused more than 70 fatalities and thousands of smaller landslides in Guerrero State (Aranza Rodríguez, 2014), indicate that the study area is highly susceptible to landslides. Hazard warnings of this type of event could be forecasted long before a large event occurs or at the very least geotechnical analysis can be performed on susceptible slopes that are at high risk of failure.

Human activity, e.g. deforestation, steepening of slopes, slope under-cutting, loading of slopes due to the construction of houses, changes in the local relative relief due to the creation of artificial slopes or by grading with construction machinery, etc. might disturb slope equilibrium and lead to the occurrence of mass movement processes with time (e.g. Alexander, 1992; Schuster, 1996). In La Pintada, landslide activity has a long history. The portion of the village significantly destroyed by the landslide on 16 September 2013 was built on the colluvium material from an older historic slide. Moreover, evidence of even older paleo-events has been observed. However, although landslide activity in this area has been proven, the government rebuilt houses and reconstructed the village in the exact pathway of the 2013 landslide, which is now known as La Nueva Pintada. Although the local community is aware of the relationship between vegetation and landslide activity, efforts toward reforestation of the slope have not been introduced by official authorities. Thus, the slope without vegetation still might present a high landslide hazard.

Earthquakes are a major trigger for landslide events (Malamud, Turcotte, Guzzetti, & Reichenbach, 2004). Examples from around the world prove that large earthquakes can produce numerous landslide features, e.g. the 7.9 Wenchuan earthquake (2008) in China (Gorum et al., 2011), the 7.8 Gorkha earthquake (2015) in Nepal (Kargel et al., 2015), the 7.4 Khait earthquake (1949) in Tajikistan (Evans et al., 2009) and the 6.6 Chuetsu earthquake (2004) in Japan (Wang, Sassa, & Xu, 2007). Seismic events with magnitudes > 4.0, which is the smallest earthquake magnitude known to produce landslides (Keefer, 1984), are common in the study area (Figure 2). However, for small-magnitude earthquakes, the main provoked events would mainly be represented by rock falls, rock slides, soil falls, disrupted soil slides, soil slumps and soil block slides (Keefer, 1984). Earthquakes with a magnitude > 5.5 have occurred recently in Guerrero, including the magnitude 5.8 earthquake that occurred in 2001 with epicenter located ca. 40 km from La Pintada (Pacheco & Singh, 2010; Pacheco et al., 2002).

This sector of the Mexican subduction zone could potentially experience large subduction earthquakes of Mw 8.1–8.4 (Suárez et al., 1990). Such events would produce from ca. 60,000 to more than 300,000 landslides in an area spanning nearly 90,000 km² (Table 5). Moreover, seismicity not only

Table 5. Area affected by landslides and number of landslides triggered by earthquake of a defined magnitude

Earthquake Magnitude, M	Area affected by landslides, A (km²)[a]	Number of triggered landslides, N[b]	Number of triggered landslides, N[c]	Magnitude of landslides, m_L[d]
5.5	109.65	88	28	1.45
5.8	218.78	206	68	1.83
8.1	43,651.58	139,675	62,951	4.80
8.4	87,096.36	326,949	153,461	5.19

[a]$Log_{10}(A) = M - 3.46$.

[b]$Log_{10}(N) = 1.2312 M - 4.8276$.

[c]$Log_{10}(N) = m_L$.

[d]$m_L = 1.29 M - 5.65$.

Sources of formulas: [a,b]Keefer (2002) and [c,d]Malamud et al. (2004).

triggers landslides directly after the earthquake, but multiple earthquakes can also have long-term effects on slope integrity, decreasing the slope stability by seismically inducing damage (Wolter, Gischig, Stead, & Clague, 2016). The effect of seismicity on the study area can also be amplified by deforestation processes, which occurred in the western Himalayas, where deforestation contributed significantly to landslides during and shortly after earthquakes (Kamp et al., 2008). Therefore, the effects of tropical cyclones can be amplified by the decreased long-term slope stability resulting from seismicity.

6. Conclusions

Our results suggest that the 2013 La Pintada landslide was a complex, two-stage event: a deep-seated slide (rocks, trees, and mud) from the rotational plane in the upper zone, and a semi-rotational to planar plane mechanisms produced a debris/mudflow in the lower zone.

The most important and direct factor that initiated the already creeping landslide in La Pintada was a four-day continuous and high rainfall event that was produced by Hurricane Manuel, which affected the western coast of Mexico. The 1974 landslide was also triggered by rainfall related to another extreme meteorological event, i.e. Hurricane Dolores. Thus, anomalous precipitation appears to be the most important factor that contributed to triggering the large event in the humid and warm climate of La Pintada. The lack of high and dense vegetation on the slope potentially amplified erosion and acted as a causal factor of the landslide, i.e. making the slope more prone to mass wasting processes. It is likely that the decrease of slope stability that resulted from cutting the slope to build houses on the toe slope contributed to landsliding in this region. Seismic activity, even if did not contribute directly to the initiation of the La Pintada landslide, could promote a decrease in slope stability, and it is likely that large subduction earthquakes produced numerous landslides in the past.

Could La Pintada landslide and other landslides in the area be forecasted? At La Pintada pre-existing slope conditions (topography, soil characteristics, lack of vegetation, changes in land use, presence of ground fractures and cracks, soil creep, and the history of landsliding in this area) suggested that this area was susceptible to landslides. Landslide susceptibility mapping produced after this event confirms that the slope condition and characteristics made this area highly susceptible to landslides (Gaidzik et al., 2017). Furthermore, the amount of accumulated rainfall, 4 days of consecutive rainfall produced by Hurricane Manuel, was the ultimate triggering indicator for a large landslide to occur. If an early warning system was implemented, indeed, this event could at least be identified as a landslide hazard, with a high-level threat. Most recent studies in Central America suggest that early warning systems combining operational flash flood guidance systems with landslide susceptibility mapping can be used for real-time landslide hazard assessment (Posner & Georgakakos, 2015a, 2015b, 2016).

The study area is extremely vulnerable to landslides. The major factor triggering landslide activity in this area was the anomalously high precipitation related to extreme, short recurrence, meteorological events that might trigger landslides from small debris flows and shallow soil slides to large deep-seated landslides. Other factors contributed to mass wasting in this area: (1) absence of vegetation on slopes and increased slope erosion, i.e. a preparatory causal factor of landsliding; (2) subduction earthquakes that could produce thousands of mass movements in a large area and the seismicity of small-magnitude earthquakes could contribute to decreased slope stability over time; and (3) the increasing human population. Finally, the increased intensity of extreme storms that trigger landslides has been associated to climate change. Therefore, in tropical climate areas, where significant population lives in mostly developing countries, the combination of these phenomena makes them highly vulnerable to extreme storms and landslide hazards.

Funding
This work was supported by the Consejo Nacional de Ciencia y Tecnología [grant number CONACyT-INEGI 209243].

Author details
María Teresa Ramírez-Herrera[1]
E-mails: tramirez@igg.unam.mx, maria_teresa_ramirez@yahoo.com
Krzysztof Gaidzik[1]
E-mails: gaidzik@igg.unam.mx, krzysztof.gaidzik@us.edu.pl
[1] Laboratorio Universitario de Geofísica Ambiental & Instituto de Geografía, Universidad Nacional Autónoma de México, Ciudad Universitaria, Coyoacán, 04510 Ciudad de México, México.
[‡] Current affiliation: Faculty of Earth Sciences, Department of Fundamental Geology, University of Silesia, Bedzinska 60, 41-200 Sosnowiec, Poland.

References

Alexander, D. (1992). On the causes of landslides: Human activities, perception, and natural processes. *Environmental Geology and Water Sciences, 20*, 165–179. https://doi.org/10.1007/BF01706160

Allan, R. P., & Soden, B. J. (2008). Atmospheric warming and the amplification of precipitation extremes. *Science, 321*, 1481–1484. doi:10.1126/science.1160787

Allstadt, K. (2013). Extracting source characteristics and dynamics of the August 2010 Mount Meager landslide from broadband seismograms. *Journal of Geophysical Research: Earth Surface, 118*, 1472–1490. doi:10.1002/jgrf.20110

Anderson, J. G., Singh, S. K., Espindola, J. M., & Yamamoto, J. (1989). Seismic strain release in the Mexican subduction thrust. *Physics of the Earth and Planetary Interiors, 58*, 307–322. https://doi.org/10.1016/0031-9201(89)90102-7

Aranza Rodríguez, S. J. (2014). *Estimación de zonas potenciales de deslizamiento debido a la ocurrencia de un evento natural a través de procesamiento de imágenes satelitales* (Dissertation). Mexico City: UNAM.

Bishop, A. W. (1955). The use of the slip circle in the stability analysis of slopes. *Géotechnique, 5*, 7–17. https://doi.org/10.1680/geot.1955.5.1.7

Buckham, R. C., Coe, J. A., Chavarría, M. M., Godt, J. W., Tarr, A. C., Bradley, L.-A., ... Johnson, M. L. (2001). *Landslides triggered by hurricane Mitch in Guatemala—Inventory and discussion*. Denver, CO: USGS.

Caballero, L., Macías, J. L., García-Palomo, A., Saucedo, G. R., Borselli, L., Sarocchi, D., & Sánchez, J. M. (2006). The September 8–9, 1998 rain-triggered flood events at Motozintla, Chiapas, Mexico. *Natural Hazards, 39*, 103–126. doi:10.1007/s11069-005-4987-7

Cannon, S. H., Haller, K. M., Ekstrom, I., Schweig, E. S., III, Devoli, G., Moore, D. W., ... Tarr, A. C. (2001). *Landslide response to Hurricane Mitch rainfall in seven study areas in Nicaragua*. Denver, CO: USGS.

Crozier, M. J. (2010). Deciphering the effect of climate change on landslide activity: A review. *Geomorphology, 124*, 260–267. doi:10.1016/j.geomorph.2010.04.009

Davies, T. C. (1996). Landslide research in Kenya. *Journal of African Earth Sciences, 23*, 41–549.

DeMets, C., Gordon, R. G., & Argus, D. F. (2010). Geologically current plate motions. *Geophysical Journal International, 181*, 1–80. doi:10.1111/j.1365-246X.2009.04491.x

Driessen, P., Deckers, J., Spaargaren, O., & Nachtergaele, F. (2000). *Lecture notes on the major soils of the world* (No. 94). Rome: Food and Agriculture Organization.

Evans, S. G., Roberts, N. J., Ischuk, A., Delaney, K. B., Morozova, G. S., & Tutubalina, O. (2009). Landslides triggered by the 1949 Khait earthquake, Tajikistan, and associated loss of life. *Engineering Geology, 109*, 195–212. doi:10.1016/j.enggeo.2009.08.007

Fellenius, W. (1927). *Earth stability calculations assuming friction and cohesion on circular slip surfaces*. Berlin: Ernst & Sohn.

Fischer, L., Amann, F., Moore, J., & Huggel, C. (2010). Assessment of periglacial slope stability for the 1988 Tschierva rock avalanche (Piz Morteratsch, Switzerland). *Engineering Geology, 116*, 32–43. doi:10.1016/j.enggeo.2010.07.005

Gaidzik, K., Ramírez-Herrera, M. T., Bunn, M., Leshchinsky, B. A., Olsen, M., & Regmi, N. R. (2017). Landslide manual and automated inventories, and susceptibility mapping using LIDAR in the forested mountains of Guerrero, Mexico. *Geomatics, Natural Hazards and Risk*, 1–26. doi:10.1080/19475705.2017.1292560

García Herrera, C. (2014). *Riesgo geológico de la inundación de la población de Tixtla Guerrero por efecto de las lluvias producto de los huracanes Manuel e Ingrid en septiembre de 2013 en el Estado de Guerrero* (pp. 182–185). Ciudad de México: Memorias de Convención Nacional Geológica.

Glade, T. (2003). Landslide occurrence as a response to land use change: A review of evidence from New Zealand. *Catena, 51*, 297–314. doi:10.1016/S0341-8162(02)00170-4

Gorum, T., Fan, X. B., van Westen, C. J., Huang, R. Q., Xu, Q., Tang, C., & Wang, G. (2011). Distribution pattern of earthquake-induced landslides triggered by the 12 May 2008 Wenchuan earthquake. *Geomorphology, 133*, 152–167. doi:10.1016/j.geomorph.2010.12.030

Haque, U., Blum, P., da Silva, P. F., Andersen, P., Pilz, J., Chalov, S. R., ... Keellings, D. (2016). Fatal landslides in Europe. *Landslides, 13*, 1545–1554. doi:10.1007/s10346-016-0689-3

Harp, E. L., Hagaman, K. W., Held, M. D., & McKenna, J. P. (2002). *Digital inventory of landslides and related deposits in Honduras triggered by Hurricane Mitch*. Denver, CO: USGS.

HURDAT2. (2016). *Norteast and North Central Pacific Hurricane Database: 1949–2015*. Hurricane Research Division. National Oceanic and Atmospheric Administration. Retrieved August 6, 2016, from https://www.nhc.noaa.gov/data/

Instituto de Geografía de la UNAM. (2007) *Nuevo Atlas Nacional de México*. Author. Retrieved June 10, 2016, from https://www.igeograf.unam.mx/web/sigg/publicaciones/atlas/anm-2007/anm-2007.php

Intergovernmental Panel on Climate Change. (2012). *Managing the risks of extreme events and disasters to advance climate change adaptation: A special report of working groups I and II of the Intergovernmental Panel on Climate Change (IPCC)* (582 pp.). (C. B. Field, V. Barros, T. F. Stocker, Q. Sahe, D. J. Dokken, K. L. Ebi, ... P. M. Midgley Eds.). Cambridge: Cambridge University Press. Retrieved August 6, 2016, from https://www.ipcc.ch/pdf/special-reports/srex/SREX_Full_Report.pdf

Intergovernmental Panel on Climate Change. (2013). *Approved summary for policy makers in Climate change 2013: The physical science basis* (pp. 3–30). (T. F. Stocker, D. Qin, G-F. Plattner, M. M. B. Tignor, S. K. Allen, J. Boschung, ... P. M. Midgley Eds.). Cambridge: Cambridge University Press. Retrieved August 6, 2016, from https://www.ipcc.ch/pdf/assessment-report/ar5/wg1/WG1AR5_SummaryVolume_FINAL.pdf

IUSS. (2015) *World reference base for soil resources 2014.* Rome: Food and Agriculture Organization of the United Nations. Retrieved August 6, 2016, from https://www.fao.org/3/a-i3794e.pdf

James, M. R., & Robson, S. (2012). Straightforward reconstruction of 3D surfaces and topography with a camera: Accuracy and geoscience application. *Journal of Geophysical Research, 117*(F3). doi:10.1029/2011JF002289

Janbu, N. (1954). *Application of composite slip surface for stability analysis.* In European conference on stability analysis, Stockholm.

Janbu, N. (1973). Slope stability computations. In R. C. Hirschfeld & S. J. Poulos (Ed.), *Embankment dam engineering* (pp. 47–86). New York, NY: Wiley.

Johnson, K., Nissen, E., Saripalli, S., Arrowsmith, J. R., McGarey, P., Scharer, K., … Blisniuk, K. (2014). Rapid mapping of ultrafine fault zone topography with structure from motion. *Geosphere, 10*, 969–986. doi:10.1130/GES01017.1

Kamp, U., Growley, B. J., Khattak, G. A., & Owen, L. A. (2008). GIS-based landslide susceptibility mapping for the 2005 Kashmir earthquake region. *Geomorphology, 101*, 631–642. doi:10.1016/j.geomorph.2008.03.003

Kargel, J. S., Leonard, G. J., Shugar, D. H., Haritashya, U. K., Bevington, A., Fielding, E. J., … Young, N. (2015). Geomorphic and geologic controls of geohazards induced by Nepal's 2015 Gorkha earthquake. *Science, 351*, aac8353. doi:10.1126/science.aac8353

Keefer, D. K. (1984). Landslides caused by earthquakes. *Geological Society of America Bulletin, 95*, 406–421. https://doi.org/10.1130/0016-7606(1984)95<406:LCBE>2.0.CO;2

Keefer, D. K. (2002). Investigating landslides caused by earthquakes–a historical review. *Surveys in Geophysics, 23*, 473–510. doi:10.1023/A:1021274710840

Knapen, A., Kitutu, M. G., Poesen, J., Breugelmans, W., Deckers, J., & Muwanga, A. (2006). Landslides in a densely populated county at the footslopes of Mount Elgon (Uganda): Characteristics and causal factors. *Geomorphology, 73*, 149–165. doi:10.1016/j.geomorph.2005.07.004

Knutson, T. R., McBride, J. L., Chan, J., Emanuel, K., Holland, G., Landsea, C., … Sugi, M. (2010). Tropical cyclones and climate change. *Nature Geoscience, 3*, 157–163. doi:10.1038/ngeo779

Kostoglodov, V., & Ponce, L. (1994). Relationship between subduction and seismicity in the Mexican part of the Middle America Trench. *Journal of Geophysical Research: Solid Earth, 99*, 729–742. https://doi.org/10.1029/93JB01556

Larsen, M. C., & Simon, A. A. (1993). rainfall intensity-duration threshold for landslides in a humid-tropical environment, Puerto Rico. *Geografiska Annaler. Series A, Physical Geography, 75*, 13–23. https://doi.org/10.2307/521049

Larsen, M. C., & Torres-Sanchez, A. J. (1992). Landslides triggered by hurricane Hugo in eastern Puerto Rico, September 1989. *Caribbean Journal of Science, 28*, 113–125.

Lawrence, M. B. (1997) *Preliminary report: Hurricane Pauline: 5–10 October 1997.* National Hurricane Center. Retrieved July 27, 2016, from https://www.nhc.noaa.gov/data/tcr/EP181997_Pauline.pdf

Malamud, B. D., Turcotte, D. L., Guzzetti, F., & Reichenbach, P. (2004). Landslide inventories and their statistical properties. *Earth Surface Processes and Landforms, 29*, 687–711. doi:10.1002/esp.1064

Matias Ramírez, L. G. (1998). Algunos efectos de la precipitación del huracán Paulina en Acapulco, Guerrero. *Investigaciones geográficas, 37*, 7–19.

Morgenstern, N. R., & Price, V. E. (1965). The analysis of the stability of general slip surfaces. *Géotechnique, 15*, 79–93. https://doi.org/10.1680/geot.1965.15.1.79

Morgenstern, N. R., & Price, V. E. (1967). A numerical method for solving the equations of stability of general slip surfaces. *The Computer Journal, 9*, 388–393. https://doi.org/10.1093/comjnl/9.4.388

Mugagga, F., Kakembo, V., & Buyinza, M. (2012). Land use changes on the slopes of Mount Elgon and the implications for the occurrence of landslides. *Catena, 90*, 39–46. doi:10.1016/j.catena.2011.11.004

Nyssen, J., Moeyersons, J., Poesen, J., Deckers, J., & Mitiku, H. (2003). The environmental significance of the remobilization of ancient mass movements in the Atbara-Tekeze headwaters near Hagere Selam, Tigray, Northern Ethiopia. *Geomorphology, 49*, 303–322. doi:10.1016/S0169-555X(02)00192-7

Pacheco, J. F., & Singh, S. K. (2010). Seismicity and state of stress in Guerrero segment of the Mexican subduction zone. *Journal of Geophysical Research: Solid Earth, 115*, B01303. doi:10.1029/2009JB006453

Pacheco, J. F., Iglesias, A., & Singh, S. K. (2002). The 8 October Coyuca, Guerrero, Mexico earthquake (Mw 5.9): A normal fault in the expected compressional environment. *Seismological Research Letters, 73*, 263.

Pasch R. J., & Zelinsky D. A. (2014) *Tropical cyclone report: Hurricane Manuel: September 13–19, 2013.* United States National Oceanic and Atmospheric Administration's National Hurricane Center. Retrieved May 5, 2016, from https://www.nhc.noaa.gov/data/tcr/EP132013_Manuel.pdf

Petley, D. (2012). Global patterns of loss of life from landslides. *Geology, 40*, 927–930. doi:10.1130/G33217.1

Phillips, C. J., & Watson, A. (1994). *Structural tree root research in New Zealand: A review* (Landcare Research Science series 7). Lincoln, NE: Manaaki Whenua Press.

Posner, A. J., & Georgakakos, K. P. (2015a). Normalized landslide index method for susceptibility map development in El Salvador. *Natural Hazards, 79*, 1825–1845. doi:10.1007/s11069-015-1930-4

Posner, A. J., & Georgakakos, K. P. (2015b). Soil moisture and precipitation thresholds for real-time landslide prediction in El Salvador. *Landslides, 12*, 1179–1196. doi:10.1007/s10346-015-0618-x

Posner, A. J., & Georgakakos, K. P. (2016). An early warning system for landslide danger. *EOS, 97.* doi:10.1029/2016EO062323

Ramírez-Herrera, M. T., Corona, N., & Suárez, G. A. (2016). Review of Great Magnitude Earthquakes and associated tsunamis along the Guerrero, Mexico Pacific Coast: A multiproxy approach. In M. Chavez, M. Ghil, & J. Urrutia-Fucugauchi (Eds.), *Extreme events: Observations, modeling, and economics* (pp. 165–176). Hoboken, NJ: Wiley. doi:10.1002/9781119157052.ch13

Ravanel, L., & Deline, P. (2011). Climate influence on rockfalls in high-Alpine steep rockwalls: The north side of the Aiguilles de Chamonix (Mont Blanc massif) since the end of the 'Little Ice Age'. *The Holocene, 21*, 357–365. doi:10.1177/0959683610374887

Restrepo, C., & Alvarez, N. (2006). Landslides and Their Contribution to Land-cover Change in the Mountains of Mexico and Central America. *Biotropica, 38*, 446–457. doi:10.1111/j.1744-7429.2006.00178.x

Scawthorn, C., Schneider, P. J., & Schauer, B. A. (2006). Natural hazards—The Multihazard approach. *Natural Hazards Review, 7*, 39–39. doi:10.1061/(ASCE)1527-6988

Schuster, R. L. (1996). Socio-economic significance of landslides. In A. K. Turner, & R. L. Schuster (Eds.), *Landslides: Investigation and mitigation special report 247* (pp. 12–35). Washington, DC: National Academy Press.

Servicio Sismológico Nacional. (2017). *Earthquake database.* Retrieved from https://www.ssn.unam.mx

Snavely, N., Seitz, S. M., & Szeliski, R. (2008). Modeling the world from internet photo collections. *International Journal of Computer Vision, 80*, 189–210. doi:10.1007/s11263-007-0107-3

Spencer, E. A. (1967). method of analysis of the stability of embankments assuming parallel interslice forces. *Géotechnique, 17*, 11–26. https://doi.org/10.1680/geot.1967.17.1.11

Stoffel, M., Tiranti, D., & Huggel, C. (2014). Climate change impascts on mass movements–Case studies from the European Alps. *Science of The Total Environment, 493*, 1255–1266. doi:10.1016/j.scitotenv.2014.02.102

Suárez, G., Monfret, T., Wittlinger, G., & David, C. (1990). Geometry of subduction and depth of the seismogenic zone in the Guerrero gap, Mexico. *Nature, 345*, 336. https://doi.org/10.1038/345336a0

UNEP. (2012). *UNEP year book. Emerging issues in our global environment 2012*. Retrieved June 10, 2016, from https://www.unep.org/yearbook/2012/pdfs/UYB_2012_FULLREPORT.pdf

UNISDR. (2012). *Number of climate-related disasters, 1980–2011—Graphic*. Retrieved June 10, 2016, from https://www.preventionweb.net/english/professional/statistics

Vanacker, V., Vanderschaeghe, M., Govers, G., Willems, E., Poesen, J., Deckers, J., & De Bievre, B. (2003). Linking hydrological, infinite slope stability and land-use change models through GIS for assessing the impact of deforestation on slope stability in high Andean watersheds. *Geomorphology, 52*, 299–315. doi:10.1016/S0169-555X(02)00263-5

Verruijt, A. (2012). *Soil Mechanics*. Delf University of Technology. Retrieved May 5, 2016, from https://geo.verruijt.net/software/SoilMechBook2012.pdf

Wang, H. B., Sassa, K., & Xu, W. Y. (2007). Analysis of a spatial distribution of landslides triggered by the 2004 Chuetsu earthquakes of Niigata Prefecture, Japan. *Natural Hazards, 41*, 43–60. doi:10.1007/s11069-006-9009-x

Wartman, J., Montgomery, D. R., Anderson, S. A., Keaton, J. R., Benoît, J., dela Chapelle, J., & Gilbert R. (2016). The 22 March 2014 Oso landslide, Washington, USA. *Geomorphology, 253*, 275–288. doi:10.1016/j.geomorph.2015.10.022

Wolter, A., Gischig, V., Stead, D., & Clague, J. J. (2016). Investigation of geomorphic and seismic effects on the 1959 Madison Canyon, Montana, Landslide using an integrated field, engineering geomorphology mapping, and numerical modelling approach. *Rock Mechanics and Rock Engineering, 49*, 2479–2501. doi:10.1007/s00603-015-0889-5

A feasible way to increase carbon sequestration by adding dolomite and K-feldspar to soil

Leilei Xiao[1,2§], Qibiao Sun[1§], Huatao Yuan[1], Xiaoxiao Li[1], Yue Chu[1], Yulong Ruan[3], Changmei Lu[1] and Bin Lian[1*]

*Corresponding author: Bin Lian, Jiangsu Key Laboratory for Microbes and Functional Genomics, Jiangsu Engineering and Technology Research Center for Microbiology, College of Life Sciences, Nanjing Normal University, Nanjing 210023, China

E-mail: bin2368@vip.163.com

Reviewing editor: Craig O'Neill, Macquarie University, Australia

Abstract: In recent years, many researchers have explored various possible ways to slow down the increase in atmospheric CO_2 concentration as this process poses a serious threat to mankind's survival. Mineral weathering is one possible way. Silicate weathering, for example, causes net carbon sequestration and carbonate weathering occurs relatively rapidly. In this study, dolomite and K-feldspar were added to soil to investigate if these minerals can increase carbon sequestration and also improve the available potassium content. The carbon content of amaranth, the organic and inorganic carbon content of the soil, two kinds of enzymes (polyphenol oxidase and urease), and the available potassium content were all tested. The experimental results show that the minerals accelerate the fixation of organic and inorganic carbon in the soil and also promote amaranth growth. Moreover, the available potassium content was increased when K-feldspar was added. Taken together, adding moderate amounts of carbonate and silicate minerals into the soil is found to be an attemptable way of accelerating CO_2 fixation and improving the potassium content of soil.

Subjects: Agriculture & Environmental Sciences; Environmental Sciences; Soil Sciences

Keywords: soil; mixed mineral; mineral weathering; carbon fixation

1. Introduction

It is widely accepted that human activity has caused the atmospheric CO_2 concentration to rise continually. The average global atmospheric CO_2 concentration reached 395.31 ± 0.10 ppm in 2014 (Le Quéré et al., 2014). Methods of reducing the rate of atmospheric CO_2 enrichment, and investigating what kind of practices are feasible and effective in blocking the trend of increasing atmosphere greenhouse gas (GHG) concentrations, have been attracting more and more attention (Liping &

ABOUT THE AUTHORS

The research of my group is concentrating in the following aspects: effects of microbes on the carbonate weathering and carbon sequestration; microbe–carbonate–silicate mineral interactions and their effects on the carbon migration and conversion; process and mechanisms in the microbe–mineral interaction. The research reported in this paper provides a possibility to slow down CO_2 release and possibly fix atmospheric CO_2 through adding silicate minerals in agricultural practice, which relates to the issues of how to deal with the global change.

PUBLIC INTEREST STATEMENT

Atmospheric CO_2 concentration increases continually. Seeking utilizable methods for blocking the trend attracts more and more attention. China is an agricultural country possessing rich mineral resources. Agricultural operations affect the carbon cycle through uptake, fixation, emission, and transfer of carbon among different pools. Based on these, we added dolomite and K-feldspar into soil to investigate if these minerals can increase carbon sequestration and also improve the available potassium content. Our research results are positive. It provides a possibility to slow down the momentum of elevated atmospheric CO_2 concentration through the practice of agriculture.

Erda, 2001). Notwithstanding the temporary nature of the release of anthropogenic CO_2, it should be noted that large quantities of carbon are also exuded through the roots of plants in the form of organic matter which degrade to gaseous form and ultimately return to the atmosphere (Ryan, Delhaize, & Jones, 2001). This process acts as a dominant conveyor in the global carbon cycle, accounting for ~120 Gt Ca^{-1} which clearly overshadows the 6 Gt Ca^{-1} produced by anthropogenic activities (Renforth, Manning, & Lopez-Capel, 2009). Many GHGs are sequestered by the agricultural and terrestrial ecosystems, and plant biomass and soil, for example, are the major sinks of atmospheric CO_2 (Liping & Erda, 2001). Therefore, carbon sequestration in soil is something that we should value (Lal, 2004).

Schlesinger and Andrews (2000) showed that the CO_2 released from the soil should be considered to be one of the largest sources of flux in the global carbon cycle, and can thus have a large effect on the atmospheric CO_2 concentration. Plants and the soil are the principal parts of the agricultural and terrestrial ecosystems. Thus, it is highly appropriate to consider the importance of the coupled plant–soil system in carbon capture, and to develop ways to enhance these natural processes (Renforth et al., 2009). For example, adopting restorative land use and using recommended management practices with agricultural soils can mitigate the negative impact of elevated atmospheric CO_2 concentrations. In temperate climates, soil organic carbon (SOC) is always a major carbon sink in agricultural soil (Smith, 2004). By adopting reasonable practices, the global SOC sequestration potential is about 0.9 ± 0.3 Pg Ca^{-1}, which offsets one-fourth to one-third of the annual increase in atmospheric CO_2 (Lal, 2004). In arid climates, the role of soil as a carbon sink is often associated with the accumulation of soil *inorganic* carbon (SIC). The maximum capacity of SIC capture technology has been shown to be limited by the availability of Ca-rich minerals (Renforth et al., 2009). Seeking cost-effective means of increasing SOC and SIC is therefore very significant, although this is a long and arduous process.

It has been demonstrated that weathering occupies an important position in present and future carbon cycling, and involvement of microbes can accelerate the weathering process of carbonate (Burford, Fomina, & Gadd, 2003; Xiao et al., 2014) and silicate (Xiao, Lian, Dong, & Liu, 2016; Xiao, Lian, Hao, Liu, & Wang, 2015). However, many researchers ignore the accumulation of weathering products in the soil and thus underestimate their contribution to the global climate (Goudie & Viles, 2012). It is also worth considering methods that can reduce atmospheric CO_2 by reacting silicate minerals to form carbonate minerals (Lackner, 2003; Manning & Renforth, 2013; Seifritz, 1990). This typically involves dissolution of the silicate minerals and subsequent precipitation of stable carbonate minerals (Power, Harrison, Dipple, & Southam, 2013). Manning and Renforth (2013) showed that the pedogenic carbonate should now be considered as a consequence of reactions between plant root exudates and calcium liberated by the silicate dissolution. However, it should be noted that the rate of silicate weathering is limited due to the slow kinetics of the CO_2–silicate reaction process (Oelkers, Gislason, & Matter, 2008). In general, the process is over an order of magnitude slower than that of the carbonate process (Mortatti & Probst, 2003; Wu, Xu, Yang, & Yin, 2008). According to calculations, it will take more than one million years to stabilize the atmospheric CO_2 level through silicate weathering (Goudie & Viles, 2012). Adding 1–2 tons of crushed olivine (grain size < 300 micron) to one hectare of soil will last approximately 30 years in a temperate climate (Schuiling & Krijgsman, 2006). Therefore, it appears to be difficult to ease the increase in atmospheric CO_2 by only employing silicate weathering in the short term. The uptake of CO_2 by atmospheric or soil respiration by carbonate rock dissolution has an important effect on the global carbon cycle and serves as one of the most important sinks (Cao et al., 2012). Liu, Dreybrodt, and Wang (2010) showed that dissolved inorganic carbon is an important but previously underestimated sink for atmospheric CO_2. This contribution to carbon sequestration reaches up to 0.8242 Pg Ca^{-1}, which amounts to 10.4% of the total anthropogenic CO_2 emission (Liu et al., 2010). Dissolution of calcite and dolomite can transform soil-generated CO_2 into alkaline form (HCO_3^-, CO_3^{2-}) (Macpherson et al., 2008).

Agricultural operations affect the carbon cycle through uptake, fixation, emission, and transfer of carbon among different pools (Lal, 2004). It should be feasible to change the carbon content distribution among these pools by anthropogenic manipulation. Recently, several studies (Fan et al., 2014; Mahmoodabadi & Heydarpour, 2014) have shown that application of manure is generally considered to increase carbon sequestration, although Schlesinger (1999) pointed out that net carbon sequestration does occur. Schuiling and Krijgsman (2006) mentioned that spreading finely powdered olivine on farmland could be extensively used to fix CO_2. In this study, we use soil as a substrate to explore if carbonate and silicate minerals, when added to the soil, can cause an increase in net carbon sequestration.

2. Materials and methods

2.1. Minerals
Dolomite, $CaMg(CO_3)_2$, for the study was provided by the Institute of Geochemistry, Chinese Academy of Sciences (Guiyang, China). Analysis using X-ray diffraction (XRD) showed that the samples were doped with small amounts of calcite and sanidine. Analysis of the K-feldspar ($KAlSi_3O_8$) used using XRD showed that quartz, muscovite, and clinochlore were present as impurities. The two kinds of mineral were both crushed and specific-sized particles (100–200 mesh) used in the study.

2.2. Plants
The study is mainly aimed at investigating whether carbon sequestration can be increased by adding carbonate and silicate to the soil. At the same time, we also explored if using K-feldspar was able to provide potassium for plant growth. Therefore, amaranth is used in this study because it is a common vegetable that is planted widely in china and has a strong ability to become enriched in K ions.

2.3. Summary of the experimental method
The experimental pot employed is illustrated in Figure S1. XRD analysis of the soil showed that it is composed of quartz, muscovite, albite ($Na(Si_3Al)O_8$), orthoclase ($KAlSi_3O_8$), and kaolinite-1A ($Al_2Si_2O_5(OH)_4$). Each pot contains 1,500 g of soil and 500 g of mineral powder. Our primary purpose is to detect the effectiveness of mineral, so the addition was relative large. The different amounts added to each pot are shown in Table1. Every treatment has three replicates. To increase the permeability of the soil, 450 g glass beads were added to the pots. The amaranth was watered timely according to the soil moisture. Water (500 ml) was regularly added to the pot every six days, and the soil infiltration water (SIW) collected the next day. The experiment was continued about two months.

2.3.1. Determination of the organic carbon content of amaranth
Whole amaranth plants were collected and dried overnight at 105°C. The dry weight was then measured. The percentage of carbon present was measured using an elemental analyzer (Elementar Vario MACRO, Germany).

2.3.2. Determination of soil parameters
At the end of the experiment, the soil moisture, pH, and activity of polyphenol oxidase and urease were determined. Portions of the naturally air-dried soil samples were used to determine the soil's organic carbon and microbial biomass.

Table 1. The different amounts of the two types of mineral added to each device

Number	Composition			Abbreviation
	Soil (g)	Dolomite (g)	K-feldspar (g)	
1	2,000	–	–	all-s
2	1,500	500	–	d-s
3	1,500	375	125	d-p-s
4	1,500	125	375	p-d-s
5	1,500	–	500	p-s

To measure the soil's moisture content, an aliquot of moist soil (about 5 g) was dried at 105°C. After 5 h, the sample was placed in a desiccator to cool for 30 min. Samples were dried and weighed repeatedly until the weight no longer decreased. Soil pH was determined using CO_2-free deionized water and a 1:1 (w/v) soil-to-water ratio. Samples were shaken for 15 min, left to settle for 30 min, and then the pH measured as in previous reports (Fierer & Jackson, 2006).

Polyphenol oxidase is known to play an important role in carbon cycling in the soil (Sinsabaugh, 1994). The analysis mirrored the work of Carney, Hungate, Drake, and Megonigal (2007). Briefly, (1) litterbags were removed as possible and approximately 2 g of wet weight placed in a blender mini-jar; (2) acetate buffer (60 ml, 50 mM, pH 5) was added and the mixture blended on "whip" for 1 min; (3) homogenate (0.750 ml) was mixed with an equal volume of substrate in a 2-ml tube—tubes were placed in a shaker and incubated for 2 h; (4) the reaction mixture was centrifuged for 2 min at 10,000 g and the absorbance of the supernatant at 460 nm immediately measured on a microplate reader.

Previous work has shown that urease activity has a positive correlation with organic carbon and total nitrogen (Zantua, Dumenil, & Bremner, 1977). For this test, toluene (1 ml) was mixed with 5 g of natural air-dried soil sample for 15 min. Then, urea solution (10 ml, 5%) and citrate buffer (20 ml, 0.96 M, pH 6.7) were added. Meanwhile, as a control, a repeat experiment was performed using an equal volume of distilled water instead of the urea solution. After incubation for 24 h at 37°C, the solution was centrifuged (4,000 g, 10 min). Supernatant (1 ml) was mixed with sodium phenoxide solution (4 ml, 2.7 M) and sodium hypochlorite solution (3 ml, 0.9%) in a 50-ml volumetric flask. After 20 min, the reaction solution was diluted to 50 ml and the absorbance at 460 nm measured. Urease activity was expressed according to the number of milligrams of NH_3–N in 1 g of soil.

For SOC determination, dry soil (2 g) was added to HCl solution (40 ml, 5%), and the mixture blended about 10 min until gas is no longer generated. The tubes were then spun at 8,000 g for 5 min. After this, the samples were rinsed three times with ultrapure water. After drying at 105°C, the residual solids were weighed and the carbon content tested using the elemental analyzer (Elementar Vario MACRO, Germany).

The microbial biomass in the soil was measured using the method outlined by Vance, Brookes, and Jenkinson (1987) involving chloroform fumigation and extraction. Moist soil was fumigated in a sealed desiccator using ethanol-free chloroform for 24 h at 25°C. Water (20 ml) and the same amount of NaOH (1 M) were placed in the bottom to trap any evolved CO_2. Non-fumigated soil was used as a control. Fumigated and non-fumigated soils were subjected to extraction using 0.5 M K_2SO_4 solution for 30 min using an "end-over-end" shaker at 350 rpm and centrifuged. The supernatant was filtered through a 0.45-μm membrane. The filtrate was tested using a total organic carbon analyzer (Shimadzu TOC-VCSN, Japan). The soil microbial biomass carbon (B_C) was estimated using $B_C = E_C/k_{EC}$, where E_C = [organic carbon extracted by K_2SO_4 from fumigated soil–carbon extracted by K_2SO_4 from non-fumigated soil] (Wu, Joergensen, Pommerening, Chaussod, & Brookes, 1990). The value of the k_{EC} parameter (the proportion of the extracted microbial biomass carbon evolved as organic carbon) was taken to be 0.45, following the work of Wu et al. (1990).

The available potassium content of the soil was measured. Briefly, a portion (0.5 g) of crushed dry sample (50–80 mesh) and ammonium acetate solution (50 ml, 1 M) were mixed in an extraction bottle (flask) following the method of Zhu, Lian, Yang, Liu, and Zhu (2013). Then, the bottle was stoppered and oscillated for 30 min. The mixture was filtered using a filter paper, and the filtrate collected for testing using a full-spectrum, direct-reading plasma emission spectrometer (Thermo Fisher Scientific, UK).

2.3.3. Determination of SIW parameters
The pH and temperature of the SIW measured using a pH meter (S20 SevenEasy, Mettler-Toledo). The concentrations of certain cations (K^+, Na^+, Ca^{2+}, and Mg^{2+}) were determined using an atomic

absorption analyzer (AA900F, PerkinElmer, US). Anion concentrations (Cl^- and SO_4^{2-}) were determined using ion chromatography (DIONEX ICS-90, US). An acid–base titration method was used to measure the content of the bicarbonate in the aqueous solution according to the published literature with a little modification (Verma, 2004; Zangen, 1962). The SIW was filtered using a 0.45-μm microporous membrane and 20 ml titrated with a standardized HCl solution. The above parameters (pH, water temperature, ion concentrations) were imported into appropriate software (MINTEQ) to calculate the saturation index of the calcite, and the disordered-dolomite and ordered-dolomite.

2.4. Statistical analysis
StatSoft's STATISTICA 6.0 software was used to analyze the data. The significance of the differences between the treatments was tested separately using one-way ANOVA tests followed by Fisher LSD tests for mean comparisons. All analyses were performed in triplicate. The data shown correspond to the means (along with the standard deviation) of at least three independent experiments.

3. Results

3.1. Amaranth carbon content
For the average carbon content, dolomite or/and K-feldspar is conducive to plant growth (see Figure 1). The minerals thus improve organic carbon fixation. From a statistical point of view, only adding dolomite (d-s) or increasing the K-feldspar content (p-d-s and p-s) were beneficial to the formation of organic carbon sinks. However, there was no difference between the all-s and d-p-s treatments.

3.2. Soil parameters
Soil moisture content was not significantly different (statistically) among the five different kinds of treatment (ranging from 11.6% to 16.8%). Dolomite significantly increased the soil pH, but K-feldspar did not (Figure 2(a)). The amount of dolomite added and the elevation of the soil pH were positively correlated. As can be seen from Figure 2(b) (compared with all-s treatment), the increase in SOC per gram of moist soil was significantly boosted after adding a large amount of dolomite (p-s and d-p-s). However, adding a large amount of K-feldspar did not have this effect (p-d-s and p-s). This may be due to the dissolution of the dolomite, which can consume gaseous or liquid CO_2. In contrast, K-feldspar does not have this ability over a relatively short period of time.

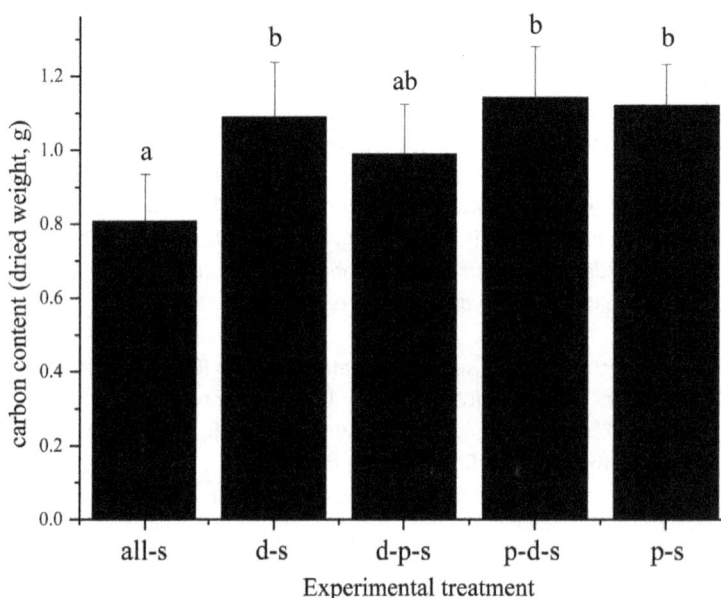

Figure 1. Amaranth carbon content after different minerals are added.

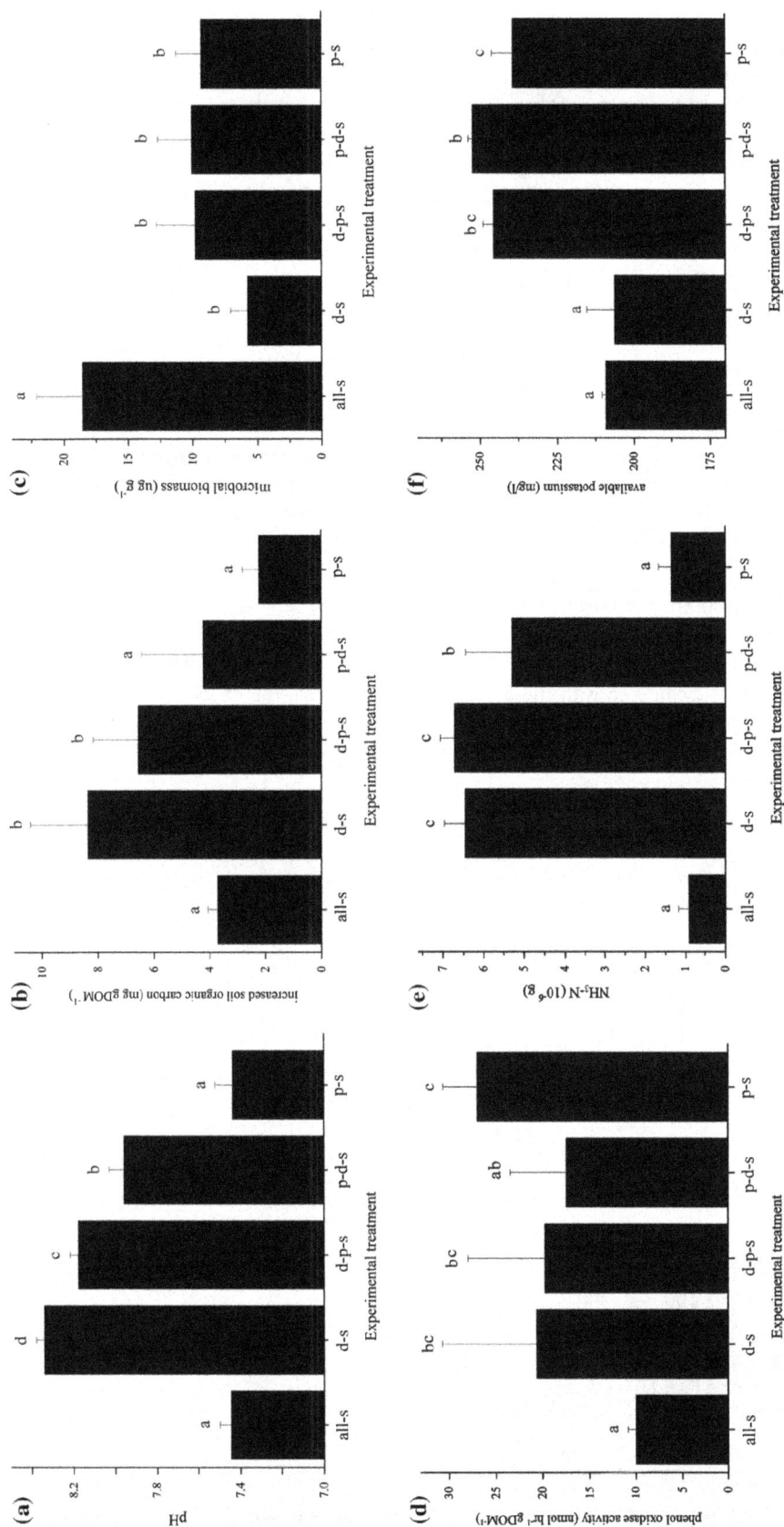

Figure 2. The effect of adding mixed minerals (dolomite and feldspar) to the soil on several soil parameters: (a) soil pH, (b) increased SOC, (c) microbial biomass, (d) polyphenol oxidase, (e) urease, and (f) available potassium.

To our surprise, the mineral (dolomite or feldspar) caused the soil microbial biomass to decline a certain amount over the experimental period. The effect of the added minerals on microbial survival in the microenvironment may be key here. Compare Figure 2(b) and (c). The microbial biomass and increase in organic carbon show some negative correlation. However, different minerals had no significant effect on soil microbial biomass. From Figure 2(a)–(c), the increase in SOC may be mainly pH dependent and little to do with the total number of micro-organisms. The minerals made the activity of the polyphenol oxidase (except for all-p) and urease (except for p-d-s) increase (Figure 2(d) and (e)). There was no significant difference between dolomite and K-feldspar with respect to polyphenol oxidase activity. Nevertheless, dolomite alone induced an increase in soil urease activity. As was expected, the amount of available potassium rose to a certain extent after K-feldspar was added (see Figure 2(f)). However, excessive addition did not significantly increase the available potassium content.

3.3. SIW parameters

3.3.1. pH
Both dolomite and K-feldspar promoted an increase in the pH of the SIW (Figure 3(a)). The impact of these two minerals on the pH of the SIW was little different. In the initial stages of the trial, the average pH in the sample with dolomite added was the highest. As the experiment progressed, a mixture of the minerals was most propitious to enhancing the SIW's pH.

3.3.2. The concentration of HCO_3^-
The addition of minerals had a great effect on HCO_3^- concentration (Figure 3(b)). No matter which mineral (dolomite or feldspar) was added, there was a significant difference at each of the eight sampling times (compared with the all-s treatment). Overall, the change in HCO_3^- concentration showed a sudden increase, followed by a rapid decrease, and then it gently changed. Dolomite alone was more influential with respect to HCO_3^- concentration compared to K-feldspar. However, adding

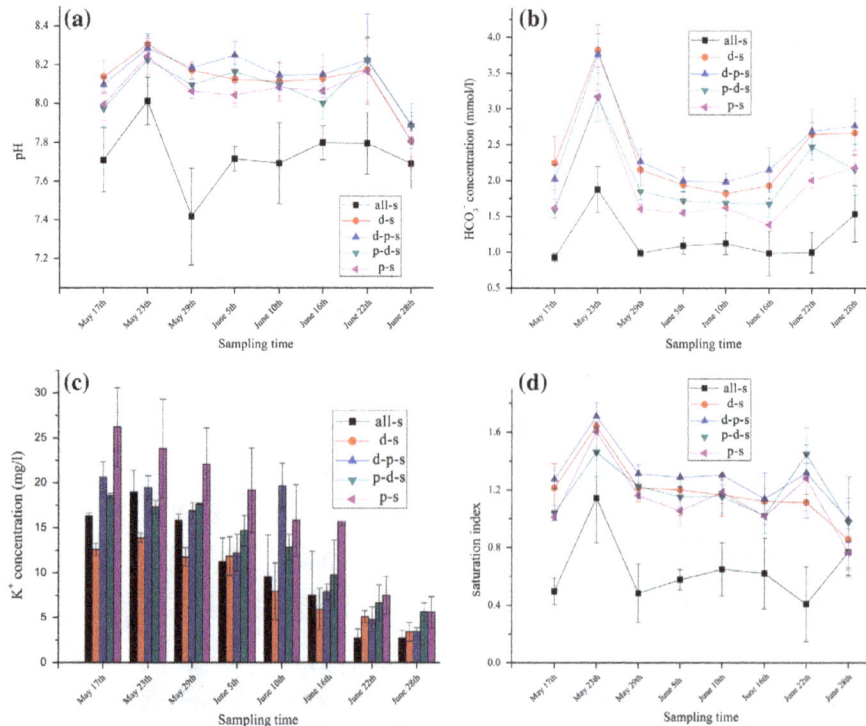

Figure 3. The effect of adding mixed minerals (dolomite and feldspar) to the soil on several of the parameters of the soil filtrate: (a) pH, (b) HCO_3^- concentration, (c) K+ concentration, and (d) calcite saturation index.

more dolomite did not cause more HCO_3^- to be produced. If standard deviation is ignored, then the HCO_3^- produced was the most in the d-s treatment at the first two sampling times, yet mixed minerals are more conducive to generating HCO_3^- in the long term.

3.3.3. K^+ concentration

Overall, the K^+ concentration gradually decreased (see Figure 3(c)). Only the average K^+ concentration in the p-s sample always exceeded that in all-s. Using mixed minerals or only K-feldspar did not significantly increase the concentration of K^+ in soil in the beginning of this experiment. The ability of K-feldspar to release K^+ can be seen in the latter part of the experiment (22 June and 28 June). Thus, it is feasible that, for long-term farming, fertilizers containing K-feldspar could be used to continue to provide K^+.

3.3.4. Saturation index calculation

MINTEQ software was used to calculate the saturation index of the carbonate using the following parameters: water temperature, pH, and the concentrations of K^+, Na^+, Ca^{2+}, Mg^{2+}, Cl^-, SO_4^{2-}, and HCO_3^- ions. Three kinds of carbonate, calcite, ordered-dolomite, and disordered-dolomite, were selected. Interestingly, the saturation indices of these three carbonates were almost the same (Figures 3(d) and S2). In the case of the calcite saturation index (Figure 3(d)), no matter which minerals were added, the calcite saturation index increased significantly. Comparing the effects of the different minerals on the saturation index, it seems mixed minerals (dolomite and K-feldspar) are more conducive to increasing the saturation index.

4. Discussion

The atmospheric CO_2 concentration continues to rise; thus researchers are constantly looking at a variety of solutions. Some of the research has shown that only a minor increase in the natural uptake is required to compensate for the extra anthropogenic CO_2 emission (Oelkers et al., 2008; Salek, Kleerebezem, Jonkers, Witkamp, & van Loosdrecht, 2013). Up until the last decade, mineral weathering has been a subject of some concern. However, this is the most important way in which nature keeps the CO_2–levels stable (Schuiling & Krijgsman, 2006). This oversight may be because people have not found a practical way to accelerate weathering (Schuiling & Krijgsman, 2006) or they have ignored the natural regulation effect. In this study, we have investigated the impact on CO_2 fixation of adding mixed dolomite–K-feldspar power to the soil. The results suggest that artificially adding suitable amount of dolomite and K-feldspar to the soil is likely to increase the amount of SOC. This is similar to the finding of previous reports investigating the potential of artificial soils (i.e. made by adding demolition waste or basic slag to soil). These were used to capture some of the transferred carbon as geologically stable $CaCO_3$ (Renforth et al., 2009). Artificial soils were also prepared by blending compost with dolerite and basalt quarry fines to be used for the purpose of CO_2 capture (Manning, Renforth, Lopez-Capel, Robertson, & Ghazireh, 2013).

   The significantly increased pH value of the soil implies that adding dolomite to the soil is possible to mitigate the effects of acid rain (Allen & Brent, 2010; Teir, Eloneva, Fogelholm, & Zevenhoven, 2006) and/or alleviates soil acidification caused by agricultural fertilizer. Although the addition of the minerals made the total microbial biomass decrease, the increased polyphenol oxidase and urease activity suggests that adding minerals may promote the reproduction and activity of the microbes in the soil which are associated with the carbon and nitrogen cycles.

   Manning and Renforth (2013) studied sequestration of atmospheric CO_2 through coupled plant–mineral reactions in urban soils. They showed that the rate-limiting factor seems to be the availability of Ca, not carbon (Manning & Renforth, 2013). Our results showed that the increase in SOC was doubled by dolomite addition, while K-feldspar addition did not do this. Both dolomite (d-s) and K-feldspar (p-f-s and p-s) improved the total carbon content of the amaranth. This means that the mineral may act as a regulatory component of the soil. After adding K-feldspar, the availability of K was significantly increased. Potassium is the most abundant cation in plants, comprising up to 10% of a plant's dry weight (Leigh & Wynjones, 1984). In China, soluble potassium is a scarce resource—there is only a 35% self-supply, so most is sourced via imports (Sun et al., 2013). Adding K-bearing minerals to soil is an attemptable way to improve the potassium content of the soil.

The following reaction occurs when dolomite is added to the soil:

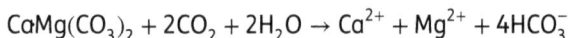

$$CaMg(CO_3)_2 + 2CO_2 + 2H_2O \rightarrow Ca^{2+} + Mg^{2+} + 4HCO_3^-$$

The experimental results showed that the amount of HCO_3^- generated was in some way related to the amount of dolomite added. In other words, the more dolomite added meant that more inorganic carbon produced (to some extent). However, the amount of Ca^{2+} in the d-s sample was lower than in the d-p-s-treated sample in the latter part of the experiment. Precipitation of carbonate mineral is controlled by the saturation state of the soil solution which itself depends on the activities of the dissolved species (cation and bicarbonate). A higher saturation index is beneficial to the formation of carbonate precipitation. From the saturation index of calcite or dolomite (Figures 3(d) and S2), we see that the saturation index of the minerals added is always higher than the all-s sample. This indicated that the addition of minerals can accelerate the fixation of gaseous carbon into the form of relatively stable inorganic carbon, e.g. HCO_3^- and CO_3^{2-} as well. Moreover, others minerals, such as anorthite, can be tested as well. Further researches are still needed to explore if the long-term minerals addition can cause soil desertification. Anyway, these results show some prospects for mineral application in the farming.

5. Conclusions
The addition of minerals was likely to accelerate the fixation of organic and inorganic carbon in the soil. Moreover, the available K content in the soil was increased when K-feldspar was added. Taken together, adding moderate amounts of carbonate and silicate minerals into soil is an attemptable way to accelerate CO_2 fixation and improve the soil's K content. However, further research is needed to improve the application of mixed minerals so as to increase soil quality and carbon sequestration.

Funding
This work was jointly supported by National Natural Science Foundation of China [Grant number: 41373078].

Author details
Leilei Xiao[1,2]
E-mail: ai-yanzi@163.com
Qibiao Sun[1]
E-mail: sunqibiao001@163.com
Huatao Yuan[1]
E-mail: 1164650987@qq.com
Xiaoxiao Li[1]
E-mail: lanlin717@foxmail.com
Yue Chu[1]
E-mail: 466393948@qq.com
Yulong Ruan[3]
E-mail: ryl880035@163.com
Changmei Lu[1]
E-mail: luchangmei@njnu.edu.cn
Bin Lian[1]
E-mail: bin2368@vip.163.com

[1] Jiangsu Key Laboratory for Microbes and Functional Genomics, Jiangsu Engineering and Technology Research Center for Microbiology, College of Life Sciences, Nanjing Normal University, Nanjing 210023, China.
[2] Key Laboratory of Coastal Biology and Utilization, Yantai Institute of Coastal Zone Research, Chinese Academy of Sciences, Yantai 264003, China.
[3] Key Laboratory of Karst Environment and Geological Hazard Prevention, Ministry of Education, Guizhou University, Guiyang 550003, China.
[§] Co-first author.

Authors' contributions
BL, CL, and LX designed the experiments; BL and LX wrote the main manuscript text; HY, LX, QS, XL, YC, and YR carried out the experiments, and LX prepared all figures. All authors reviewed the manuscript.

References
Allen, D. J., & Brent, G. F. (2010). Sequestering CO_2 by mineral carbonation: Stability against acid rain exposure. *Environmental Science & Technology, 44,* 2735–2739.
Burford, E., Fomina, M., & Gadd, G. (2003). Fungal involvement in bioweathering and biotransformation of rocks and minerals. *Mineralogical Magazine, 67,* 1127–1155. http://dx.doi.org/10.1180/0026461036760154
Cao, J. H., Yuan, D. X., Chris, G., Huang, F., Yang, H., & Lu, Q. (2012). Carbon fluxes and sinks: The consumption of

atmospheric and soil CO$_2$ by carbonate rock dissolution. *Acta Geologica Sinica-English Edition, 86*, 963–972.

Carney, K. M., Hungate, B. A., Drake, B. G., & Megonigal, J. P. (2007). Altered soil microbial community at elevated CO$_2$ leads to loss of soil carbon. *Proceedings of the National Academy of Sciences, 104*, 4990–4995. http://dx.doi.org/10.1073/pnas.0610045104

Fan, J., Ding, W., Xiang, J., Qin, S., Zhang, J., & Ziadi, N. (2014). Carbon sequestration in an intensively cultivated sandy loam soil in the North China Plain as affected by compost and inorganic fertilizer application. *Geoderma, 230*, 22–28. http://dx.doi.org/10.1016/j.geoderma.2014.03.027

Fierer, N., & Jackson, R. B. (2006). The diversity and biogeography of soil bacterial communities. *Proceedings of the National Academy of Sciences, 103*, 626–631. http://dx.doi.org/10.1073/pnas.0507535103

Goudie, A. S., & Viles, H. A. (2012). Weathering and the global carbon cycle: Geomorphological perspectives. *Earth-Science Reviews, 113*, 59–71. http://dx.doi.org/10.1016/j.earscirev.2012.03.005

Lackner, K. S. (2003). Climate change: A guide to CO2 Sequestration. *Science, 300*, 1677–1678. http://dx.doi.org/10.1126/science.1079033

Lal, R. (2004). Soil carbon sequestration to mitigate climate change. *Geoderma, 123*, 1–22. http://dx.doi.org/10.1016/j.geoderma.2004.01.032

Le Quéré, C., Moriarty, R., Andrew, R. M., Peters, G. P., Ciais, P., Friedlingstein, P., ... Zeng, N. (2014). Global carbon budget 2014. *Earth system science data discussions, 7*, 521–610. http://dx.doi.org/10.5194/essdd-7-521-2014

Leigh, R. A., & Wynjones, R. G. (1984). A hypothesis relating critical potassium concentrations for growth to the distribution and functions of this ion in the plant cell. *New Phytologist, 97*, 1–13. http://dx.doi.org/10.1111/nph.1984.97.issue-1

Liping, G., & Erda, L. (2001). Carbon sink in cropland soils and the emission of greenhouse gases from paddy soils: A review of work in China. *Chemosphere-Global Change Science, 3*, 413–418. http://dx.doi.org/10.1016/S1465-9972(01)00019-8

Liu, Z. H., Dreybrodt, W., & Wang, H. J. (2010). A new direction in effective accounting for the atmospheric CO$_2$ budget: Considering the combined action of carbonate dissolution, the global water cycle and photosynthetic uptake of DIC by aquatic organisms. *Earth-Science Reviews, 99*, 162–172. http://dx.doi.org/10.1016/j.earscirev.2010.03.001

Macpherson, G. L., Roberts, J. A., Blair, J. M., Townsend, M. A., Fowle, D. A., & Beisner, K. R. (2008). Increasing shallow groundwater CO$_2$ and limestone weathering, Konza Prairie, USA. *Geochimica et Cosmochimica Acta, 72*, 5581–5599. http://dx.doi.org/10.1016/j.gca.2008.09.004

Mahmoodabadi, M., & Heydarpour, E. (2014). Sequestration of organic carbon influenced by the application of straw residue and farmyard manure in two different soils. *International Agrophysics, 28*, 169–176.

Manning, D. A. C., & Renforth, P. (2013). Passive sequestration of atmospheric CO$_2$ through coupled plant-mineral reactions in urban soils. *Environmental Science & Technology, 47*, 135–141.

Manning, D. A. C., Renforth, P., Lopez-Capel, E., Robertson, S., & Ghazireh, N. (2013). Carbonate precipitation in artificial soils produced from basaltic quarry fines and composts: An opportunity for passive carbon sequestration. *International Journal of Greenhouse Gas Control, 17*, 309–317. http://dx.doi.org/10.1016/j.ijggc.2013.05.012

Mortatti, J., & Probst, J.-L. (2003). Silicate rock weathering and atmospheric/soil CO$_2$ uptake in the Amazon basin estimated from river water geochemistry: Seasonal and spatial variations. *Chemical Geology, 197*, 177–196. http://dx.doi.org/10.1016/S0009-2541(02)00349-2

Oelkers, E. H., Gislason, S. R., & Matter, J. (2008). Mineral carbonation of CO$_2$. *Elements, 4*, 333–337. http://dx.doi.org/10.2113/gselements.4.5.333

Power, I. M., Harrison, A. L., Dipple, G. M., & Southam, G. (2013). Carbon sequestration via carbonic anhydrase facilitated magnesium carbonate precipitation. *International Journal of Greenhouse Gas Control, 16*, 145–155. http://dx.doi.org/10.1016/j.ijggc.2013.03.011

Renforth, P., Manning, D. A. C., & Lopez-Capel, E. (2009). Carbonate precipitation in artificial soils as a sink for atmospheric carbon dioxide. *Applied Geochemistry, 24*, 1757–1764. http://dx.doi.org/10.1016/j.apgeochem.2009.05.005

Ryan, P. R., Delhaize, E., & Jones, D. L. (2001). Function and mechanism of organic anion exudation from plant roots. *Annual Review of Plant Physiology and Plant Molecular Biology, 52*, 527–560. http://dx.doi.org/10.1146/annurev.arplant.52.1.527

Salek, S. S., Kleerebezem, R., Jonkers, H. M., Witkamp, G.-J., & van Loosdrecht, M. C. M. (2013). Mineral CO$_2$ sequestration by environmental biotechnological processes. *Trends in Biotechnology, 31*, 139–146. http://dx.doi.org/10.1016/j.tibtech.2013.01.005

Schlesinger, W. H. (1999). Carbon and agriculture: Carbon sequestration in soils. *Science, 284*, 2095. http://dx.doi.org/10.1126/science.284.5423.2095

Schlesinger, W. H., & Andrews, J. A. (2000). Soil respiration and the global carbon cycle. *Biogeochemistry, 48*, 7–20. http://dx.doi.org/10.1023/A:1006247623877

Schuiling, R. D., & Krijgsman, P. (2006). Enhanced weathering: An effective and cheap tool to sequester CO$_2$. *Climatic Change, 74*, 349–354. http://dx.doi.org/10.1007/s10584-005-3485-y

Seifritz, W. (1990). CO$_2$ disposal by means of silicates. *Nature, 345*, 486. http://dx.doi.org/10.1038/345486b0

Sinsabaugh, R. S. (1994). Enzymic analysis of microbial pattern and process. *Biology and Fertility of Soils, 17*, 69–74. http://dx.doi.org/10.1007/BF00418675

Smith, P. (2004). Soils as carbon sinks: the global context. *Soil Use and Management, 20*, 212–218. http://dx.doi.org/10.1079/SUM2004233

Sun, L., Xiao, L., Xiao, B., Wang, W., Pan, C., Wang, S., & Lian, B. (2013). Differences in the gene expressive quantities of carbonic anhydrase and cysteine synthase in the weathering of potassium-bearing minerals by *Aspergillus niger*. *Science China Earth Sciences, 56*, 2135–2140. http://dx.doi.org/10.1007/s11430-013-4704-4

Teir, S., Eloneva, S., Fogelholm, C. J., & Zevenhoven, R. (2006). Stability of calcium carbonate and magnesium carbonate in rainwater and nitric acid solutions. *Energy Conversion and Management, 47*, 3059–3068. http://dx.doi.org/10.1016/j.enconman.2006.03.021

Vance, E., Brookes, P., & Jenkinson, D. (1987). An extraction method for measuring soil microbial biomass C. *Soil Biology and Biochemistry, 19*, 703–707. http://dx.doi.org/10.1016/0038-0717(87)90052-6

Verma, M. P. (2004). A revised analytical method for HCO$_3^-$ and CO$_3^{2-}$ determinations in geothermal waters: An assessment of IAGC and IAEA interlaboratory comparisons. *Geostandards and Geoanalytical Research, 28*, 391–409. http://dx.doi.org/10.1111/ggr.2004.28.issue-3

Wu, J., Joergensen, R., Pommerening, B., Chaussod, R., & Brookes, P. (1990). Measurement of soil microbial biomass C by fumigation-extraction—an automated procedure.

Soil Biology and Biochemistry, 22, 1167–1169. http://dx.doi.org/10.1016/0038-0717(90)90046-3

Wu, W., Xu, S., Yang, J., & Yin, H. (2008). Silicate weathering and CO_2 consumption deduced from the seven Chinese rivers originating in the Qinghai-Tibet Plateau. *Chemical Geology, 249*, 307–320. http://dx.doi.org/10.1016/j.chemgeo.2008.01.025

Xiao, L., Hao, J., Wang, W., Lian, B., Shang, G., Yang, Y., ... Wang, S. (2014). The up-regulation of carbonic anhydrase genes of bacillus mucilaginosus under Soluble Ca^{2+} deficiency and the heterologously expressed enzyme promotes calcite dissolution. *Geomicrobiology Journal, 31*, 632–641. http://dx.doi.org/10.1080/01490451.2014.884195

Xiao, L., Lian, B., Hao, J., Liu, C., & Wang, S. (2015). Effect of carbonic anhydrase on silicate weathering and carbonate formation at present day CO_2 concentrations compared to primordial values. *Scientific Reports, 5*, 7733. http://dx.doi.org/10.1038/srep07733

Xiao, L., Lian, B., Dong, C., & Liu, F. (2016). The selective expression of carbonic anhydrase genes of Aspergillus nidulans in response to changes in mineral nutrition and CO_2 concentration. *MicrobiologyOpen, 5*, 60–69. http://dx.doi.org/10.1002/mbo3.2016.5.issue-1

Zangen, M. (1962). Titration of carbonate-bicarbonate leach solutions. *Journal of Applied Chemistry, 12*, 92–96.

Zantua, M. I., Dumenil, L. C., & Bremner, J. M. (1977). Relationships between soil urease activity and other soil properties1. *Soil Science Society of America Journal, 41*, 350–352. http://dx.doi.org/10.2136/sssaj1977.036159950041000 20036x

Zhu, X., Lian, B., Yang, X., Liu, C., & Zhu, L. (2013). Biotransformation of earthworm activity on potassium-bearing mineral powder. *Journal of Earth Science, 24*, 65–74. http://dx.doi.org/10.1007/s12583-013-0313-6

Regression models for intrinsic constants of reconstituted clays

Farzad Habibbeygi[1]*, Hamid Nikraz[1] and Bill K Koul[2]

*Corresponding author: Farzad Habibbeygi, Civil Engineering, Curtin University, Australia
E-mail: farzad.habibbeygi@postgrad. curtin.edu.au
Reviewing editor: Giulio Iovine, IRPI, Consiglio Nazionale delle Ricerche, Italy

Abstract: In this study, four models were developed to predict intrinsic constants based on some simple physical parameters as well as clay mineralogy of a reconstituted clay sample. The effect of each predictor on the response was evaluated for each individual clay mineralogy. According to the results, it appears that the void ratio at liquid limit has the greatest effect on clays with a considerable amount of smectite, while the effect of the initial void ratio of such clays is the least amongst other clay minerals. The accuracy of the predictive model increases with the inclusion of clay mineralogy as an input parameter. R^2 increases from 0.978 to 0.99 and from 0.831 to 0.896 for intrinsic parameters of e^*_{100} and e^*_{1000}, respectively. A simplified method is also presented to determine the virgin compression line of reconstituted clays using the initial void ratio, the void ratio at the liquid limit, and its clay mineralogy.

Subjects: Earth Sciences; Earth Systems Science; Geology - Earth Sciences; Civil, Environmental and Geotechnical Engineering

Keywords: reconstituted clay; consolidation; intrinsic constant; compressibility; void ratio; model; mineralogy

1. Introduction

Dredged slurry sedimentation is currently used frequently in land reclamations as a sustainable solution to manage the huge quantities of slurries dredged from lakes and rivers every year. The deposited slurries can be found all over the world where reclamation is considered to be an essential step in developing new urban areas. Over the past seven decades, researchers and geotechnical engineers across the world have been interested in better comprehending the compression behaviour of dredged slurries at high initial water content due to the potential effect of their settlement on future coastal development projects.

ABOUT THE AUTHOR

Farzad Habibbeygi is a PhD student at School of Civil and Mechanical Engineering, Faculty of Science and Engineering, Curtin university and a member of Engineers of Australia. His research areas of interest are mechanical and compression behaviour of expansive clays, numerical modelling and artificial neural network. He has also published several research papers in international journals in his area of interest.

PUBLIC INTEREST STATEMENT

This paper introduces four predictive models for intrinsic constants based on some simple geotechnical parameters as well as predominant clay mineral of a reconstituted clay sample. The effect of each input parameter on the response of the models was evaluated for each individual clay mineralogy. According to the results, the void ratio at liquid limit has the greatest effect on clays with a considerable amount of smectite, while the effect of the initial void ratio of such clays is the least amongst other clay minerals.

During the dredging process, the structure of a clayey material is broken down such that the sedimented dredged clay can be considered as a reconstituted soil. A reconstituted soil is defined as a type of remoulded sample which is prepared at a water content equal to or greater than its liquid limit following Burland's (1990) procedure. During the past seven decades, many researchers have performed a considerably large number of consolidation tests on remoulded/reconstituted samples to investigate the compressibility of clays (Butterfield, 1979; Habibbeygi, Nikraz, & Verheyde, 2017; Hong & Tsuchida, 1999; Hong, 2006; Mesri & Olson, 1971; Mitchell & Soga, 1976; Nagaraj & Miura, 2001; Nagaraj & Murthy, 1983; Sridharan & Gurtug, 2005; Sridharan & Nagaraj, 2000). However, a relatively accurate prediction of the compressibility behaviour of dredged slurries can be derived by performing modified oedometer (consolidation) tests on these types of materials; unfortunately, consolidation tests on reconstituted clays are always time-consuming, costly and cumbersome. The determination of compression index is considered to be a relatively expensive and time-consuming test in most geotechnical projects, especially when undertaken on clays with high initial water content. Considering all these difficulties and limitations in assessing the compressibility of reconstituted clays at high initial water content, geotechnical engineers estimating the volumetric behaviour of reconstituted clays often tend to use empirical correlations instead of performing one-dimensional consolidation tests. Accordingly, having some reliable empirical equations to predict the compression behaviour of such soils is beneficial. In this case, compressibility can be estimated by understanding some simple physical characteristics of the soil sample under study (i.e. initial water content, liquid and plastic limits, natural void ratio), which can be determined conveniently in the laboratory.

To the authors' best knowledge, some research has previously been conducted to consider the effect of initial water content and soil properties on the compression behaviour of reconstituted clays (Cerato & Lutenegger, 2004; Kootahi & Moradi, 2016; Lee, Hong, Kim, & Lee, 2015; Lei, Wang, Chen, Huang, & Han, 2015; Takashi, 2015; Xu & Yin, 2015; Zeng, Hong, Cai, & Han, 2011), but only a few studies have taken into account the influence of clay mineralogy (Habibbeygi, Nikraz, & Chegenizadeh, 2017; Xu & Yin, 2015). In this paper, the intrinsic concept was used as a basic frame of reference for interpreting and evaluating the compressibility of reconstituted clays. Furthermore, the effect of clay mineralogy on compression behaviour has been assessed using a broad range of geotechnical data from the literature. Eventually, four series of practical relationships were developed to estimate the intrinsic constants and compressibility of a reconstituted clay based on some simple physical parameters, as well as the clay's mineralogy.

2. Existing correlations for predicting intrinsic constants
Burland (1990) proposed a unique framework for normalising the compression behaviour of reconstituted clays by introducing the intrinsic concept in his 40th Rankine lecture. The intrinsic concept has since been consistently used worldwide to explain the behaviour of clays (Al Haj & Standing, 2015; Habibbeygi et al., 2017; Hong, Lin, Zeng, Cui, & Cai, 2012; Hong, Yin, & Cui, 2010; Horpibulsuk, Liu, Zhuang, & Hong, 2016). Intrinsic compression behaviour can be interpreted by two constants of compressibility—compression index (C_c^*) and e_{100}^*—and a void ratio invariant, named the void index (I_v). According to Burland, compression index (C_c^*) can be defined as the subtraction of two definite void ratios (Equation 1). The void index at each state (I_v) can also be expressed with the related void ratio and the intrinsic constants as follows (Burland, 1990):

$$C_c^* = e_{100}^* - e_{1000}^* \tag{1}$$

$$I_v = \frac{e - e_{100}^*}{C_c^*} \tag{2}$$

where e_{100}^* and e_{1000}^* are the void ratios of the reconstituted clay at the effective vertical consolidation stresses (σ_v') at 100 kPa and 1000 kPa, respectively. To distinguish the reconstituted invariants from those of natural clays, the symbol (*) has been used with compression parameters in the equations related to a reconstituted clay.

Burland (1990) stated that the compression behaviour of reconstituted clays can be generalised by using the normalised invariant of the void index and, in fact, a unique line can interpret the compressibility of all different clays independent of their natural state. However, it has been recently documented that the initial water content (w_0) of reconstituted clays has a considerable effect on the intrinsic compression line (ICL) at low stress levels (Habibbeygi et al., 2017; Hong et al., 2010; Xu, Gao, Yin, Yang, & Ni, 2014), and the ICL is almost unique for a broad range of initial water content for a "medium to high" stress range when I_v is plotted against σ'_v in a semi-log scale.

Burland (1990) also proposed an empirical equation for calculating the ICL indirectly. The value of void index (I_v) to estimate the inherent compressibility of reconstituted clays can be calculated as follows:

$$I_v = 2.45 - 1.285\left(\log \sigma'_v\right) + 0.015\left(\log \sigma'_v\right)^3 \tag{3}$$

Burland (1990) also suggested a method for calculating the ICL indirectly in the absence of one-dimensional consolidometer test results, based on the regression analyses on a various range of clays (Equations 4 and 5). The use of the following equations must be limited to the liquid limit (w_L) from 25% to 160% (i.e. e_L in the range of 0.6–4.5) (Burland, 1990):

$$e^*_{100} = 0.109 + 0.679e_L - 0.089e_L{}^2 + 0.016e_L{}^3 \tag{4}$$

$$C^*_c = 0.256e_L - 0.04 \tag{5}$$

where e_L is the void ratio of reconstituted clays at liquid limit.

Yin and Miao (2013) proposed a modification to Burland's relationships by considering the influence of initial water content on the intrinsic constants of reconstituted clays. Forty-two samples from three different sampling sites in China were collected in their study. The geotechnical parameters of the investigated soils can be summarised as follows:

The liquid limit varied between 61% and 91%; The plastic limit was limited to a narrow range of 30% to 38%; and the initial water content varied between 43% and 180%.

The modified equations for estimating the intrinsic constants (C^*_c and e^*_{100}) in terms of w_L, and w_0 are as follows:

$$e^*_{100} = 1.13w_L + 0.39w_0/w_L - 0.084 \tag{6}$$

$$C^*_c = 0.91w_L + 0.25w_0/w_L - 0.461 \tag{7}$$

where w_0 and w_L are the initial water content and the liquid limit of the studied reconstituted clay, respectively.

Zeng, Hong, and Cui (2015) performed 48 consolidometer tests on natural clays, as well as reconstituted kaolinite clays, with different initial water content to assess the inherent compression behaviour of such soils. They used the consolidation test results of their study, as well as the data of earlier research, to suggest equations for estimating the intrinsic constants. They suggested two groups of equations based on the range of e_L. Equations (8) and (9) were developed for the e_L range of 0.76–2.7, while Equations (10) and (11) were suggested to be used for a broader range of e_L of 0.66 to 5.72. Density of solid particles of the studied soils varied from 2.65 g/cm³ to

2.75 g/cm^3. The initial water content (w_0) ranged from 22.3% to 163.3% and the liquid limit from 28.1% to 100%:

$$e_{100}^* = 0.357 + 0.171e_0 + 0.223e_L \quad (For\ e_L = 0.76 - 2.7) \tag{8}$$

$$C_c^* = -0.069 + 0.109e_0 + 0.152e_L \tag{9}$$

$$e_{100}^* = 0.223 + 0.261e_0 + 0.282e_L - 0.018e_0^2 - 0.05e_L^2 + 0.015e_L^3 \ (For\, e_L = 0.66 - 5.72) \tag{10}$$

$$C_c^* = -0.064 + 0.153e_0 + 0.11e_L - 0.006e_0^2 \tag{11}$$

where e_0 and e_L are the void ratios at initial state and the liquid limit of the investigated reconstituted clay, respectively.

Xu and Yin (2015) investigated the intrinsic compression curves of clays with three different minerals (kaolinite, illite and smectite). The consolidation tests results of their work on the reconstituted samples, with initial water content ranging from 1 to 2 times their respective liquid limits, revealed that the effects of the initial water content on the intrinsic constants of reconstituted clays depended on the type of the clay minerals. For example, their study demonstrated that w_0 has less impact on e_{100}^* of the predominant montmorillonite clay than on other secondary clay minerals.

Habibbeygi et al. (2017) performed eight series of consolidation tests on reconstituted samples to investigate the impact of mineralogy on the compressibility of expansive clays. The initial water content of their study ranged from $0.67w_L$ to $1.33w_L$, and vertical consolidation stress varied from as low as 1 kPa to as high as 1,600 kPa to consider a broad range of consolidation stress. Their studies depicted that the initial water content had a considerable impact on the intrinsic constants for the studied soil. In fact, e_{100}^* and C_c^* increased with increases in the initial water content. Furthermore, the intrinsic constants of the studied clay with smectite as the predominant clay mineral were higher than the estimated values from existing empirical equations.

In summary, it appears that there is a considerable influence of clay mineralogy on the intrinsic constants of a reconstituted clay. While some experimental equations have been proposed to estimate the intrinsic constants of reconstituted clays (e_{100}^* and C_c^*) by past researchers (Burland, 1990; Yin & Miao, 2013; Zeng et al., 2015), none of these equations has taken into account the effect of clay mineralogy on the intrinsic constants. Given past research, the main objective of this paper has been to investigate the effect of clay mineralogy on the value of intrinsic constants, and to suggest some equations for estimating these constants and compressibility based on simple, measureable, physical geotechnical parameters considering the influence of clay mineralogy.

3. Estimating the intrinsic constants
To develop some empirical equations that are able to predict values of the intrinsic constants under different circumstances, a large series of 1D consolidation test data was collected in this study to include not only diverse initial states and physical properties, but also the variety of clay mineralogy in the predicting relationships. The collected data were then used to derive some regression models based on the availability of the input data. Finally, the proposed model, including the clay mineralogy, was compared to the existing models, and the influence of each parameter in the model on the response was evaluated for various clay minerals.

4. Experimental data of reconstituted clays
A broad range of experimental data of 94 consolidation tests on various reconstituted/remoulded clays, with different mineralogy and initial water contents, was used in this study. A summary of the geotechnical properties, initial state, and references to these tests is tabulated in Table 1.

Table 1. Geotechnical properties of consolidometer tests

Soil description	w_o	G_s	w_L	w_P	Chief clay mineral	Clay %	Stress range	No. of tests	Reference
Black soil	88	2.72	60	30	Illite-Smectite	48	Up to 8 MPa	1	Al Haj and Standing (2015)
Red soil	138	2.78	87	33	Illite-Smectite	50	Up to 8 MPa	1	Habibbeygi et al. (2017)
Baldivis clay	55–109	2.6	82	35	Smectite	68	1–1600 kPa	8	
Atchafalaya	101–177	2.8	101	35	Montmorillonite	66.6	Up to 2 MPa	8	Cerato and Lutenegger (2004)
Boston blue clay	45–79	2.8	45	23	Illite	48.4			
Kaolinite	42–74	2.68	42	26	Kaolinite	36.2			
Kaolin	45–113	2.69	55.3	25.2	Kaolinite	62	0.5–1600 kPa	9	Xu and Yin (2015)
Illite	45–104	2.7	52.7	29.6	Illite	40	0.5–1600 kPa	10	Xu and Yin (2015)
Montmorillonite	190–529	2.72	258.9	42.8	Smectite	84	0.5–1600 kPa	8	Xu and Yin (2015)
Lianyungang clay	50–146	2.71	74	33		23	0.5–1600 kPa	14	Hong et al. (2010)
Baimahu clay	64–180	2.65	91	38		20	0.5–1600 kPa	14	Hong et al. (2010)
Kemen clay	43–122	2.67	61	30		19	0.5–1600 kPa	14	Hong et al. (2010)
Lin-Gang Clay1 (undisturbed)	46–124	2.8	56.4	26.5	Illite	55.4	2–827 kPa	4	Lei et al. (2015)
Central Fishing Port Clay (undisturbed)	22.4	2.8	39.8	15	Illite	46		1	
Qing-Fang Clay1 (undisturbed)	43.4	2.8	40.8	22.1	Illite	38.1		1	
Qing-Fang Clay2 (remoulded)	43.3							1	

Density of soil particles ranged from 2.57 g/cm^3 to 2.80 g/cm^3 and the initial water content varied from 22.4% to 528.7%. The liquid and plastic limits of the studied soils were in the range of 39.8% to 258.9% and 15% to 42.8%, respectively. As can be seen from Table 1, the data include three different clay minerals—kaolinite, illite and smectite.

Figure 1 illustrates the compression curves of all these clays in the form of e vs. $\log \sigma_v'$ relationship. The initial consolidation stress in the tests was as low as 0.5 kPa and reached high stress levels of 1,600 kPa to 4,000 kPa. As expected, most curves are inverse S-shaped, with a distinct remoulded yield stress. Figure 1 shows two sets of curves, separated approximately at e = 5, as

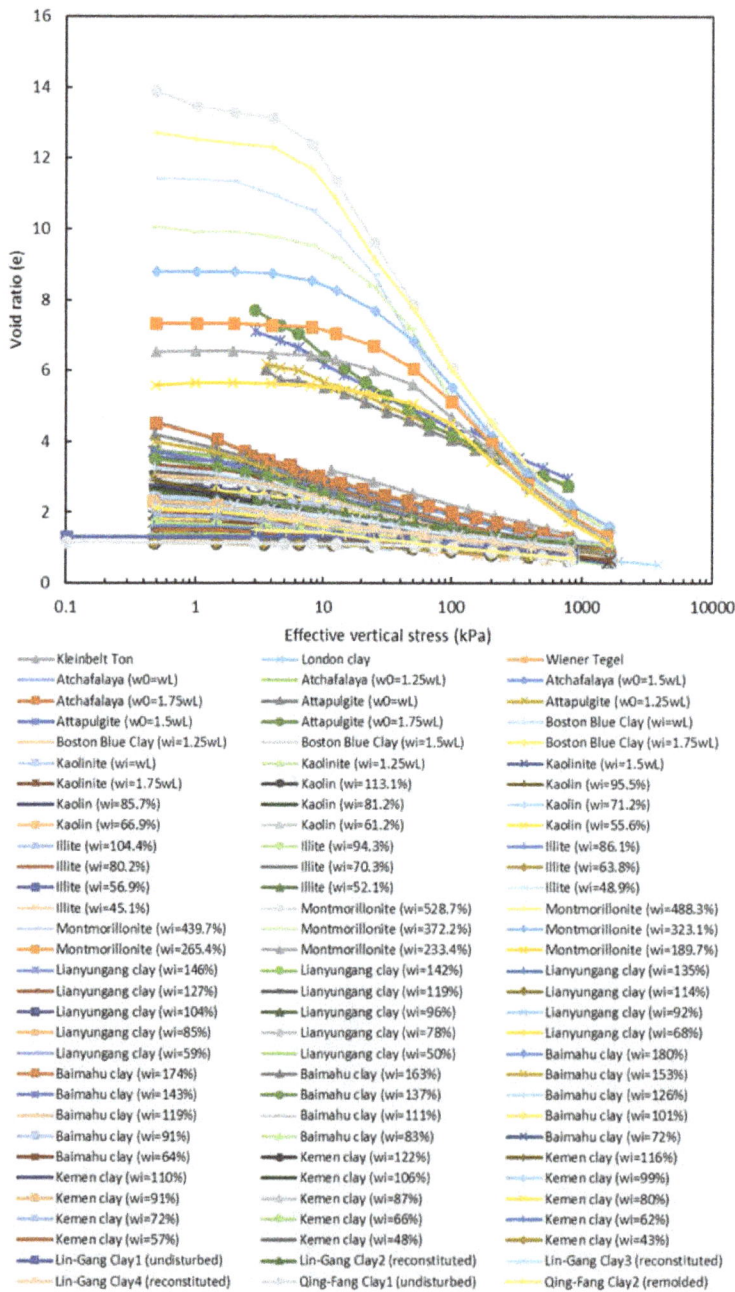

Figure 1. Compression curves of the studied clays in e- $\log \sigma_v'$ space (references for experimen-tal data are listed in Table 1).

there were no samples available at this void ratio. Remoulded yield stress of a reconstituted clay, which is similar to pre-consolidation stress of a natural clay, is a particular stress beyond which the inherent structure of a reconstituted clay breaks down, and the compressibility (the slope of virgin compression line) increases abruptly.

The compression curves of the studied soils are replotted in a normalised plane, using the void index in Figure 2 (I_v vs. $\log \sigma'_v$). However, there is a considerable disparity between normalised compression curves of reconstituted clays at low stress level; Burland's equation can express these well when the vertical consolidation stress is higher than the vertical consolidation stress at the remoulded yield stress. A new polynomial equation (Equation 12)

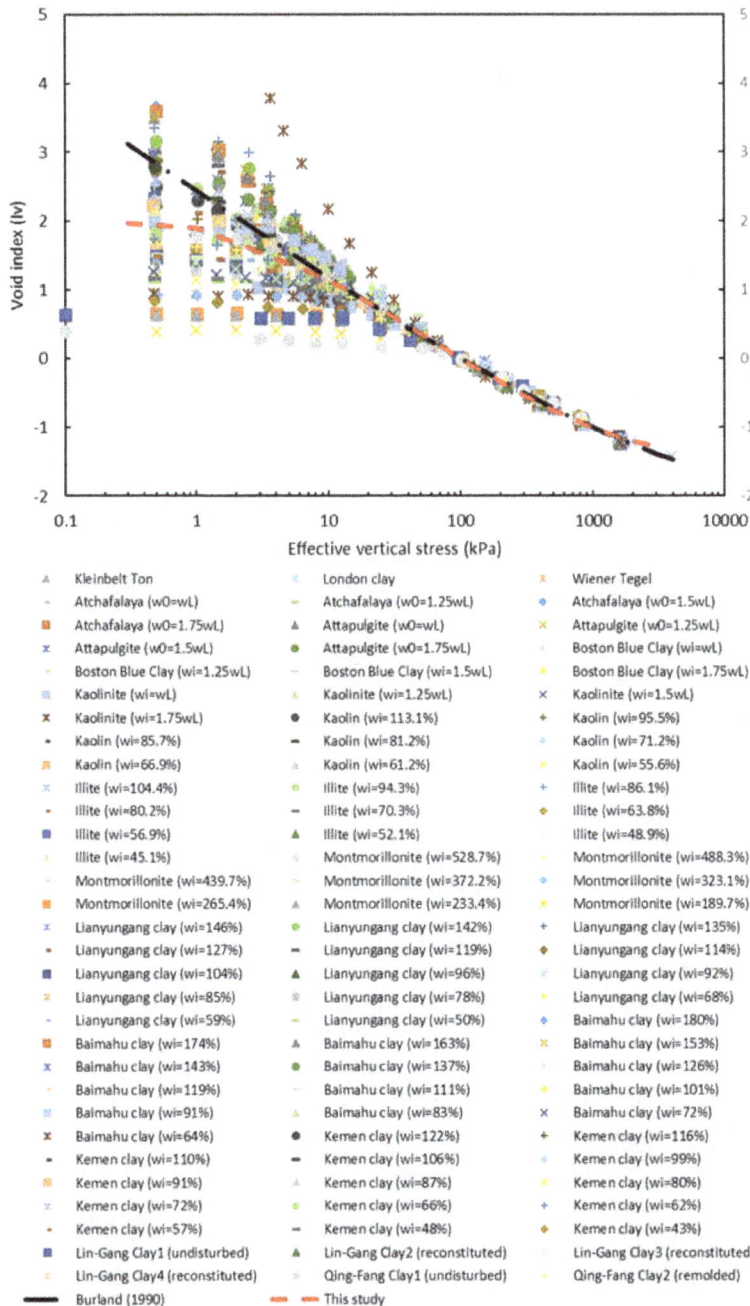

Kleinbelt Ton	London clay	Wiener Tegel
Atchafalaya (w0=wL)	Atchafalaya (w0=1.25wL)	Atchafalaya (w0=1.5wL)
Atchafalaya (w0=1.75wL)	Attapulgite (w0=wL)	Attapulgite (w0=1.25wL)
Attapulgite (w0=1.5wL)	Attapulgite (w0=1.75wL)	Boston Blue Clay (wi=wL)
Boston Blue Clay (wi=1.25wL)	Boston Blue Clay (wi=1.5wL)	Boston Blue Clay (wi=1.75wL)
Kaolinite (wi=wL)	Kaolinite (wi=1.25wL)	Kaolinite (wi=1.5wL)
Kaolinite (wi=1.75wL)	Kaolin (wi=113.1%)	Kaolin (wi=95.5%)
Kaolin (wi=85.7%)	Kaolin (wi=81.2%)	Kaolin (wi=71.2%)
Kaolin (wi=66.9%)	Kaolin (wi=61.2%)	Kaolin (wi=55.6%)
Illite (wi=104.4%)	Illite (wi=94.3%)	Illite (wi=86.1%)
Illite (wi=80.2%)	Illite (wi=70.3%)	Illite (wi=63.8%)
Illite (wi=56.9%)	Illite (wi=52.1%)	Illite (wi=48.9%)
Illite (wi=45.1%)	Montmorillonite (wi=528.7%)	Montmorillonite (wi=488.3%)
Montmorillonite (wi=439.7%)	Montmorillonite (wi=372.2%)	Montmorillonite (wi=323.1%)
Montmorillonite (wi=265.4%)	Montmorillonite (wi=233.4%)	Montmorillonite (wi=189.7%)
Lianyungang clay (wi=146%)	Lianyungang clay (wi=142%)	Lianyungang clay (wi=135%)
Lianyungang clay (wi=127%)	Lianyungang clay (wi=119%)	Lianyungang clay (wi=114%)
Lianyungang clay (wi=104%)	Lianyungang clay (wi=96%)	Lianyungang clay (wi=92%)
Lianyungang clay (wi=85%)	Lianyungang clay (wi=78%)	Lianyungang clay (wi=68%)
Lianyungang clay (wi=59%)	Lianyungang clay (wi=50%)	Baimahu clay (wi=180%)
Baimahu clay (wi=174%)	Baimahu clay (wi=163%)	Baimahu clay (wi=153%)
Baimahu clay (wi=143%)	Baimahu clay (wi=137%)	Baimahu clay (wi=126%)
Baimahu clay (wi=119%)	Baimahu clay (wi=111%)	Baimahu clay (wi=101%)
Baimahu clay (wi=91%)	Baimahu clay (wi=83%)	Baimahu clay (wi=72%)
Baimahu clay (wi=64%)	Kemen clay (wi=122%)	Kemen clay (wi=116%)
Kemen clay (wi=110%)	Kemen clay (wi=106%)	Kemen clay (wi=99%)
Kemen clay (wi=91%)	Kemen clay (wi=87%)	Kemen clay (wi=80%)
Kemen clay (wi=72%)	Kemen clay (wi=66%)	Kemen clay (wi=62%)
Kemen clay (wi=57%)	Kemen clay (wi=48%)	Kemen clay (wi=43%)
Lin-Gang Clay1 (undisturbed)	Lin-Gang Clay2 (reconstituted)	Lin-Gang Clay3 (reconstituted)
Lin-Gang Clay4 (reconstituted)	Qing-Fang Clay1 (undisturbed)	Qing-Fang Clay2 (remolded)
Burland (1990)	This study	

Figure 2. Normalised compres-sion curves of the studied clays in I_v - $\log \sigma'_v$ space (references for experimental data are listed in Table 1).

has also been fitted to the data with a reasonable correlation coefficient, as high as 0.93, to refine the existing ICL equations:

$$I_v = 1.906 - 0.392\left(\log \sigma_v'\right) - 0.457\left(\log \sigma_v'\right)^2 + 0.088\left(\log \sigma_v'\right)^3 \tag{12}$$

The refined equation, as well as the Burland's equation, is plotted in Figure 2 for comparison. It is noted that Burland's empirical equation has been developed for the reconstituted soils, with e_L lower than 4.44, while the proposed equation used data with e_L up to 7.04. However, Burland's equation and the refined equation are close to each other for the medium level of effective vertical stress (50–100 kPa), and even overlap for some stresses in this range. The modified equation becomes slightly more concave upwards than Burland's equation for consolidation stresses higher than 100 kPa. As shown in Figure 2, there is a disparity between the results of this modified equation (i.e. central fitting of scatter plot) and the experimental data, especially for the low-stress level (i.e. less than 20 kPa). The initial soil sample condition, clay mineralogy, non-linear behaviour of clays, change in the soil permeability, the coefficient of consolidation during consolidation procedure and secondary consolidation are some of the factors that affect the compression curves.

5. Regression models

Following the introduction of the intrinsic concept by Burland (1990), the void index has been used herein to normalise the compression curves and to interpret the compressibility of reconstituted clays. Based on Burland's work, the compressibility of a reconstituted clay can also be estimated by knowing the values of e_{100}^* and e_{1000}^*. These two void ratios can also be used to determine the intrinsic compression index (I_v) and to predict the compression behaviour of a reconstituted clay.

Two simple geotechnical parameters, i.e. the void ratio at initial condition (e_0) and the void ratio at liquid limit (e_L), along with the type of clay mineralogy (CLM), were adopted as predictors to obtain the empirical equations for estimating the intrinsic constants in this study. As all input parameters may not be available, the following four regression models have been developed based on the degree of availability of geotechnical parameters of the investigated soil:

Model 1: If only the void ratio at the liquid limit (e_L) is available.

Model 2: If both the void ratio at the liquid limit (e_L) and the initial void ratio (e_0) are known.

Model 3: If, in addition to the void ratios of (e_0 and e_L), clay mineralogy (CLM) is also available.

Model 4: A simplified form of Model 3, which is easier to use without losing much accuracy from Model 3.

5.1. Model 1

Three different regression models were developed to estimate e_{100}^* and e_{1000}^* depending on the information obtainable for the studied reconstituted clay. The first model, which is the simplest one, uses only one parameter, e_L, to estimate the intrinsic parameters. The void ratio at the liquid limit can be calculated by assuming saturated condition for the clayey soil, using the relationship $e_L = G_s.w_L$, where G_s is the specific gravity, and w_L is the liquid limit of the studied clay. The equations for this model can be summarised as follows:

$$e_{100}^* = 1.167 - 0.334e_L - 0.242e_L^2 - 0.016e_L^3, \ \left(R^2 = 0.978\right) \tag{13}$$

$$e_{1000}^* = 2.093 - 1.712e_L - 0.637e_L^2 - 0.057e_L^3, \ \left(R^2 = 0.831\right) \tag{14}$$

$$C_c^* = -0.926 - 2.046e_L - 0.879e_L^2 - 0.073e_L^3 \tag{15}$$

The relationship of e_L and the intrinsic constants predicted from this model are illustrated in Figure 3. The values of intrinsic constants increase continuously when e_L increases in this model, which is in good agreement with the observed data and the results of past research (Burland, 1990; Hong et al., 2010).

5.2. Model 2

In Model 2, the effect of initial state has been applied to the model by adding the initial void ratio into the predicting relationships. The following equations estimate the intrinsic parameters of a reconstituted clay when both initial state (e_0) and e_L are accessible:

$$e_{100}^* = 0.813 + 0.182e_0 - 0.112e_L - 0.001e_0{}^2 + 0.075e_L{}^2 , \ (R^2 = 0.987) \tag{16}$$

$$e_{1000}^* = 0.777 + 0.101e_0 - 0.166e_L - 0.003e_0{}^2 + 0.033e_L{}^2, \ (R^2 = 0.877) \tag{17}$$

$$C_c^* = 0.036 + 0.081e_0 + 0.054e_L + 0.002e_0{}^2 + 0.042e_L{}^2 \tag{18}$$

5.3. Model 3

In the third model, which is the most sophisticated model to predict intrinsic constants, in addition to initial state parameters, e_0 and e_L, the clay mineralogy is also included. Clay mineralogy of the investigated soil was assumed to be represented by its most predominant clay mineral. Clay was implied into the regression analysis by a dummy parameter named CLM. Intrinsic equations and the values of dummy parameters of CLM_1 to CLM_3, depending on the type of principal clay mineral, are presented as follows:

$$e_{100}^* = CLM_1 + 0.2e_0 - 0.208e_L - 0.002e_0{}^2 + 0.081e_L{}^2, \ (R^2 = 0.990) \tag{19}$$

$$e_{1000}^* = CLM_2 + 0.111e_0 - 0.209e_L - 0.003e_0{}^2 + 0.035e_L{}^2, \ (R^2 = 0.896) \tag{20}$$

$$C_c^* = CLM_3 + 0.081e_0 + 0.054e_L + 0.002e_0{}^2 + 0.042e_L{}^2 \tag{21}$$

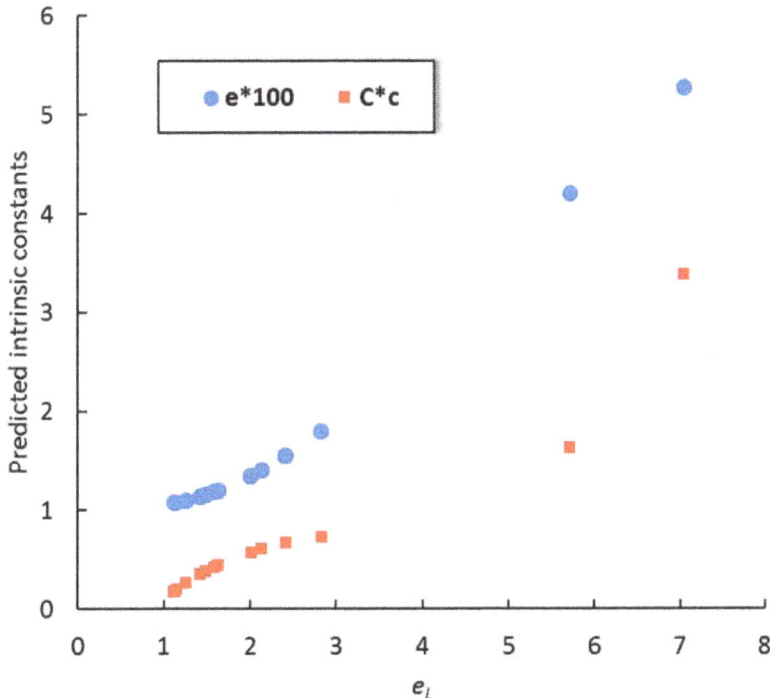

Figure 3. Plot of intrinsic con-stants e_{100}^* and C_c^* against e_L in Model 1.

$$CLM_1 = \begin{cases} 0.956 - \textit{for chief clay mineral of kaolinite} \\ 0.887 - \textit{for chief clay mineral of illite} \\ 1.094 - \textit{for chief clay mineral of smectite} \end{cases}$$

$$CLM_2 = \begin{cases} 0.862 - \textit{for chief clay mineral of kaolinite} \\ 0.796 - \textit{for chief clay mineral of illite} \\ 0.924 - \textit{for chief clay mineral of smectite} \end{cases}$$

$$CLM_3 = \begin{cases} 0.094 - \textit{for chief clay mineral of kaolinite} \\ 0.091 - \textit{for chief clay mineral of illite} \\ 0.170 - \textit{for chief clay mineral of smectite} \end{cases}$$

It is much easier to use the Wilkinson notation to express the regression equations (Wilkinson & Rogers, 1973), as it is a very powerful notation in presenting a complex regression model. The notation is used to present a model in terms of a response and predictors in the form of a simple equation with a table of predictor coefficients. In this notation, '1' stands for constants and each predictor is separated with a symbol of "+". If there is a powered predictor in the equation, it means that all lower order terms of the predictor are included in the equation, unless otherwise noted. Thus, substituting Equations (19)–(21) yields

$$e_{100}^*, \ e_{1000}^*, \ C_c^* \sim 1 + CLM + e_0 + e_L + e_0^2 + e_L^2 \tag{22}$$

where CLM is the categorical invariant of clay mineralogy and can be expressed in this model by two dummy variables, CLM_K and CLM_M. The coefficients of predictors (e_0, e_L, e_0^2, e_L^2, and CLM) for this relationship are tabulated in Table 2.

5.4. Model 4

Model 4 is a simplified version of Model 3. The model still considers the clay mineralogy, initial state, and properties of the studied soil, yet it retains its simplicity. Despite the fact the model is simple, it still has a good accuracy in predicting the intrinsic parameters of e_{100}^* and e_{1000}^*. The correlation coefficients of R^2 are 0.99 and 0.85 for e_{100}^* and e_{1000}^*, respectively,

$$e_{100}^* = 1 + e_0 + e_0 : CLM + e_L : CLM + e_L^2 + e_0^2, \ (R^2 = 0.990) \tag{23}$$

$$e_{1000}^* = 1 + e_L + e_L : CLM, \ (R^2 = 0.840) \tag{24}$$

Wilkinson notation has been used to express Equations (23) and (24). A symbol of ":" means only the product of two predictors without the lower order terms of the predictor. The coefficient of each predictor and dummy variables are presented in Table 3.

Table 2. Predictor coefficients of Model 3

Coefficients	e_{100}^*	e_{1000}^*	C_c^*
Constant	0.887	0.796	0.092
e_0	0.200	0.111	0.088
e_L	−0.208	−0.209	0.001
CLM_K	0.069	0.066	0.003
CLM_M	0.207	0.128	0.080
e_0^2	−0.002	−0.003	0.002
e_L^2	0.081	0.035	0.046

CLM_K and CLM_M are 1.0 when the principal clay of the studied soil is Kaolinite and Smectite, respectively, otherwise they are zero.

Table 3. Predictor coefficients of Model 4			
Coefficients	e^*_{100}	**Coefficients**	e^*_{1000}
Constants	0.696	Constants	0.502
e_0	0.157	e_L	0.147
$e_0 : CLM_K$	0.139	$e_L : CLM_K$	0.092
$e_0 : CLM_{MM}$	−0.096	$e_L : CLM_{MM}$	0.059
$e_L : CLM_K$	−0.110		
$e_L : CLM_{MM}$	0.215		
e_0^2	0.005		
e_L^2	0.040		

6. Comparison and discussion

A comparison of the ability of the proposed models to predict intrinsic constants has been presented in this section. Figure 4 demonstrates the values of predicted e^*_{100} and C^*_c against the related measured values obtained from the experimental data. It can be seen that the estimated values of e^*_{100} and C^*_c are in good agreement with the observed values. Moreover, Model 4 can estimate the parameters accurately even though it has relatively lesser order of terms of predictors than Model 3. A summary of the results is presented in Table 4. As can be noted, the accuracy of the models increases when the clay mineralogy is considered as an input parameter (Models 3 and 4).

The dataset reported by Habibbeygi et al. (2017) were used to compare the different empirical equations existing in the literature. The e^*_{100} and C^*_c values of Baldivis clay, estimated for different conditions of initial water content, are plotted in Figure 5. The geotechnical and physical parameters for the studied clay are presented in Table 1. According to Burland's equation, intrinsic constants are independent of the initial state (for example, e_0 or w_0). Zeng et al. (2015) modified Burland's equation to consider the effect of the initial state. The values estimated by equations suggested by Zeng et al. (2015) increase with an increase in the value of w_0, in a trend similar to the observed values. However, both these equations underestimate the intrinsic constants of the investigated clays with a considerable amount of smectite in its mineralogy. On the other hand, as depicted in Figure 5, the estimated values of intrinsic constants using the proposed model of this study are higher than the predicted values from other methods, and are close to the observed values. Thus, the predicted intrinsic constants from the proposed model are in good agreement with the measured constants for the investigated clay.

The effect of each predictor on the response of the proposed model has been examined for Model 4. Figure 6 presents the influence of change in each predictor on e^*_{100}. As can be seen from Figure 6(a), the change in e_0 has the greatest effect on e^*_{100} among other predictors. It shows that changing e_0 from 0.63 to about 14 increases e^*_{100} by about 3. Moreover, it also shows that changing e_L from 1.11 to 7 raises e^*_{100} by about 2. Figure 6(b) shows the interaction plot for observing the effect of changing one predictor while keeping others fixed. For example, the increase in e_L (from 1.11 to 7.0) is much more effective on the growth of e^*_{100} when the principal mineral of clay is smectite, with an increase of 3, whereas the increase of e^*_{100} is just one (1) when the principal clay mineral is illite. Interaction graphs for e_0 and e_L are plotted in Figure 6(c,d) for comparison. The increase in both e_0 and e_L lead to an increase in e^*_{100}, but with different rates for different clay minerals. For example, the rate of increase of e^*_{100} with e_0 for smectite is the lowest. On the other hand, the rate of increase of e^*_{100} with the increase of e_L is the highest in smectite. In Model 4, e^*_{1000} is only a function of e_L and its mineralogy. The effects of the variation of these predictors are plotted in Figure 7. While the increase in e_L leads to an increase in e^*_{1000}, the variation is the least for illite, with a growth of about 0.8.

The non-linear compressibility can be explained by the rheological behaviour. Rheological behaviour accounts for the secondary consolidation of clays. The rheological behaviour exists in the primary consolidation and cannot be neglected.

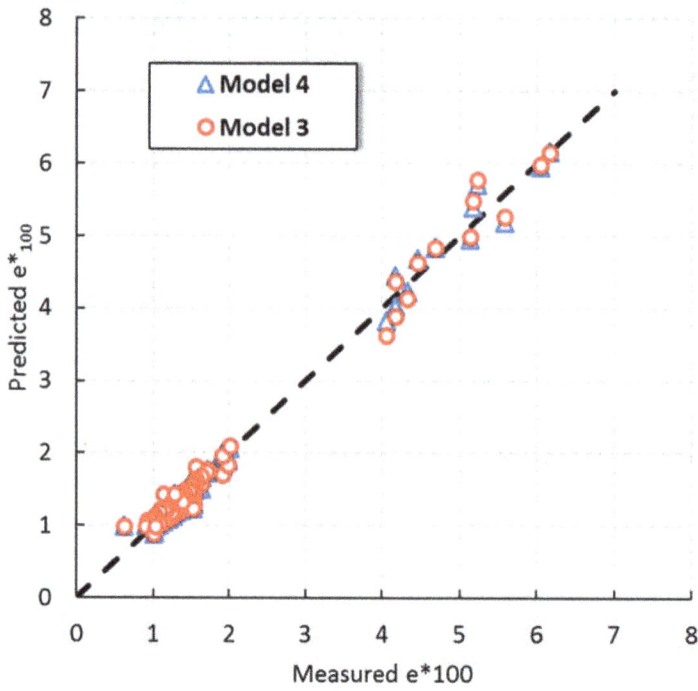

(a)

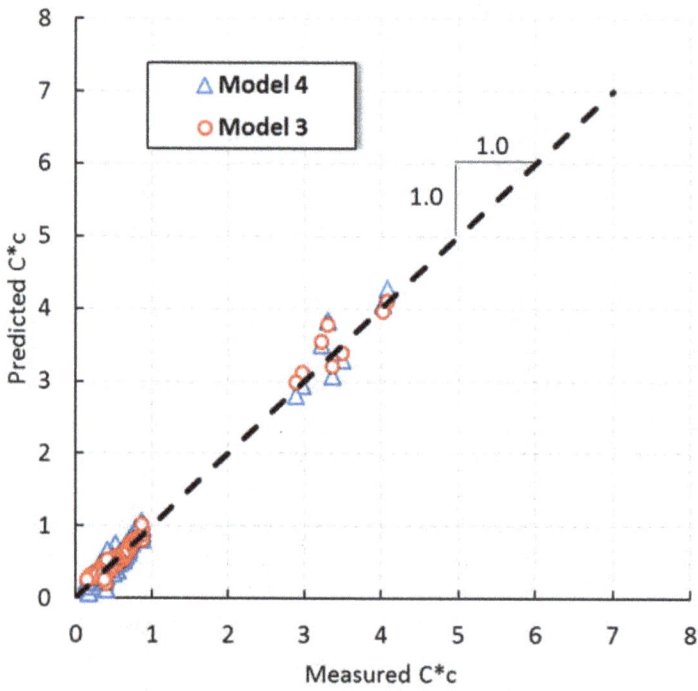

(b)

Figure 4. Comparison of pre-dicted intrinsic constants and measured ones: (a) e^*_{100} (b) C_c^*.

Table 4. R^2 of the proposed models for intrinsic parameters (e^*_{100} and e^*_{1000})

Model	R^2 for e^*_{100}	R^2 for e^*_{1000}	Input parameters
Model 1	0.978	0.831	e_L
Model 2	0.987	0.877	e_0 and e_L
Model 3	0.990	0.896	e_0, e_L and CLM
Model 4 (simplified)	0.990	0.840	e_0, e_L and CLM

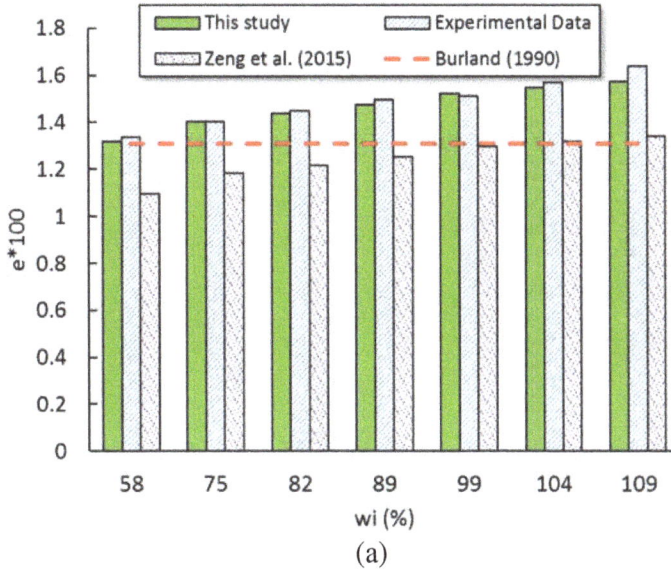

(a)

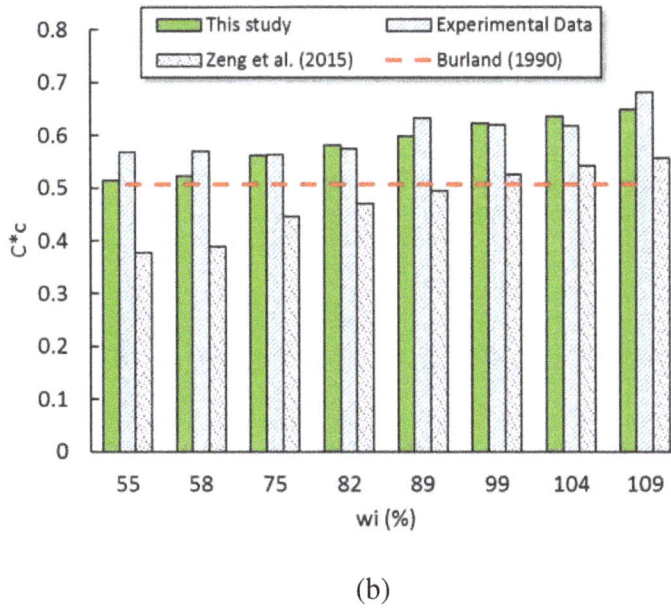

(b)

Figure 5. Comparison of pre-dicted intrinsic constants by different methods for Baldivis clay: (a) e^*_{100} **(b)** C^*_c.

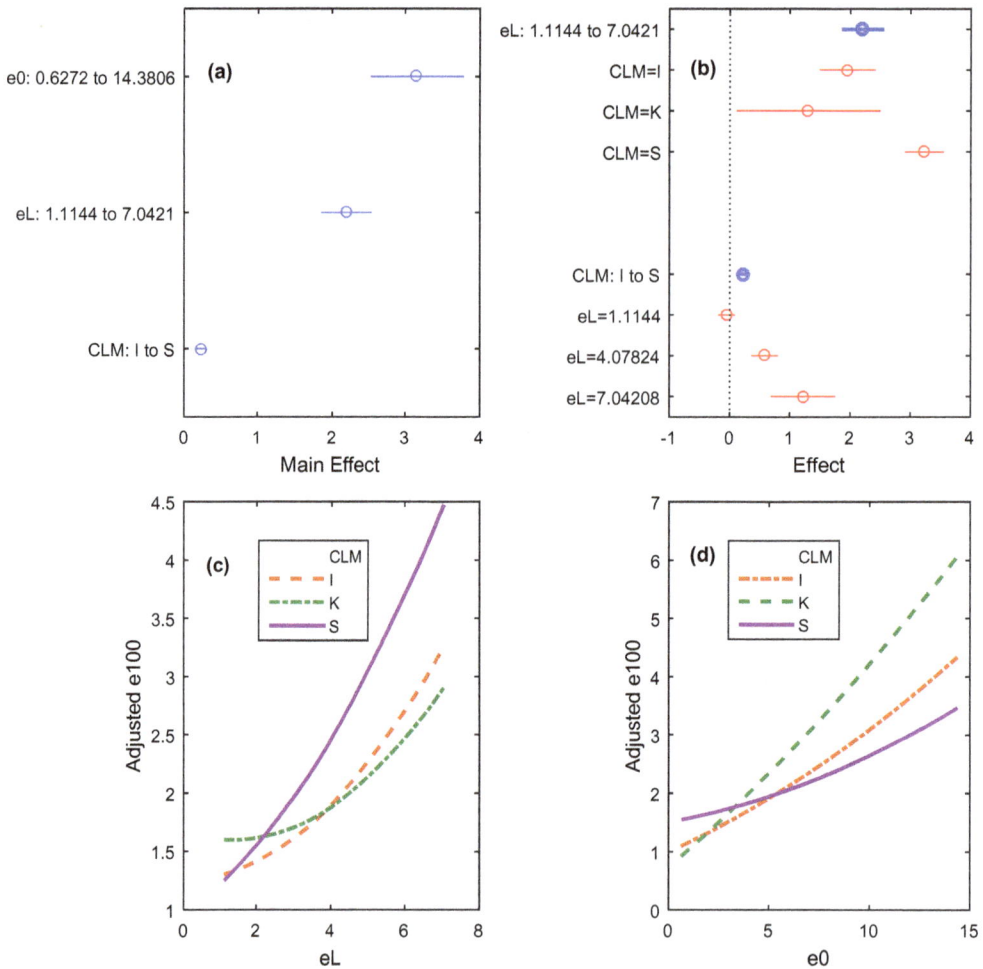

Figure 6. The effect of predic-tors on e^*_{100} (I: illite, K: kaolo-nite, S: smectite): (a) predictor effect, (b) predictor effect with CLM, (c) e_L effect and (d) e_0 effect.

7. Virgin compression line estimation

An advantage of using the void index for normalisation is that it allows estimation of compression curves without performing consolidation tests. In other words, the virgin compression line (VCL) of a reconstituted clay can be determined by substituting the modified equation of ICL (Equation 12) into the void index definition equation (Equation 2) as follows:

$$e = C_c^* \left(1.906 - 0.392x - 0.457x^2 + 0.088x^3\right) + e^*_{100} \ For \, \sigma'_v \geq \sigma'_{yr} \tag{25}$$

where $x = \log \sigma'_v$; σ'_v is in kPa, and σ'_{yr} is the remoulded yield stress of a reconstituted clay. This equation can be used for a broad range of e_L (1.1–7.0) and the initial state of e_0 (0.63 to 14.38).

Depending on the availability of information about the studied soil, intrinsic constants of e^*_{100} and e^*_{1000} can be calculated from any of the four models—Models 1 to 4. In cases where the three main physical parameters of clay mineralogy, the initial void ratio and the void ratio at liquid limit are available, Equations (19) and (21) can be used. Otherwise, if clay mineralogy is not known with any degree of accuracy, either groups of equations—Equations (13)–(15) or Equations (16)–(18)—can be used for two different conditions of accessibility to the initial state. Moreover, the void ratio at liquid limit can always be computed by knowing the specific gravity of the investigated soil and the liquid limit. Therefore, the VCL of a reconstituted clay can be computed indirectly, without performing time-consuming laboratory 1D consolidation tests, by knowing simple physical geotechnical parameters.

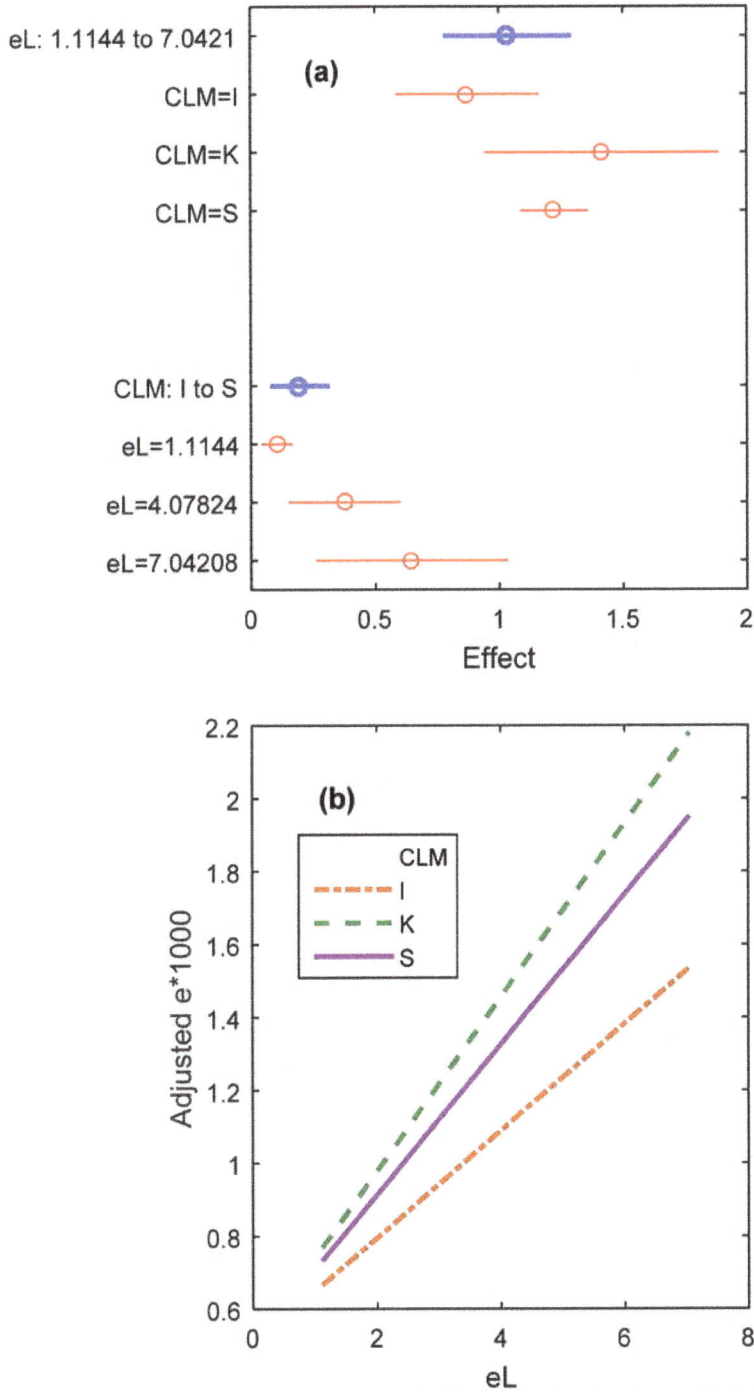

Figure 7. The effect of predic-tors on e^*_{1000} (I: illite, K: kaolo-nite, S: smectite): (a) predictor effect and (b) e_L effect.

To sum up, the steps are as follows:

- First determine the initial state, including clay mineralogy and the liquid limit of the studied soil;
- Secondly, the intrinsic parameters should be estimated by using one of the four proposed models—Models 1 to 4—based on the degree of understanding of the initial state; and
- Finally, the VCL can be determined by Equation (25).

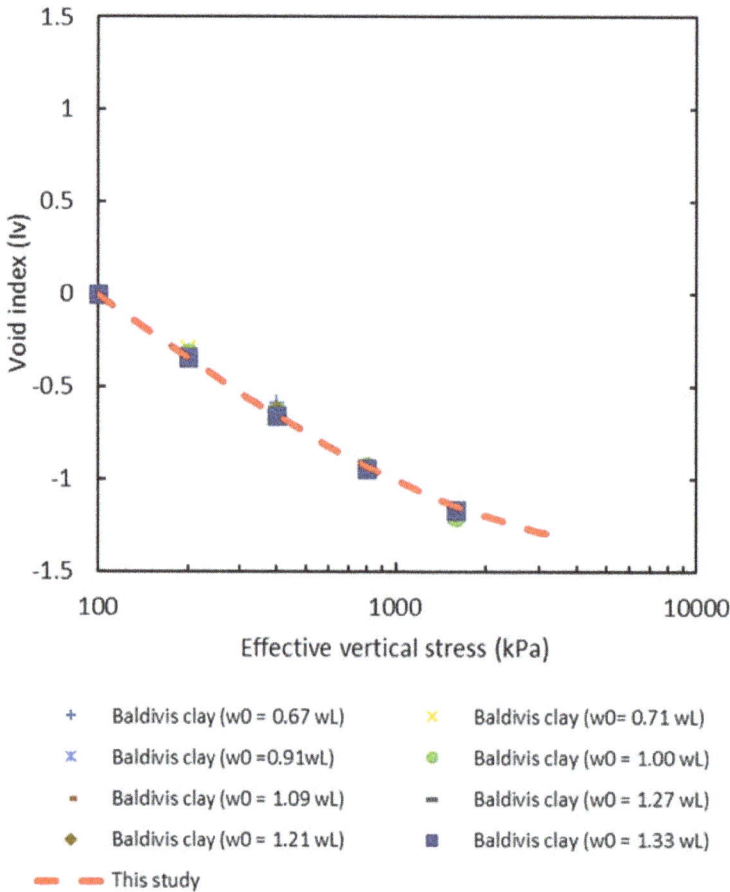

Figure 8. intrinsic compression line of Baldivis clay in I_v - log σ'_v space (reference for experimen-tal data is listed in Table 1).

The ICL calculated by this method has been plotted for Baldivis clay in Figure 8 for consolidation stress greater than 100 kPa (i.e. stress range higher than the remoulded yield stress of Baldivis clay).

Figure 8 identifies that the refined equation of the void index can satisfactorily predict the void index of the studied reconstituted clay for stresses higher than the remoulded yield stress. Predicted values of the void ratio of Baldivis clay using this method are illustrated in Figure 9 for models 3 and 4. As can be seen, there is good agreement between the measured values and predicted values for both models. The estimated void ratios are in the range of ± 10% of the observed value for the studied soil.

8. Conclusions

Based on the intrinsic concept, the compression behaviour of a reconstituted clay can be esti-mated if the intrinsic constants are known. Four regression models are proposed for estimating the intrinsic constants based on the degree of available data for the studied clay. The proposed models use the maximum advantage of the available data so that when enough information is obtainable, accuracy increases significantly. The results show that the clay mineralogy of a reconstituted clay has a considerable impact on the values of intrinsic constants. The effect of each predictor on the response of the suggested model is also investigated in this paper. The results show that, however, e^*_{100} increases with an increase in e_L for all types of clay minerals, but the increase is greatest in clays with smectite in their mineralogy. On the other hand, it seems that e_0 has a lesser effect on e^*_{100} for a clay with smectite as its principal mineral as compared to other clay minerals. The refined equation of the void index is also used to predict the compression behaviour of

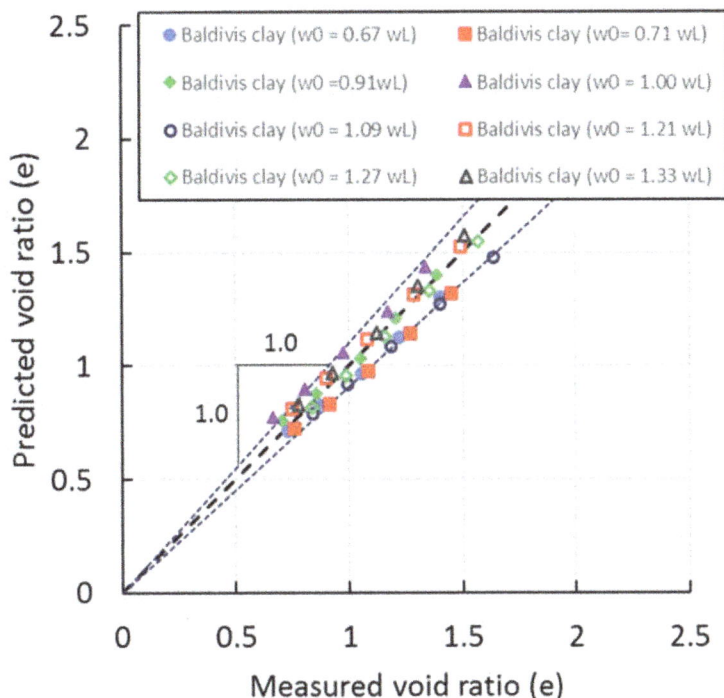

(a)

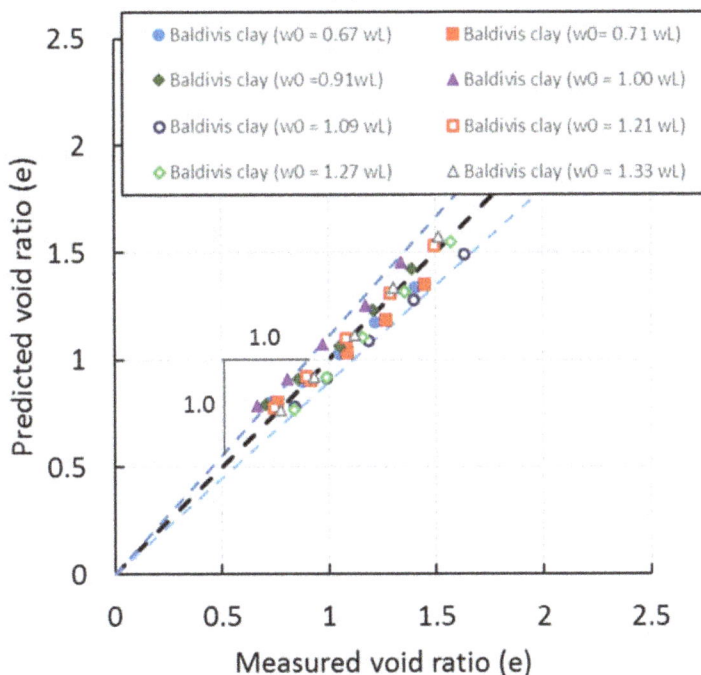

(b)

Figure 9. Comparison of esti-mated and observed void ratios for Baldivis clay (a) Model 3 and (b) Model 4.

a reconstituted clay from Baldivis, Western Australia. The results show that the predicted void ratios are in good agreement with measured values from consolidation tests for a wide range of initial water content. It is recommended that further investigation on the natural clays be carried out through a laboratory testing programme targeted to study the accuracy of the proposed models for a broader range of initial conditions and the clay mineralogy. Remote sensing can be used as a more accurate method for identifying the soil characteristics and the models can accordingly be modified based on the test results.

Acknowledgements
The first author would like to acknowledge the contribution of an Australian Government Research Training Program Scholarship in supporting this research.

Author details
Farzad Habibbeygi[1]
E-mail: farzad.habibbeygi@postgrad.curtin.edu.au
ORCID ID: http://orcid.org/0000-0002-5231-3397
Hamid Nikraz[1]
E-mail: H.Nikraz@curtin.edu.au
Bill K Koul[2]
E-mail: bkk@4dg.com.au
[1] Department of Civil Engineering, Science and Engineering, Curtin University, Perth, Australia.
[2] 4DGeotechnics Pty Ltd, West Leederville, Perth, WA, Australia.

References
Al Haj, K. M. A., & Standing, J. R. (2015). Mechanical properties of two expansive clay soils from Sudan. *Geotechnique, 65*(4), 258–273. doi:10.1680/geot.14.P.139

Burland, J. B. (1990). On the compressibility and shear strength of natural clays. *Geotechnique, 40*(3), 329–378. doi:10.1680/geot.1990.40.3.329

Butterfield, R. (1979). A natural compression law for soils (an advance on e–Log p'). *Geotechnique, 29*(4), 469–480. doi:10.1680/geot.1979.29.4.469

Cerato, A. B., & Lutenegger, A. J. (2004). Determining intrinsic compressibility of fine-grained soils. *Journal of Geotechnical and Geoenvironmental Engineering, 130*(8), 872–877. doi:10.1061/(ASCE)1090-0241-(2004)130:8(872)

Habibbeygi, F., Nikraz, H., & Chegenizadeh, A. (2017). Intrinsic compression characteristics of an expansive clay from Western Australia. *International Journal of GEOMATE, 12*(29), 140–147. doi:10.21660/2017.29.20455

Habibbeygi, F., Nikraz, H., & Verheyde, F. (2017). Determination of the compression index of reconstituted clays using intrinsic concept and normalized void ratio. *International Journal of GEOMATE, 13*(39), 54–60. doi:10.21660/2017.39.98271

Hong, Z., & Tsuchida, T. (1999). On compression characteristics of Ariake clays. *Canadian Geotechnical Journal, 36*(5), 807–814. doi:10.1139/t99-058

Hong, Z. S. (2006). Void ratio-suction behavior of remolded ariake clays. *Geotechnical Testing Journal, 30*(3), 234–239.

Hong, Z. S., Lin, C., Zeng, L. L., Cui, Y. J., & Cai, Y. Q. (2012). Compression behaviour of natural and reconstituted clays. *Geotechnique, 62*(4), 291–301. doi:10.1680/geot.10.P.046

Hong, Z. S., Yin, J., & Cui, Y. J. (2010). Compression behaviour of reconstituted soils at high initial water contents. *Geotechnique, 60*(9), 691–700. doi:10.1680/geot.09.P.059

Horpibulsuk, S., Liu, M., Zhuang, Z., & Hong, Z.-S. (2016). Complete compression curves of reconstituted clays. *International Journal of Geomechanics, 16*, 06016005. doi:10.1061/(ASCE)GM.1943-5622.0000663

Kootahi, K., & Moradi, G. (2016). Evaluation of compression index of marine fine-grained soils by means of index tests. *Marine Georesources & Geotechnology.* doi:10.1080/1064119x.2016.1213775

Lee, C., Hong, S.-J., Kim, D., & Lee, W. (2015). Assessment of compression index of Busan and Incheon clays with sedimentation state. *Marine Georesources & Geotechnology, 33*(1), 23–32. doi:10.1080/1064119x.2013.764947

Lei, H., Wang, X., Chen, L., Huang, M., & Han, J. (2015). Compression characteristics of ultra-soft clays subjected to simulated staged preloading. *KSCE Journal of Civil Engineering.* doi:10.1007/s12205-015-0343-y

Mesri, G., & Olson, R. E. (1971). Consolidation characteristics of montmorillonite. *Geotechnique, 21*(4), 341–352. doi:10.1680/geot.1971.21.4.341

Mitchell, J. K., & Soga, K. (1976). *Fundamentals of soil behavior* (Vol. 422). New York, NY: University of California. Berkeley, John Wiley & Sons Inc.

Nagaraj, T., & Miura, N. (2001). *Soft clay behaviour analysis and assessment.* Rotterdam, Netherlands: A. A. Balkema.

Nagaraj, T., & Murthy, S. (1983). Rationalization of Skempton's compressibility equation. *Geotechnique, 33*(4), 433–443. doi:10.1680/geot.1983.33.4.433

Sridharan, A., & Gurtug, Y. (2005). Compressibility characteristics of soils. *Geotechnical & Geological Engineering, 23*(5), 615–634. doi:10.1007/s10706-004-9112-2

Sridharan, A., & Nagaraj, H. (2000). Compressibility behaviour of remoulded, fine-grained soils and correlation with index properties. *Canadian Geotechnical Journal, 37*(3), 712–722. doi:10.1139/t99-128

Takashi, T. (2015). e-log σv' relationship for marine clays considering initial water content to evaluate soil structure. *Marine Georesources & Geotechnology.* doi:10.1080/1064119x.2015.1113577

Wilkinson, G., & Rogers, C. (1973). Symbolic description of factorial models for analysis of variance. *Applied Statistics, 22*(3), 392–399.

Xu, G., Gao, Y., Yin, J., Yang, R., & Ni, J. (2014). Compression behavior of dredged slurries at high water contents. *Marine Georesources & Geotechnology, 33*(2), 99–108. doi:10.1080/1064119x.2013.805287

Xu, G.-Z., & Yin, J. (2015). Compression behavior of secondary clay minerals at high initial water contents. *Marine Georesources & Geotechnology, 34*(8), 721–728. doi:10.1080/1064119x.2015.1080333

Yin, J., & Miao, Y. (2013). Intrinsic compression behavior of remolded and reconstituted clays-reappraisal. *Open Journal of Civil Engineering, 03*(03), 8–12. doi:10.4236/ojce.2013.33B002

Zeng, -L.-L., Hong, Z.-S., Cai, Y.-Q., & Han, J. (2011). Change of hydraulic conductivity during compression of undisturbed and remolded clays. *Applied Clay Science, 51*(1–2), 86–93. doi:10.1016/j.clay.2010.11.005

Zeng, -L.-L., Hong, Z.-S., & Cui, Y.-J. (2015). Determining the virgin compression lines of reconstituted clays at different initial water contents. *Canadian Geotechnical Journal, 52*(9), 1408–1415. doi:10.1139/cgj-2014-0172

Strike-slip deformation in the Inkisi Formation, Brazzaville, Republic of Congo

Timothée Miyouna[1]*, Hardy Medry Dieu-Veill Nkodia[1], Olivier Florent Essouli[1], Moussa Dabo[2], Florent Boudzoumou[1,3] and Damien Delvaux[4]

*Corresponding author: Timothée Miyouna, Marien NGOUABI University, PO Box 69, Brazzaville, Republic of Congo
E-mail: miyounatim@yahoo.fr
Reviewing editor: Robin John Armit, School of Earth, Atmosphere and Environment, Monash University, Australia

Abstract: Evidence of strike-slip deformation in the Inkisi Formation was over-looked for a long time. After controversial characterization, this paper demonstrates that the Inkisi Formation underwent at least two phases of strike-slip deformation accompanied with a compressive component, which created faults that help determine paleostress. Field observations permitted to characterize NW–SE trends sinistral strike-slip faults systems and NE–SW oriented dextral strike-slip faults systems. The strike-slip faults are associated with flower structures in profile view and with damage zones along tips, wall and linking zones in plan view. Both faults systems (sinistral and dextral) initiated from joints, which show similar orientations with faults. Many kinematic indicators of slip sense or extension have enabled to determine the stress stages and the evolution of the structures. The first tectonic phase has a horizontal maximum principal compressive stress σ_1 of 319 ± 21,1/03, which probably have a potential correlation with far-field stress propagation which occurs during the subduction of Gondwana south margin in the Permo-Trias. The second tectonic phase with a slightly inclined maximum principal compressive stress σ_1 of 264°± 22,3/12 potentially results from the intraplate stress propagation, due to the opening of the Southern Atlantic Ocean.

Subjects: Earth Sciences; Geology – Earth Sciences; Sedimentology & Stratigraphy

Keywords: strike-slip faults; joints; damage zones; plumose structures; flower structure; paleostress; Inkisi formation; Congo

1. Introduction

Characterization of strike-slip deformation has increased worldwide, as their fault zone characteristics has been proven to have a significant control on fluid flow and earthquakes initiation and termination. These studies evolved with proposition of several examples and models of fault growth and propagation (Kim et al., 2000; Martel, 1990; Olson & Pollard, 1991; Pollard & Aydin, 1988; Segall & Pollard, 1983). These models include characterization of different types of

ABOUT THE AUTHOR

Timothée Miyouna is an assistant professor at Marien NGOUABI University of Brazzaville, where he teaches mineralogy and petrology of sedimentary rocks. He extensively worked in gold mineralization in West Africa Belt. Actually, he is working in basin researches and tectonic deformations in the Paleozoic sandstones of Inkisi.

PUBLIC INTEREST STATEMENT

The study of the tectonic structures that affected the Inkisi sandstone of Paleozoic age, in the southern Brazzaville has a paramount importance. A large part of the city is built on these sandstones that carry fractures that reach the topographic surface. This constitutes a large danger to the buildings and the stability of the geotechnical structures. This study can also be a good guideline in groundwater exploration.

structures that occur (Christie-Blick & Biddle, 1985; Kim, Peacock, & Sanderson, 2004; Slyverster, 1988) both in plan view and in cross-section around strike-slip faults. Particularly, identification of damage zones structures around strike-slip faults has been subject to classification by Kim and al. (2004). This classification, proven to be reliable, has been applied to the less studied Inkisi Formation in Republic of Congo in order to identify sets of fractures that affect this Formation.

Previous studies on the Inkisi Formation have first revealed two sets of orthogonal non-described fractures (Dadet, 1969). Then these fractures were subject of two controversial considerations according to their characterization. They were first considered as faults by Cornet and Pourret (1982) without specifying their classification types, and then reconsidered as synsedimentary hydroplastic faults by Alvarez, Maurin, and Vicat (1995). The latter characterization implies that the Inkisi Formation has not undergone a tectonic event. Nonetheless, in Angola, the Inkisi Formation is overlain by Karoo deposits Formation of Permian age (Oesterlen, 1976), which are widespread in the southern Sahara Desert of Africa (Tack et al., 2001), but Karoo deposits are deformed in most part of Africa in the Congo basin and in rift segments of Tanganyika–Rukwa–Malawi (Catuneanu et al., 2005; Catuneanu, 2004; Daly, Lawrence, Dirmu-Tshiband, & Matouana, 1992; Daly, Lawrence, Kimun'a, & Binga, 1991; Delvaux, 2001a, 2001b; Delvaux, Kervyn, Macheyeki, & Temu, 2012). Thus, it necessarily implies that the underlying unit of Inkisi Formation would be deformed also.

This paper firstly presents a characterization of joints and strike-slip faults by description of different types of damage zones around faults according to the classification of Kim et al. (2004) and their profile view description. Secondly, we present the different fracture systems and determine paleostress with the Win-Tensor program, using the stress inversion method (Delvaux, 1993). At the end, we prove that strike-slip faults nucleate from joints.

2. Geological sitting

The Inkisi Formation extends from Republic of Congo to Angola via Democratic Republic of Congo (Figure 1). It outcrops in the southern part of Brazzville along the Congo river. It overlays the Congo craton, which is of Archean age, where it is separated by an angular unconformity with underlying units of West Congolian Group, a foreland part of the Panafrican West Congo Orogen (from 625Ma to 490Ma) (Dadet, 1969; Alkmim et al., 2006).

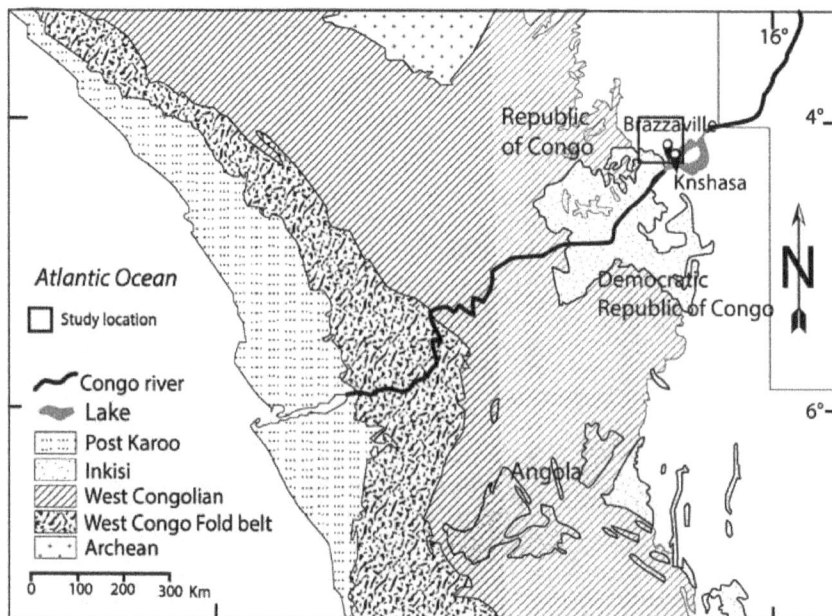

Figure 1. Location of the stu-died area (black square) and the distribution of the Inkisi Formation over the Central Africa. Redwrawn after Dadet (1969).

The Inkisi Formation is also part of the "redbeds" sequence in the Congo basin of 1000 m of thickness, where it is correlated with Banalia Group in the Lindi Basin in RDC (Figure 2) and the Banio group in Central Africa Republic (Delpomdor & Préat 2015; Kadima et al., 2011; Tack et al., 2008). These three correlated units show arkoses of same types of sedimentary structures.

The Inkisi Formation is considered as an individual lithological unit unrelated to Pan-African orogeny (Tack, Wingate, Liégeois, Fernandez-Alonso, & Deblond, 2001). The age of the Inkisi Formation is not well constrained. However, it is suggested of pre-Karoo age (320Ma) (Tack et al., 2008), because in Angola the Inkisi Formation is overlain by Karoo deposits of Permian age (Oesterlen, 1979). Affaton et al. (2016) in Congo Brazzaville suggested a deposition U-Pb age, obtained through detrital zircon between 500 and 800 Ma. However, in Democratic Republic of Congo, Frimmel et al. (2006) suggested U-Pb ages between 558 and 851 Ma. Therefore, the Inkisi Formation seems to be of earlier Paleozoic age.

Age (MA)		Units of West Congo regions		
		SuperGroup	Group	Lithology
~320				Karoo deposits
558 to 851 (Frimmel et al., 2006)	Early Paleozoic		Inkisi	Sandstones, intercalated conglomerates.
				Tectonic uncoformity (Pan-African orogeny ~635MA)
566±46 (Straathof, 2011)	Late Neoproterozoic	West-Congo SuperGroup	Mpioka	Shales, siltstones, sandstones conglomerate at base.
575 (Poidevin, 2007)			Schisto-Calcaire	Limestones, dolomites, oolites, cherts, and intercalations of shales and marns
635 (Mickala et al., 2014)			Upper Diamictite	Glacial diamictite
1113 to 901 (Dianzenza-Ndefi, 1983)	Late Neoproterozoic		Bouenzian	Cross-Bedded sandstones, clays, and conglomerates
			Lower Diamictite	Glacial diamictite
910 to 1000 (Tait et al., 2011)	Proterozoic			West Congo Orogen
2500 (Dianzenza-Ndefi, 1983)	Archean			Congo Craton

Figure 2. Lithostratigraphic synthesis of Neoproterozoic to Early Paleozoic period of geo-logical units that overlies the Congo Craton, in Republic of Congo. Compiled after Dadet (1969), Dianzenza-Ndefi (1983), Boudzoumou, (1986), Frimmel et al. (2006), Poidevin (2007), Straathof (2011), Tait et al. (2011), Mickala et al. (2014).

Alvarez et al. (1995) after a thorough facies description concluded that the Inkisi Formation corresponds to a large deltaic body, but these assertions still debatable as the pro-delta is currently missing in the sequence. However, a well-defined subdivision of Inkisi Formation which agrees with our field observations was done by Boudzoumou (1986). He suggested that the Inkisi Formation is of fluvial origin, and it is made up of repeatedly three terms: (i) coarse sandstones associated with quartzite elliptical pebbles, which sometimes show in its lower part conglomerates; (ii) coarse sandstone with trough cross-bedding; and (iii) alternating fine sandstones to very fine sandstone with horizontal laminations.

The studied area is located in south west of Brazzaville, in the Republic of Congo (Figure 1). Fractures have been recorded in the artisanal quarries of Brossete and Kombé along the Congo river.

3. Methodology
The methodology was developed from literature on features of joints and strike-slip faults. The collection of data were done on the field. Then data analyses were computed in programs.

3.1. Field data collection
The collection proceeds first by searching of geological structures. Once found, geographic coordinate of the station was taken. Then we identified faults and joints: firstly by a geometric analysis, by recording their orientation, their architecture, their connections and secondly by a kinematic analysis by looking offsets features (pebbles, etc.), secondary structures along the traces of structures and by looking for slip sense of slip indicators on the surfaces of faults. After a characterization, a sketch of these structures was produced and recorded measurements of structures (by Topochaix compass) were put in the sketch. At the end, photographs of field evidence on identified structures were taken and their number reported on the sketch.

3.2. Data analysis
Data were first recorded in a summary table of measurements with Excel program. Then we produced histograms from strikes measurements. The Win-Tensor program, version: 5.8.8 was used to produce rose diagrams, stereograms and to determine paleostress (Delvaux, 1993). The determination of paleostress from the Win-Tensor program used the stress inversion method (Angelier, 1994, 1989) which complies with the procedure described *in* Delvaux and Sperner (2003). The stress inversion method helps reconstitute four parameters of the reduced tensors: σ_1; σ_2; σ_3 (where $\sigma_1 \geq \sigma_2 \geq \sigma_3$) and the ratio $R = (\sigma_2 - \sigma_3)/(\sigma_1 - \sigma_3)$ (where $0 < R < 1$), which determine the intensity of σ_2 relative to σ_1; and σ_3 . The program determines the index regime called R' based on the ratio R and on a vertical continuous scale from 0 (radial extension) to 3 (constriction), with $R' = R$, for an extension regime (0 to 1); $R' = 2 - R$ for a strike-slip regime (1 to 2); $R' = 2 + R$ for a thrust regime (2 to 3). The reduced tensor is first estimated by Right dihedral method. They are more precisely determined with an iterative rotational stress optimization. The program assesses the quality of the obtained tensors from two qualities ranking parameter QR. The first, the QRw, which is a quality rank defined as in World Stress Map project, it ranges from A (best); B (good), C (medium), D (poor) to E (worst). It is determined in function of threshold values of series of criteria (Delvaux et Sperner, 2003). The second, the QRt, is a quality rank of diversity of faults orientation and the found slip lineation. The program uses several criteria to evaluate the ranking quality, as well described in Delvaux et Sperner (2003).

Data were input manually in the program, as characterized in the field, as fault with slip line, shear plane with tension facture, shear fractures and joints (see Table A1). Additionally, we carry out a grouping into primary subsets, according to field criteria (cross-cutting principle, relative age, reactivation, neoformation, etc.). Furthermore, after processing, the program suggests an optional data separation into secondary subsets of homogeneous orientation of initially chosen subsets.

4. Field data results and faults patterns

4.1. Strike-slip faults

Strike-slip faults form two major sets of fractures that cross cut each other at varying angle (from acute to almost orthogonal). The first set is made up of sinistral strike-slip faults and has strike orientation that ranges from 335° to 019° (Figure 3b, c). The second set is made up of dextral strike-slip faults and has strike orientation ranging from 020° to 065° (Figure 4b, c). Both sets show a steep dip, ranging from 70° to 90° and their major line plunge at to 5° (Figure 3a, d; 4a, d). The polished surfaces of faults are generally filled with palygorskite and calcite (Figure 5c), accretion steps of calcites (Figure 5a) and show within some surfaces crystal fiber lineation (Figure 5b).

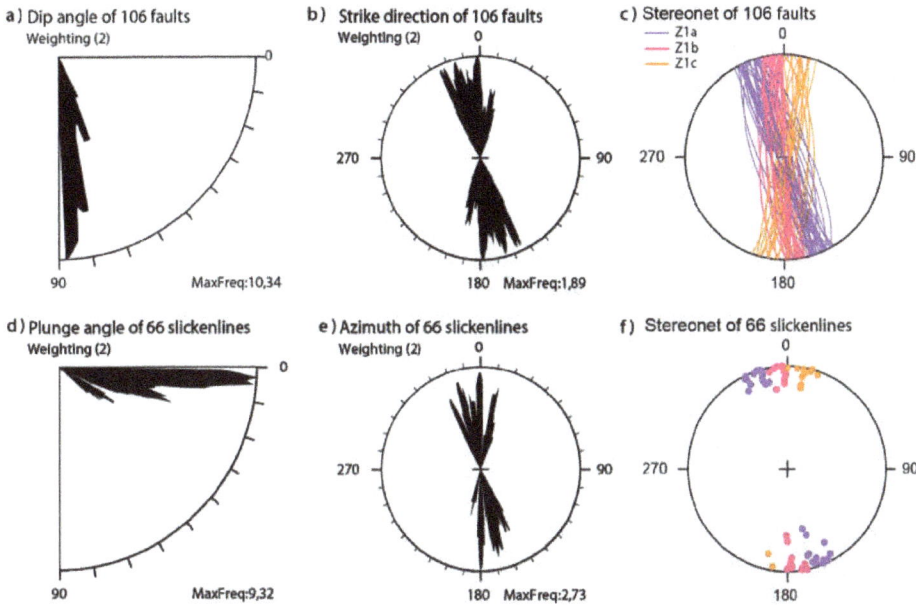

Figure 3. Rose diagrams and stereograms of sinistral strike-slip faults. Three classes of orientations Z1a, Z1b, and Z1c were distinguished.

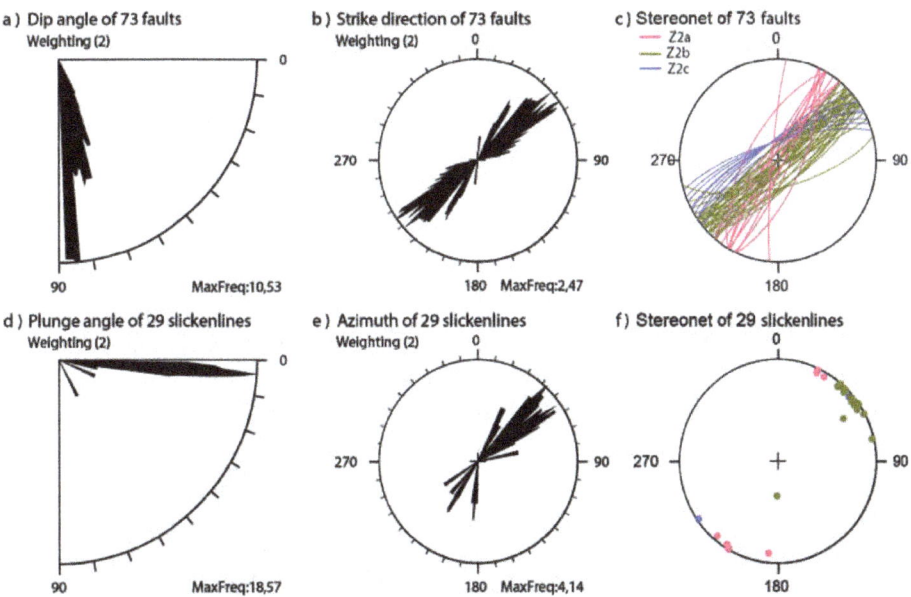

Figure 4. Rose diagrams and stereograms of dextral strike-slip faults. Three classes of orientations Z2a, Z2b, and Z2c were distinguished.

Through their traces, they showed many kinematic indicators that help inferred the sense of movement, as displaced pebbles (Figure 6), pull-apart structures, horstailing splays, etc.

Both strike-slip faults show a progression from single small faults with little displacement (7 mm) to faults zones of moderate displacement (up to 12 cm) and length (up to 400 m). In horizontal plan view, both sets of strike-slip faults (dextral and sinistral) show subparallel faults which relayed each other, associated with bends and characteristic feature of fault damage zones around their tips, linking zones and wedge zones. In cross-section view, both sets show negative and positive flower structures; however, sinistral strike-slip faults showed frequently positive flower structures, while dextral strike-slip faults showed negative structures flowers numerously. Fault damage structures are well expressed by sinistral strike-slip faults, which form compound strike-slip fault zones (Figure 7), as described by Martel (1990). Compound strike-slip faults are presumed to accommodate larger displacement than other faults (Martel, 1990). Figure 8 shows the relationship of two set fractures. The sinistral strike-slip faults are

Figure 5. Strike-slip fault sur-faces and kinematic indicators.(A) Slickensides highlighted by accretion steps of calcite, or (B) crystal fiber stretching. (C) A left-step sinistral strike-slip with palygorskite and overprint slickensides in its surface show details in (D).

Figure 6. Displaced pebbles within strike-slip faults. (A) Sinistral lateral strike-slip faults with 2 cm of displace-ment. (B) Dextral strike-slip faults with a small vertical component.

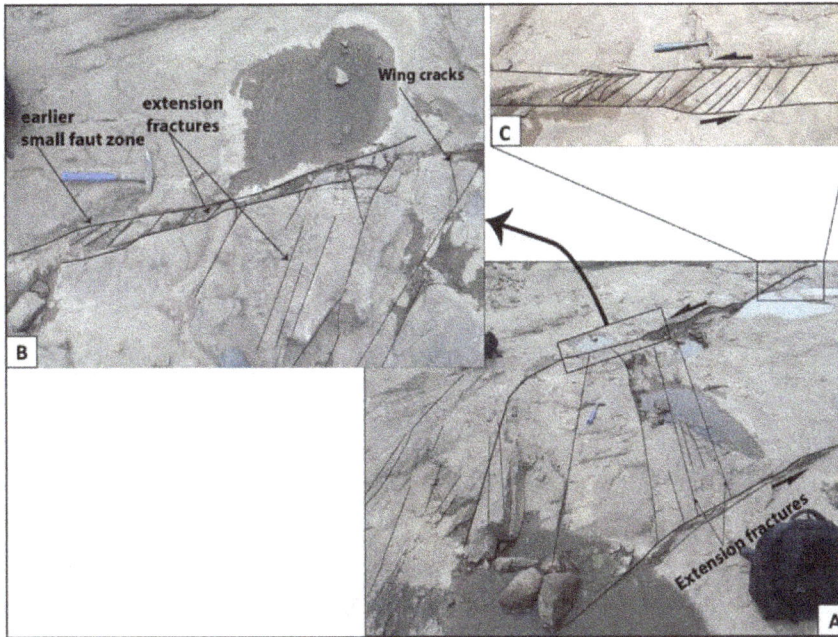

Figure 7. Compound fault zone associated with small fault zone. (A) Compound fault zone with extension fractures. (B) Small fault zone within bound-aries of compound fault zone and extension fractures show-ing wing crack terminations that overprint small fault zone.(C) Another small fault zone with extensions fractures at the tip of the compound fault zone.

crossed-cut and displaced by the dextral strike-slip faults. More than five zones of this type were found with a displacement ranging between 2cm and 5cm.

4.1.1. Tip zones damage structures

4.1.1.1. Horsetail splays and en-echelon fractures. Faults display in some places horsetail fractures at their tips (Figure 9). The angles between the main faults and the horsetail fractures range from 20° to 45°. The local orientation of the maximum compressive stress has been assumed to be parallel to horsetail fractures (320° for sinistral fault). Some horsetail fractures show a normal slip component within their traces, a change in their orientation, and wing cracks development at their intersection with the main fault. Wing cracks development within horsetail fractures tips imply that they formed from mode I loading.

En-echelon fractures (Figure 9) can be associated with extensions fractures. They are assumed to propagate away from the fault terminations and they show steep dip.

4.1.2. Linking damage zones

They are dominated by extension fractures, dilation jogs and pull-apart structures.

4.1.2.1. Extension fractures. They dominantly link faults (Figure 10) and are assumed to develop parallel to the maximum compressive stress σ_1. The angles between main faults and extensions fractures range from 40 to 55°. Some of them of wing crack at their intersection with the principal fault. A principal fault might be associated both with extensions fractures and horsetail fractures. The identification of extensions fractures is supported by the presence of wing cracks at the intersection with main fault and additionally, by range of their intersection angle (40° to 55°) with main faults, as described by Hancock (1985). They constitute the type of structures most frequently observed in the field and help greatly to determine paleostress.

Figure 8. Cross-cutting rela-tionship between sinistral strike-slip fault and dextral strike-slip fault. The sinistral strike-slip fault is displaced by dextral strike-slip fault over 2 cm of offset. (A) Cross-cutting relationship. (B) Sketch of cross-cutting relationship of sinistral fault (Z1) and dextral fault (Z2) in A. (C) and (D) Corrugation (b) and crushed rock (a) within the surface of Z2.

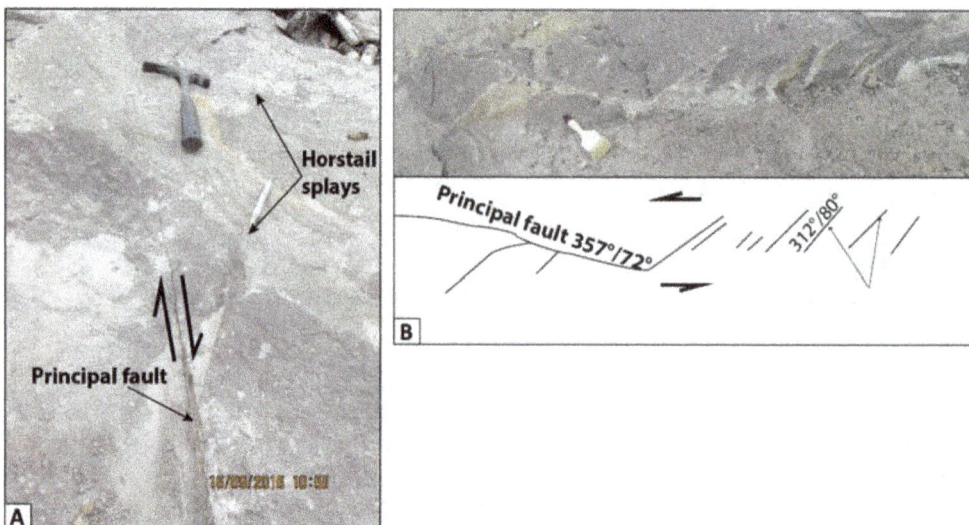

Figure 9. (A): Horsetail splays at 190/85 oriented dextral strike-slip fault tip. (B) En-echelon fractures at sinistral strike-slip fault tip.

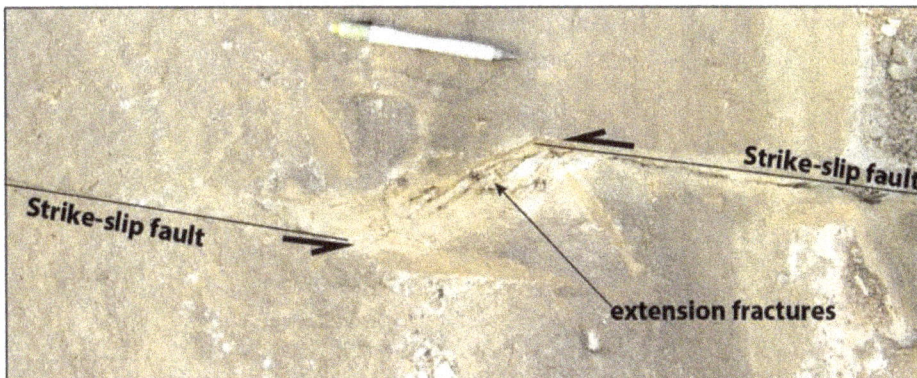

Figure 10. Extension fractures around linking zone of a sinis-tral strike-slip fault, observed in horizontal plan view.

4.1.2.2. Dilation jogs. They appear most of time within the fault traces (Figure 11). They often show lens shape with crushed material and duplex.

4.1.2.3. Pull-apart. Pull-apart structures (Figure 12) are extension fractures that opened up between two linked faults due to increasing slip in fault segment (Kim et al., 2004). They show a normal component in their interior, suggesting an opening. The geometric shape is controlled by boundary fractures.

4.1.3. Wall damage zone

Wall damage zone as linking damage has sometimes quite similar architecture. We find extension fractures intersecting two faults traces as in linking damage zone, but here no relay zones, only two bounding strike-slip faults. Extension fractures show also wing crack at their intersection with the main fault (Figure 13).

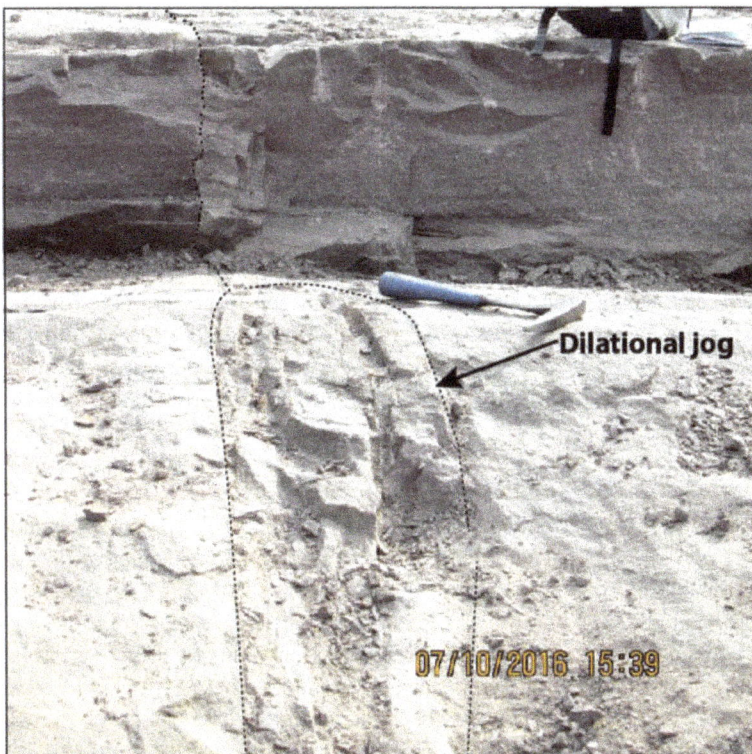

Figure 11. Dilation jog around strike-slip fault.

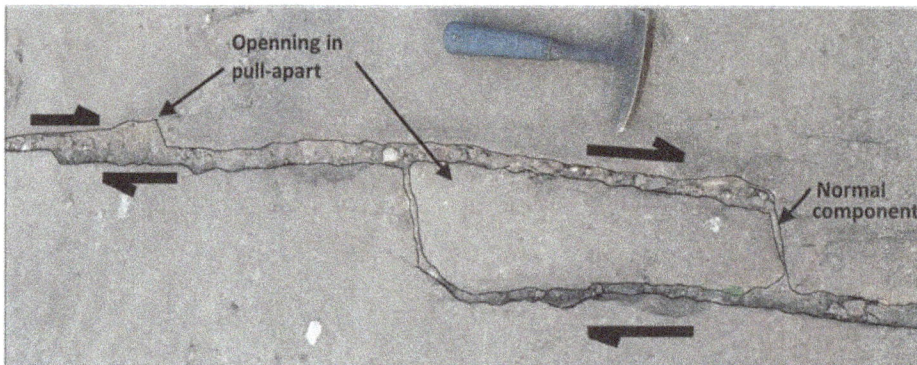

Figure 12. Pull-apart structure associated with a dextral strike-slip fault.

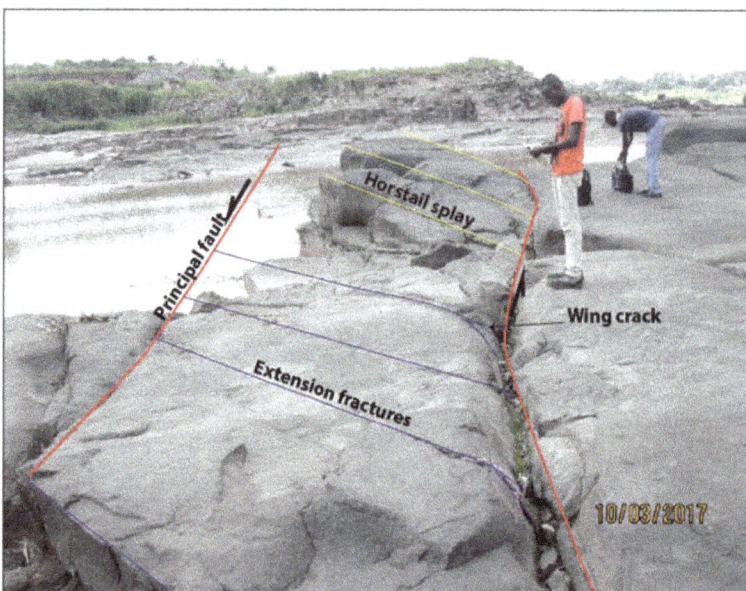

Figure 13. Strike-slip fault showing both a wall damage zone with extension fractures and horsetail splay fractures at its tips. Note wing cracks on extension fractures.

4.1.4. Flower structures

They have been first noticed as lens shaped in plan view, and then observed in profile view at least two or three plans of fractures that bifurcate from one steep fault. Bifurcated planes delimit a space which frequently comprises lots of fractures. Most of the fractures within the space show characteristics of negative flower structures (Figure 14) with normal-slip component. Flower structures have same orientation relation to strikes of both strike-slip faults systems.

4.2. Joints

Joints show parallel orientation with strike-slip faults (Figure 15b, c). They are predominantly steeply dipping (Figure 15a). Therefore, they were subdivided into two sets: one parallel to sinistral strike-slip faults (Figure 15d, e) and the other parallel to dextral strike-slip (Figure 15f). Figures 15 and 16 show rose diagrams and stereograms that support this evidence. At least 16% of joints were found with well-marked plumose structures (Figures 17, 18). Most plumose structures display straight plume axis, and seemed to originate from pebbles or micro-cracks in the rock. Some plumose structures show fine hackles to more noticeable ones within the joint planes. Apart from straight plumes, plumes with many arrest lines have been found (Figure 18a). Some plumose structures have been found with slickensides overprinting them (Figure. 18b, c) as it was described

Figure 14. Flower structures associated with a dextral strike-slip faults.

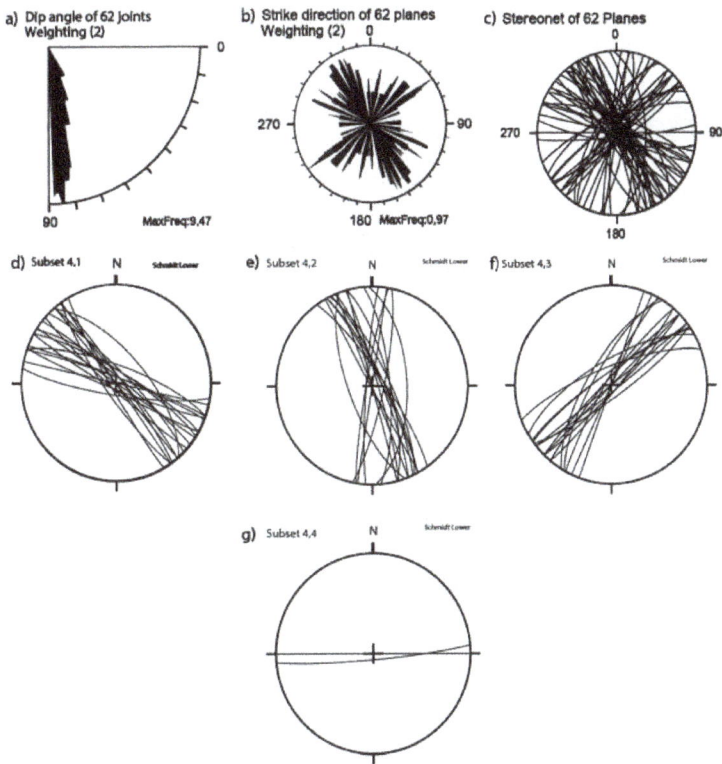

Figure 15. Rose diagrams and stereograms of joints without plumose structures (a,b,c). Distinction of five subsets (a,b, c,d,e,f).

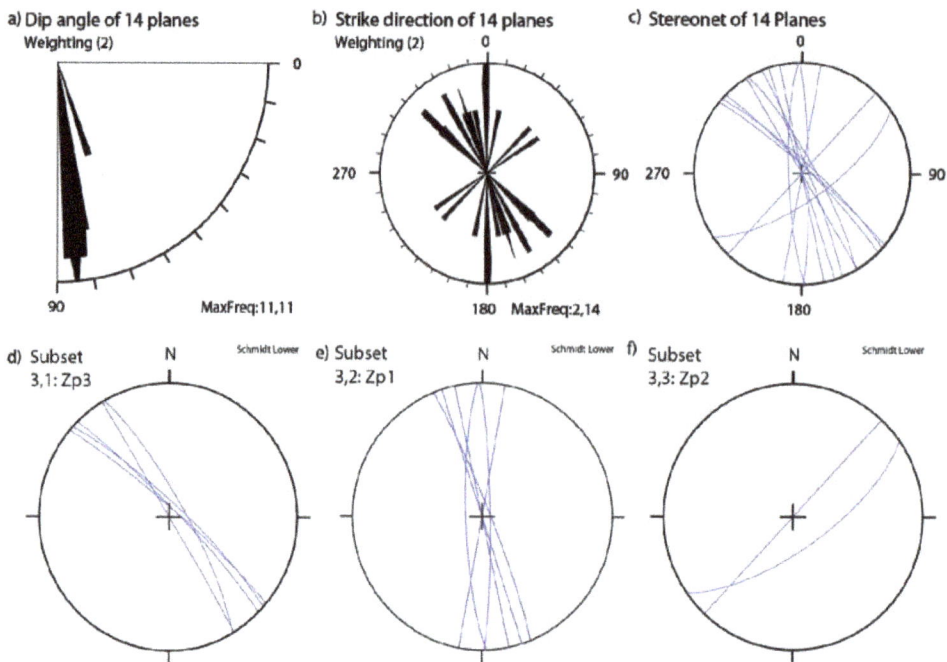

Figure 16. Rose diagrams and stereograms of joints with plumes structures (a,b,c). Distinction of different subsets (d,e,f).

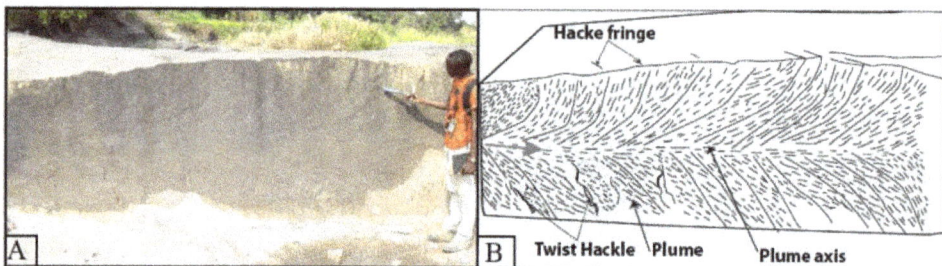

Figure 17. Plumose structures with weakly marked hackle. (B) Sketch of the photo, with the grey arrow showing the propa-gation direction.

by Barton (1983) and Segall and Pollard (1983). This evidence indicates at least two fault phases of deformation (Pollard & Aydin, 1988). Plumose joints with many arrest lines have also through their surface decimetric veins of calcite.

5. Results of stress inversion

Measurements were grouped into four (04) subsets: (1) subset of sinistral strike-slip (Z1); (2) subset of dextral strike-slip (Z2); (3) subset of parallel joints to Z1; and (4) subset of parallel joints to Z2. From the four subsets, four stages of stress have been found. Chronologically these stages have to be grouped into two tectonic stress stages, because field observations suggested the evolution of joints into faults:

- Stress stage 1 (Figure 19): This is the oldest, represented by 182 fractures (106 faults and 76 joints), with 66 striated planes of faults. It started with an extension regime ($R' = 0{,}5$) (Figure 19b) with NE–SW principal direction of extension, forming the majority of joints with plumes, the quality rank QRw and QRt of this stage are medium. The change in orientation of joints is probably due to rock defaults (Davis, Reynolds & Kluth, 2011). The passage from extension regime to strike-slip regime (Figure 19a) was achieved through a rotation of the maximum

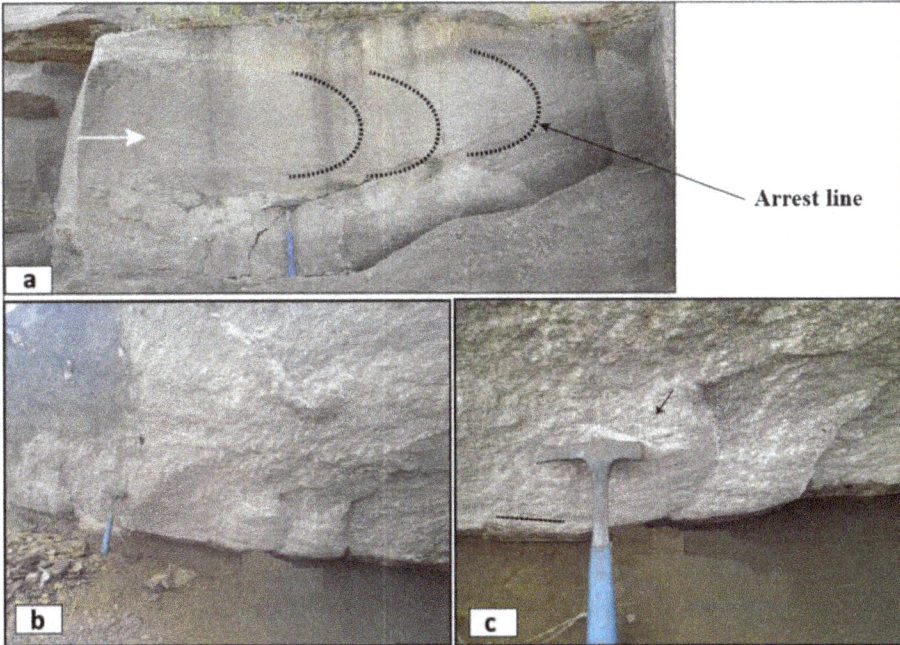

Figure 18. (a) Plumose structures with many arrest lines. (b and c) Well-marked twist hackles with overprinted slickensides.

compressive stress. The strike-slip regime, slightly compressional (R' = 1,54) has a maximum compressive stress oriented NW-SW (319 ± 21,1/03). The quality rank QRw is medium and the QRt is poor, therefore the solution is not perfectly constrained, but the orientation of the maximum compressive stress fits the range of the solution of tσ1 inferred on the field from

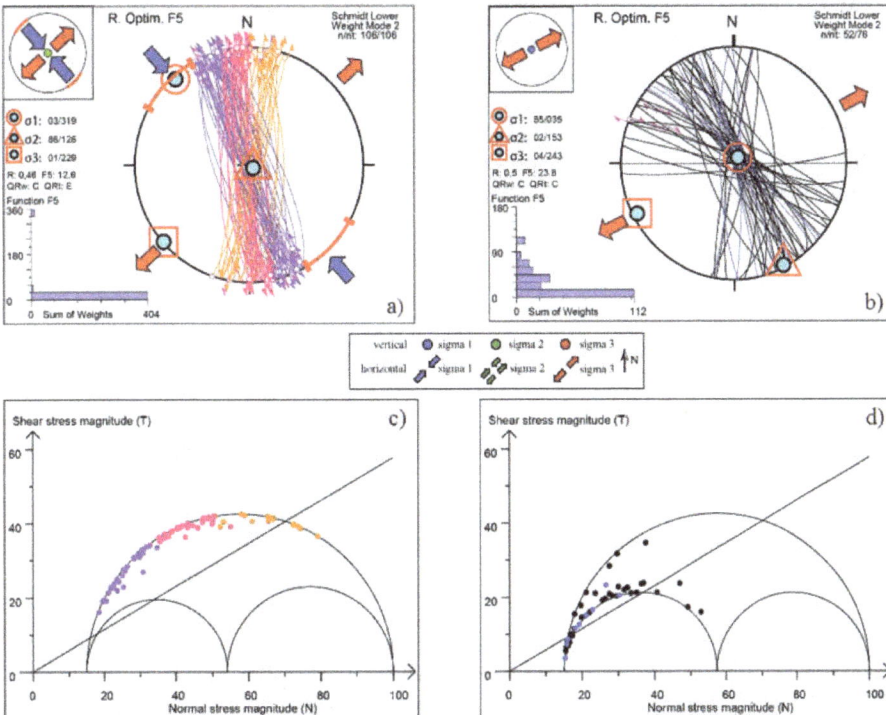

Figure 19. Fault slip data and stress inversion results. Lower-hemisphere Schmidt stereo plot of the fault-slip data subsets and corresponding stress ten-sor. (a) Subset 1 data (Z1) results. (b) Subset 3 (joints) data results. (c) Mohr diagram plots of subset 1 data. (d) Mohr diagram plots of subset 3 data.

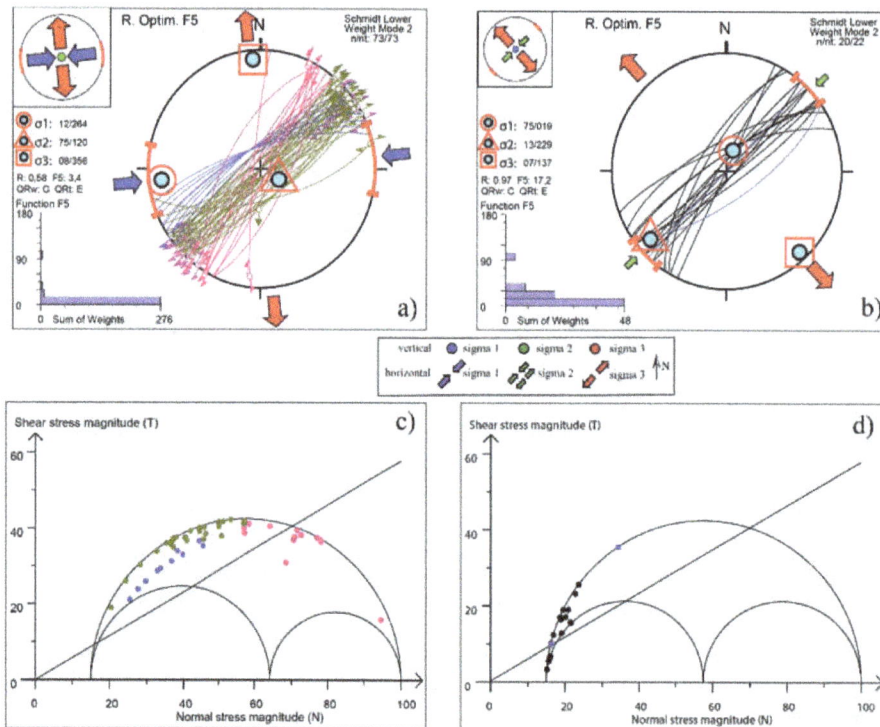

Figure 20. Fault slip data and stress inversion results. Lower-hemisphere Schmidt stereo plot of the fault-slip data subsets and corresponding stress ten-sor. (a) Subset 2 data (Z2) results. (b) Subset 4 (joints) data results. (c) Subset 2 data plot in Mohr diagram. (d) Subset 4 data in Mohr diagram.

extension fractures or horsetails fractures. The Mohr diagram show that some NW–SE to N–S faults have experienced an intense shear stress while others NNE-SSW faults have experienced an intense normal stress relative to the shear stress.

- Stress stage 2 (Figure 20): This last stage is defined by 95 fractures (73 faults and 22 joints), with 29 striated planes of faults. It started with an extension regime (Figure 20b) to slightly strike-slip ($R' = 1,42$) with a principal direction of extension oriented NW-SW of average orientation. Then, it evolved into a strike-slip regime to slightly compressional with a principal stress oriented W-E (264 ± 22,3/12) (Figure 20a). The Mohr diagram of both faults and joints confirms that evolution, of strike slip from joint.

6. Discussion

Our field observations show detail strike-slip faults and joints structures that prove that the Inkisi Formation underwent a strike-slip regime. Strike-slip faults show a large range of minor structures across their traces and their profiles, which go from horsetail fractures, extensional fractures in faults linking zones, dilation jogs, wing cracks, en-echelon fractures at their tips and flowers structures (Figure 21).

Rose diagrams and stereograms reveal that faults and joints are organized into two systems of fractures: the first one shows NW–SE to NNE–SSW trends and comprises sinistral faults and joints, the second one shows NE–SW trends and comprises dextral faults and joints too.

Observations indicate a close link between development of joints and strike-slip faults. The presence of plume structures with many arrest lines and calcite veins within their planes suggests that pore fluid pressure probably plays a role in the formation of these fractures (Bons, Elburg, & Gomez-Rivas, 2012; Secor, 1965; Tullis and Tullis J., 1986). Additionally, joints have parallel orientation with faults. On the other hand, some plume joints have overprinted slickensides on their surfaces. Moreover, faults show crystal fiber lineation (striation) and also accretion steps of calcite.

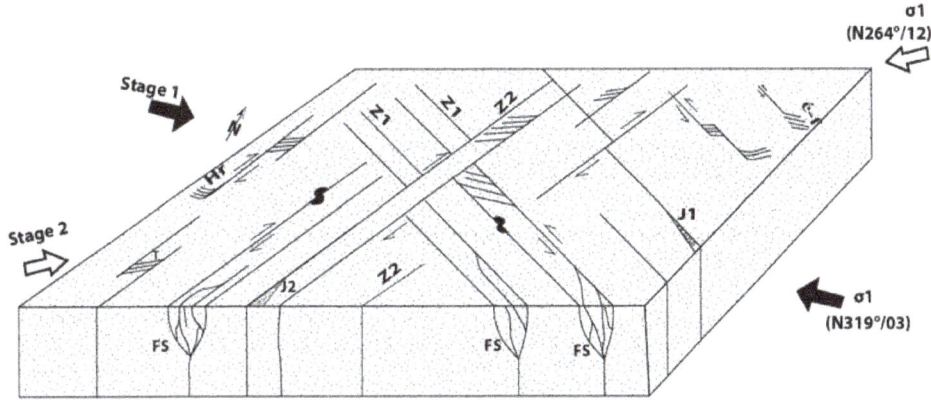

Figure 21. Schematic diagram which idealizes architecture of structures in Inkisi Formation. J1: joints parallel to Z1; J2: joints parallel to Z2; FS: Flower structures; Hr: horsetail splays; e-f: en-echelon fracture.

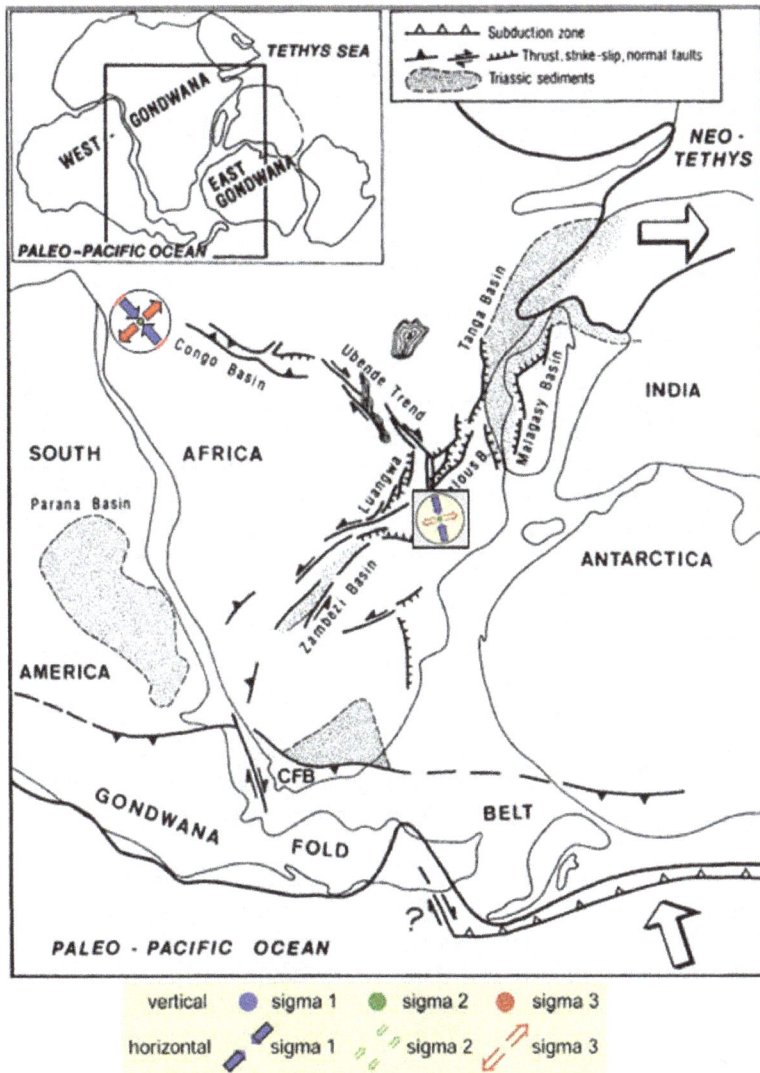

Figure 22. Paleotectonic map of the Permo-triassic transition in African part of Gondwna. The reduce tensor find in Brazzaville are in the same range of orientation as that find in the TRM. Modified by from Delvaux (2001b).

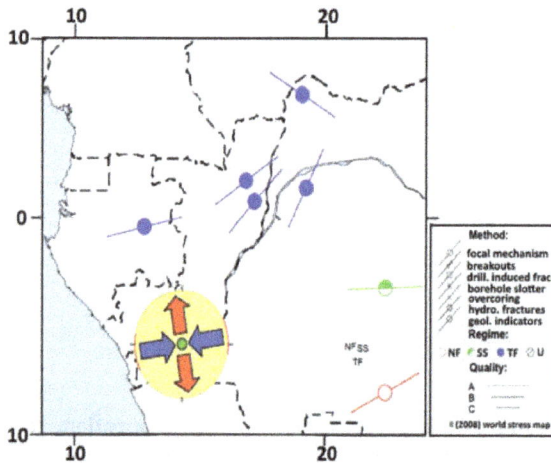

Figure 23. World stress map (2008) and second stress stage reported in the study zone. (Map from www.world-stress-map.org).

This clearly indicates that strike-slip faults nucleated from joints. The same findings have been made by Segall and Pollard (1983), Kim et al. (2004), and Barton (1983). Thus, the two stress stages found from stress inversions method support this field evidence. Both stages of tectonic strike-slip grew from pre-existing joints and some strike-slip faults form small faults with little displacement and others evolved into fault zones and compound fault zones as described in Martel (1990).

Strike-slip faults show clear examples of damage zones structures through their traces as described by Kim et al. (2004) in the classification of fault damage zones. The assumed compressive principal stress on field from horsetail fractures, en-echelon fractures, and extension fractures of N320° supports our determined maximum principal compressive paleostress.

The early stress stage is broadly constrained, because few conjugates faults planes of that stage have been observed on the field. However, these results place a range of limits for the reduced tensor solution with a maximum compressive stress of 319°± 21,1/03. Although uncertainties are quite large (±21,1°), the orientation of the maximum compressive stress obtained on the field from extension fractures gives an orientation that fits in the interval determine from Win-Tensor. This stress stage has a potential correlation with stresses that propagate from the southern subduction margin of Gondwana at the Permo-Trias which formed the Cap Fold Belt, currently preserved in South Africa. This affirmation is supported by two principal arguments. The first point is that the Inkisi Formation is overlain by Karoo deposits in Angola (Tack et al., 2001), and the Karoo deposits have been deformed in most part of Africa by the subduction that affected the southern margin of Gondwana at Permo-Trias (Catuneanu et al., 2005). Therefore, the Inkisi Formation might have been affected by this tectonic event. More importantly, the reconstruction of paleostress by Delvaux et al. (2012), in the Tanganyika-Rukwa-Malawi (TRM) rift segment in western Tanzania, located the Karoo deposits, gives three stress stages. The second stage which is well constrained, obtained from Karoo deposits of Permian age, called Namwele-Mkolomo coal field, has caused a dextral strike-slip tectonic associated with a transpression that has a subparallel orientation with our first stage (Figure 22). Nonetheless, the stress obtained in TRM is oriented NNW–SSE and cause dextral kinematic, while our orientation is NW–SE and causes sinistral kinematic. This change in kinematic sense might probably be due to a rotation of stress counterclockwise during the collision, as the stress propagated over 2,000 km in the crust (Daly et al., 1992, 1991). The second point is that in Democratic Republic of Congo, in the Congo basin stratigraphic unit, the Banalia Group are correlated to Inkisi Formation (Delpomdor & Préat, 2015; Kadima et al., 2011) and they underlay the Karoo deposits that are also affected by the subduction in the southern margin of Gondwana at Permo-Trias (Daly et al., 1992). Therefore, the underlaying units of Karoo deposits in the Congo basin must be deformed also. Thus, it proves that the Inkisi Group has been affected by the subduction at Permo-Trias at the southern margin of Gondwana.

The latest stress stage with a sub-horizontal principal compressive stress (264°± 22,3/12) is not as the early stage well constrained for same reasons, but the maximum compressive stress recorded in extension fractures fits in the same range of that calculated. This stress stage might probably result from ridge push of South Atlantic mid-ocean ridge in continental part. The world map of stress data supports this evidence with the presence of a horizontal compressive stress parallel in orientation with what we determine (Figure 23). This explanation has also been given for Northern Europe and Scandinavia, where the horizontal compressive stress is oriented NW–SE

(Fossen, 2010). This result suggests that the latest stress stage is still active but must be studied through different methods to confirm its activity.

7. Conclusion

This paper shows and proves that the Inkisi Formation has been affected by at least two strike-slip tectonic events associated both with a compression, initiated from joints. The early event produces parallel sinistral strike-slip faults and joints and the latter event produces parallel dextral strike-slip faults and joints. Both tectonic events show similar features within their structures (horstail splays, extension fractures, dilation jogs, en-echelon fractures, and flower structures), which form the basis of their identification according to Kim et al. (2004) classification of fault damage zones.

The early tectonic deformation with a horizontal principal compressive stress of 319°± 21,1/03 (NW–SE) probably results from the subduction at Gondwana southern margin in Permo-Trias. However, the latest tectonic event with a sub-horizontal principal compressive stress (264° ± 22,3/12) (W–E) is potentially generated by ridge push of South Atlantic mid-ocean ridge.

Acknowledgements
We would like specially to thank Joachim Miyouna for English corrections of the manuscript. Instructive comments of anonymous reviewers substantially improve the quality of the manuscript and were greatly appreciated. We are also grateful to Pr. El Hadji Sow from Cheikh Anta Diop University of Dakar for his advices and encouragements.

Funding
The author received no direct funding for this research.

Author details
Timothée Miyouna[1]
E-mail: miyounatim@yahoo.fr
Hardy Medry Dieu-Veill Nkodia[1]
E-mail: nkodiahardy@gmail.com
Olivier Florent Essouli[1]
E-mail: Oflorentessouli@gmail.com
Moussa Dabo[2]
E-mail: moussadabo@hotmail.com
Florent Boudzoumou[1,3]
E-mail: boudzoumouflorent@yahoo.fr
Damien Delvaux[4]
E-mail: damien.delvaux@africamuseum.be
[1] Marien NGOUABI University, Brazzaville, Republic of Congo.
[2] Cheikh Anta DIOP University, Dakar, Republic of Senegal.
[3] Geosciences Department, National Research Institute in Exact and Natural Sciences (IRSEN) of Brazzaville, Brazzaville, Republic of Congo.
[4] Africa Museum,Leuvensesteenweg 13, Tervuren, B-3080, Belgique.

Description
This study of deformations in the Inkisi sandstone Formation shows that it is affected by two major directions of NW–SE and WSW–ENE strike-slip faults. The NW SE faults have displayed sinistral movement, whereas the WSW–ENE faults that offset the previous one have dextral movement. NW–SE fractures may result from the subduction of Gondwana south margin in the Permo-Trias. The second results from the interplate stress propagation due to the opening of the southern Atlantic ocean.

References
Affaton, P, Kalsbeek, F, Boudzoumou, F, Trompette, R, Thrane, K, & Frei, R. (2016). The Pan-African West Congo belt in the Republic of Congo (Congo Brazzaville): Stratigraphy of the Mayombe and West Congo Supergroups studied by detrital zircon geochronology. *Precambrian Research, 272*, 185-202.
Alkmim, F.F, Marshak, S, Pedrosa-Soares, A.C, Peres, G.G, Cruz, S, & Whittington, A. (2006). Kinematic evolution of the Araçuaí–West Congo orogen in Brazil and Africa: nutcracker tectonics during the Neoproterozoic assembly of Gondwana. *Precambrian Research, 149*, 43–64.
Alvarez, P., Maurin, J.-C., & Vicat, J.-P. (1995). La formation de l'Inkisi (Supergroupe Ouest-Congolien) en Afrique Centrale (Congo et Bas-Zaïre): Un delta d'âge Paléozoïque comblant un bassin en extension. *Journal of African Earth Sciences, 20*(2), 119–131. doi:10.1016/0899-5362(95)00038-U
Angelier, J. (1989). From orientation to magnitudes in paleostress determinations using fault slip data. *Journal of Structural Geology, 11*(1/2), 37–50. doi:10.1016/0191-8141(89)90034-5
Angelier, J. (1994). Fault slip data an paleostress reconstruction. In I. P. Hancock (Ed.), *Continental deformation* (pp. 53–100). Oxford: Pergamon Press.

Barton, C. C. (1983). Systhematic jointing in the Cardium Sandstone along the Bow River, PhD thesis, Alberta, Canada: New Haven, Connecticut, *Yale University*, 301p.

Bons, D. P., Elburg, A. M., & Gomez-Rivas, E. (2012). A review of the formation of tectonic veins and their microstructures. *Journal of Structural Geology, 43*, 33–62. doi:10.1016/j.jsg.2012.07.005

Boudzoumou, F. (1986). La chaîne Ouest-Congolienne et son avant-pays au Congo : relations avec le Mayombien ; sedimentation des sequences d'âge Protérozoïque supérieur. Thèse de doctorat, *Université Aix-Marseille*, France. 220p.

Catuneanu, O. (2004). Retroarc foreland systems evolution through time. *Journal of African Earth Sciences, 38*, 225–242. doi:10.1016/j.jafrearsci.2004.01.004

Catuneanu, O., Wopfener, H., Eriksson, P. G., Cairncross, B., Rubidge, B. S., Smith, R. M., & Hancox, P. J. (2005). The Karoo basins of South-central Africa. *Journal of Earth Sciences, 43*, 211–253.

Christie-Blick, N., & Biddle, T. K. (1985). Deformation and basin formation along strike-slip faults. *Special Publication Society of Paleontologists and Mineralogists*, N°37, 1–34.

Cornet, P., & Pourret, G. (1982). Contrôle structural du Congo dans les formations de l'Inkisi en aval de Brazzaville. *Annales Université Brazzaville, 1976-1997* (12–13), 17–28.

Dadet, P. (1969). *Notice explicative de la carte géologique de la République du Congo au 1:500 000* (pp. 103). Orléans: Mémoire BRGM, n°70.

Daly, M. C., Lawrence, R. S., Dirmu-Tshiband, K., & Matouana, B. (1992). Tectonic evolution of the Cuvette Centrale, Zaïre. *Journal of Geological Society, London, 149*, 539–546. doi:10.1144/gsjgs.149.4.0539

Daly, M. C., Lawrence, S. R., Kimun'a, D., & Binga, M. (1991). Late Paleozoic deformation in central Africa; a result of distant collision? *Journal Southern American Earth Sciences, 6*, 33–47.

Davis, G. H, Reynolds, S. J, & Kluth, C. F. (2011). *Structural geology of rocks and regions*. John Wiley & Sons.

Delpomdor, F., & Préat, A. (2015). Overview of the Neoproterozoic sedimentary series exposed along margins of the Congo Basin. *In* M. J. de Wit, F. Guillocheau, & M. C. J. de Wit (eds.), Geology and resource potential of the Congo basin (p. 41–58). Berlin, Heidelberg: Springer Berlin Heidelberg.

Delvaux, D. (1993). *The TENSOR program for paleostress reconstruction: Examples from the east African and the Baikal rift zones* (Vol. 5, pp. 216). France: Abstract supplement N°1 to Terra Nova.

Delvaux, D. (2001a). Tectonic an paleostress evolution of the Tanganyika-Rukwa-Malawi rift segment, East Africain Rift System. In D. P. A. Ziegler, W. Cavazza, A. H. Roberston, & Crasquin-Soleau (Eds.), *Peri-Tethys Memoir 6: Peri Tethyan Rift/Wrench Basins and Passive Margins* (pp. 545–567). Paris 186: Mém. Mus. National Hist. Nat.

Delvaux, D. (2001b). Karoo rifting in western Tanziania: Precursor of Gondwana breakup. In *Contributions to Geology and Paleontology of Gondwana* (pp. 111–125). Cologne: In honour of Prof. Dr. Helmut Wopfner.

Delvaux, D., & Sperner, B. (2003). Stress tensor inversion from fault kinematic indicators and focal mechanism data: The TENSOR program. In D. Nieuwland (Ed.), *New Insights into Structural Interpretation and Modelling* (Vol. 212, pp. 75–100). Geological Society, London, Special Publications.

Delvaux, D., Kervyn, K., Macheyeki, S. A., & Temu, E. B. (2012). Geodynamic significance of the TRM segment in the East African Rift (W-Tanzania): Active tectonics and paleostress in the Ufipa plateau and Rukwa

basin. *Journal of Structural Geology, 37*, 161–180. doi:10.1016/j.jsg.2012.01.008

Dianzenza-Ndefi, H. (1983). *Les sédiments du Protérozoique supérieur et leurs frmations au Nord-Ouest de la Cuvette Congolaise (Afrique Centrale)* (Apport des datations par les méthodes Rb-Sr et K-Ar. Louis Pasteur). Strasbourg, France.

Fossen, H. (2010). *Structural Geology* (pp. 163). New York: Cambridge University Press.

Frimel, H. E., Tack, L., Basei, M. S., Nutman, A. P., & Boven, A. (2006). Provenance and chemostratigraphy of the Neoproterozoïc West congolian Group in the Democratic Republic of Congo. *Journal of African Earth Sciences, 46*(2006), 221–239. doi:10.1016/j.jafrearsci.2006.04.010

Hancock, P. L. (1985). Brittle microtectonics: Principles and practice. *Journal of Structural Geology, 7*, 437–457. doi:10.1016/0191-8141(85)90048-3

Kadima, E, Delvaux, D, Sebagenzi, S.N, Tack, L, & Kabeya, S.M. (2011). Structure and geological history of the Congo Basin: an integrated interpretation of gravity, magnetic and reflection seismic data. *Basin Research, 23*, 499-527.

Kim, Y.-S, Andrews, J.R, & Sanderson, D.J. (2000). Damage zones aroundstrike-slip fault systems and strike-slip fault evolution, Crackington, Haven, southwest England. *Geoscience Journal, 4*, 53–72.

Kim, Y.-S. P., Peacock, D. C., & Sanderson, J. (2004). Fault damage zones. *Journal of Structural Geology, 26*, 503–517. doi:10.1016/j.jsg.2003.08.002

Martel, S. J. (1990). Formation of compound strike-slip zones, Mount Abbot quadrangle, California. *Journal of Structural Geology, 12*(7), 869–882. doi:10.1016/0191-8141(90)90060-C

Mickala, O. R., Vidal, L., Boudzoumou, F., Affaton, F., Vandamme, D., Borshneck, D., ... Miche, H. (2014). Geochemical characterization of the Marinoan "Cap Carbonate" of the Niari-Nyanga Basin (Central Africa). *Precambrian research, 255*, Part 1, 357–380.

Oesterlen, M. (1976). Karoo-System und präkambrische Unterlage im nördlichen Angola-I. Stratigraphie, Tektonik und Petrographie. Geologisches Jahrbuch, Reihe B, 20, 3-55, Hannover.

Oesterlen, M. (1979). Karoo system und präkambrishe unterlage im nördlichen Angola-II. Diagenese un Sedimentologie des Karoo Systems. Geologisches Jahrbuch, Reihe B, 36, 3-41, Hannover.

Olson, E. J., & Pollard, D. D. (1991). The initiation and growth of en échelon veins. *Journal of Structural Geology, 13*(5), 595–608. doi:10.1016/0191-8141(91)90046-L

Poidevin, J. L. (2007). Stratigraphie isotopique du stron-tium et datation des formations carbonatées et glaciogéniques néoprotérozoiques du Nord et de l'Ouest du craton du Congo. *Comptes Rendus Geoscience, 339*, 259–273. doi:10.1016/j.crte.2007.02.007

Pollard, D., & Aydin, A. (1988). Progress in understanding jointing over the past century. *The Geological Society of America, 100*, 1181–1204. doi:10.1130/0016-7606 (1988)100<1181:PIUJOT>2.3.CO;2

Secor, D. J. (1965). Role of fluid pressure in jointing. *American Journal of Science, 263*, 633–646. doi:10.2475/ajs.263.8.633

Segall, P., & Pollard, D. D. (1983). Nucleation and growth strike-slip faults in granite. *Journal of Geophysics Research, 14*, 555–568. doi:10.1029/JB088iB01p00555

Straathof, G. B. (2011). *Neoproterozoic low latitude glaciations: An African perspective*. PhD, thesis, University of Edinburgh, U.K, Edinburgh, p. 263

Sylvester, A. G. (1988). Strike slip faults. *Geological Society of America Bulletin, 100*, 1666–1703. doi:10.1130/0016-7606(1988)100<1666:SSF>2.3.CO;2

Tack, L., Delvaux, D., Kadima, E., Delpomdor, F., Tahon, A., Dumont, P., ... Dewaele, S. (2008). The 1.000 m thick Redbeds sequence of the Congo River Basin (CRB): A generally overlooked testimony in Central Africa of post-Gondwana amalgamation (550 Ma) and pre-Karoo break-up (320 Ma). *22nd Colloquium of African Geology, Hammamet, Tunisia, November 4-6, 2008, Abstract book, 86-88.*

Tack, L., Wingate, M., Liégeois, J.-P., Fernandez-Alonso, M., & Deblond, A. (2001). Early Neoproterozoic magmatism (1000-910 MA) of the Zadinian and Mayumbian Groups (Bas-Congo): Onset of Rodinian rifting at western edge of the Congo craton. *Precambrian Research, 110*, 277–306. doi:10.1016/S0301-9268(01)00192-9

Tait, J., Delpomdor, F., Préat, A., Tack, L., Straathof, G., & Nkula, V. K. (2011). Neoproterozoic sequences of the West Congo and Lindi/Ubangi Supergroups in the Congo Craton, Central Africa. *Geological Society, London, Memoirs, 36*, 185–194.

Tullis, T. E., & Tullis, J. (1986). Experimental rock deformation techniques, *In* Hobbs B. E., & Heard H. C. (eds.), Mineral and rock deformation; laboratory studies; the Paterson volume: Geophysical Monograph, Monash University, Clayton, Victoria, Australia, v. 36, p. 297–324.

Appendices

Table A1. Input data table of joints and faults slip in Win-Tensor with primary subset associated. Type (1: fault with slip line; 3: faults with tension fracture, 4: plane of fracture); column slip sense (S: Sinistral; D: dextral; T: tension); column confidence level (C: certain; P: probable)

Id	Properties		Fracture plane		Slip line/tension fracture		Slip	Conf.	Weight	Activ.	Striae	Subset
	Format	Type	Dip	Dip-Direction	Plunge/Dip	Azimuth/Dip-Dir.	Sense	Level	Factor	Type	Intens.	Index
Bro-1	11	4	80	80			S	C	2,0	1		1,0
Bro-2	11	4	80	116			D	C	2,0	1		2,0
Bro-3	11	4	80	268			T	C	2,0			3,0
Bro-4	11	4	85	325			D	C	2,0	1		2,0
Bro-5	11	4	90	180			D	C	2,0	1		4,0
Bro-6	11	4	70	263			S	C	2,0	1		1,0
Bro-7	11	4	75	328			D	C	2,0	1		2,0
Bro-8	11	4	75	325			D	C	2,0	1		2,0
Bro-9	11	1	80	320	4	49	D	C	2,0	1	2	2,0
Bro-10	11	4	85	135			D	C	2,0	1		2,0
Bro-11	11	4	85	245			S	C	2,0	1		1,0
Bro-12	11	1	79	315	3	44	D	C	2,0	1	2	2,0
Bro-13	11	4	75	300			T	P	2,0			4,0
Bro-14	11	4	80	146			D	C	2,0	1		2,0
Bro-15	11	1	85	152	1	62	D	C	2,0	1	2	2,0
Bro-16	11	3	90	72	90	195		P	2,0	1	2	1,0
Bro-17	11	1	70	146	1	56	D	C	2,0	1	2	2,0
Bro-18	11	4	80	235			T	P	2,0			4,0
Bro-19	11	4	85	267			S	C	2,0	1		1,0

(Continued)

Table A1. (Continued)

Id	Properties		Fracture plane		Slip line/tension fracture		Slip	Conf.	Weight	Activ.	Striae	Subset
	Format	Type	Dip	Dip-Direction	Plunge/Dip	Azimuth/Dip-Dir.	Sense	Level	Factor	Type	Intens.	Index
Bro-20	11	3	90	72	90	194		P	2,0	1	2	1,0
Bro-21	11	4	80	235			S	P	2,0	1		4,0
Bro-22	11	3	85	267	75	58		P	2,0	1	2	1,0
Bro-23	11	3	85	267	75	52		P	2,0	1	2	1,0
Bro-24	11	1	75	133	2	43	D	C	2,0	1	2	2,0
Bro-25	11	4	85	318			D	C	2,0	1		2,0
Bro-26	11	3	85	267	75	52		P	2,0	1	2	1,0
Bro-27	11	4	75	65			S	C	2,0	1		1,0
Bro-28	11	4	90	135			D	C	2,0	1		2,0
Bro-29	11	4	90	50			T	P	2,0			4,0
Bro-30	11	1	75	145	1	55	D	C	2,0	1	2	2,0
Bro-31	11	3	85	250	65	17		P	2,0	1	2	1,0
Bro-32	11	1	85	115	5	25	D	C	2,0	2	2	2,0
Bro-33	11	3	85	262	85	220		P	2,0	1	2	1,0
Bro-34	11	3	85	262	85	210		P	2,0	1	2	1,0
Bro-35	11	3	85	73	85	195		P	2,0	1	2	1,0
Bro-36	11	4	90	299			D	P	2,0	1		2,0
Bro-37	11	3	85	73	85	210		P	2,0	1	2	1,0
Bro-38	11	3	85	74	85	220		P	2,0	1	2	1,0
Bro-39	11	4	90	252			M	C	2,0	1		1,0
Bro-40	11	4	75	332			D	C	2,0	1		2,0
Bro-41	11	1	85	268	4	178	S	C	2,0	1	2	1,0

(Continued)

Table A1. (Continued)

Id	Properties		Fracture plane		Slip line/tension fracture		Slip	Conf.	Weight	Activ.	Striae	Subset
	Format	Type	Dip	Dip-Direction	Plunge/Dip	Azimuth/Dip-Dir.	Sense	Level	Factor	Type	Intens.	Index
Bro-42	11	1	82	300	1	210	D	C	2,0	1	2	2,0
Bro-43	11	4	90	129			D	C	2,0	1		2,0
Bro-44	11	4	70	273			S	C	2,0	1		1,0
Bro-45	11	3	85	261	87	227		P	2,0	1	2	1,0
Bro-46	11	4	75	343			D	C	2,0	1		2,0
Bro-47	11	4	85	244			S	C	2,0	1		1,0
Bro-48	11	1	70	245	4	157	S	C	2,0	1	2	1,0
Bro-49	11	1	65	300	4	212	D	C	2,0	1	2	2,0
Bro-50	11	1	80	248	1	158	S	C	2,0	1	2	1,0
Bro-51	11	3	80	248	75	199		P	2,0	1	2	1,0
Bro-52	11	4	75	339			T	P	2,0			4,0
Bro-53	11	4	70	330			T	P	2,0			4,0
Bro-54	11	4	80	264			S	P	2,0	1		1,0
Bro-55	11	4	85	156			D	P	2,0	1		2,0
Bro-56	11	4	75	336			D	P	2,0	1		2,0
Bro-57	11	1	75	266	5	177	S	C	2,0	1	2	1,0
Bro-58	11	3	75	266	85	220		P	2,0	1	2	1,0
Bro-59	11	3	75	266	86	220		P	2,0	1	2	1,0
Bro-60	11	4	80	135			D	C	2,0	1		2,0
Bro-61	11	4	80	144			T	P	2,0			4,0
Bro-62	11	3	85	261	86	236		P	2,0	1	2	1,0
Bro-63	11	4	80	325			T	P	2,0			4,0

(Continued)

Table A1. (Continued)

Id	Properties		Fracture plane		Slip line/tension fracture		Slip	Conf.	Weight	Activ.	Striae	Subset
	Format	Type	Dip	Dip-Direction	Plunge/Dip	Azimuth/Dip-Dir.	Sense	Level	Factor	Type	Intens.	Index
Bro-64	11	4	80	65			T	P	2,0			4,0
Bro-65	11	4	80	326			T	P	2,0			4,0
Bro-66	11	4	60	88			T	P	2,0			4,0
Bro-67	11	4	80	326			T	P	2,0			4,0
Bro-68	11	4	85	246			M	P	2,0	1		1,0
Bro-69	11	4	85	247			S	P	2,0	1		1,0
Bro-70	11	4	80	254			S	P	2,0	1		1,0
Bro-71	11	4	80	245			S	P	2,0	1		1,0
Bro-72	11	4	75	331			D	P	2,0	1		2,0
Bro-73	11	4	80	267			S	P	2,0	1		1,0
Bro-74	11	4	78	68			S	P	2,0	1		1,0
Bro-75	11	4	75	320			D	P	2,0	1		2,0
Bro-76	11	4	80	245			S	P	2,0	1		1,0
Bro-77	11	3	80	245	80	210		P	2,0	1	2	1,0
Bro-78	11	3	85	245	85	213		P	2,0	1	2	1,0
Bro-79	11	1	85	130	5	40	D	C	2,0	1	2	2,0
Bro-80	11	4	90	130			D	C	2,0	1		2,0
Bro-81	11	4	90	78			S	C	2,0	1		1,0
Bro-82	11	3	90	78	90	36		P	2,0	1	2	1,0
Bro-83	11	3	90	78	90	38		P	2,0	1	2	1,0
Bro-84	11	1	90	75	5	345	S	C	2,0	1	2	1,0
Bro-85	11	4	90	135			D	C	2,0	1		2,0

(Continued)

Table A1. (Continued)

	Properties			Fracture plane		Slip line/tension fracture		Slip	Conf.	Weight	Activ.	Striae	Subset
Id	Format	Type		Dip	Dip-Direction	Plunge/Dip	Azimuth/Dip-Dir.	Sense	Level	Factor	Type	Intens.	Index
Bro-86	11	3		88	263	70	210		P	2,0	1	2	1,0
Bro-87	11	4		90	139			D	C	2,0	1		2,0
Bro-88	11	4		89	69			S	C	2,0	1		1,0
Bro-89	11	4		80	72			S	C	2,0	1		1,0
Bro-90	11	4		85	145			D	C	2,0	1		2,0
Bro-91	11	4		85	148			D	C	2,0	1		2,0
Bro-92	11	1		85	148	5	58	D	C	2,0	1	2	2,0
Bro-93	11	4		85	256			S	C	2,0	1		1,0
Bro-94	11	4		85	325			D	C	2,0	1		2,0
Bro-95	11	1		90	245	3	155	S	C	2,0	1	2	1,0
Bro-96	11	3		90	245	87	210	S	P	2,0	1	2	1,0
Bro-97	11	4		90	252			T	C	2,0			3,0
Bro-98	11	4		85	138			D	C	2,0	1		2,0
Bro-99	11	4		85	145			D	C	2,0	1		2,0
Bro-100	11	4		85	79			S	C	2,0	1		1,0
Bro-101	11	4		90	82			S	C	2,0	1		1,0
Bro-102	11	4		85	135			D	C	2,0	1		2,0
Bro-103	11	4		85	130			D	C	2,0	1		2,0
Bro-104	11	3		85	130	85	190		P	2,0	1	2	2,0
Bro-105	11	3		85	135	85	195		P	2,0	1	2	2,0
Bro-106	11	3		90	88	85	208		P	2,0	1	2	1,0
Bro-107	11	3		90	88	85	142		P	2,0	1	2	1,0

(Continued)

Table A1. (Continued)

Id	Properties		Fracture plane		Slip line/tension fracture		Slip	Conf.	Weight	Activ.	Striae	Subset
	Format	Type	Dip	Dip-Direction	Plunge/Dip	Azimuth/Dip-Dir.	Sense	Level	Factor	Type	Intens.	Index
Bro-108	11	4	79	258			S	C	2,0	1		1,0
Bro-109	11	4	80	144			D	C	2,0	1		2,0
Bro-110	11	4	90	322			D	C	2,0	1		2,0
Bro-111	11	3	80	75	90	222		P	2,0	1	2	1,0
Bro-112	11	3	80	265	90	230		P	2,0	1	2	1,0
Bro-113	11	3	65	250	70	220		P	2,0	1	2	1,0
Bro-114	11	1	90	130	2	40	D	C	2,0	1	2	2,0
Bro-115	11	4	75	269			S	C	2,0	1		1,0
Bro-116	11	4	90	299			D	C	2,0	1		2,0
Bro-117	11	4	70	273			S	C	2,0	1		1,0
Bro-118	11	4	85	244			S	C	2,0	1		1,0
Bro-119	11	4	90	75			S	C	2,0	1		1,0
Bro-120	11	4	75	339			D	C	2,0	1		2,0
KO-1	11	4	80	104			S	P	2,0	1		1,0
KO-2	11	4	85	94			S	P	2,0	1		1,0
KO-3	11	4	80	82			S	P	2,0	1		1,0
KO-4	11	1	85	89	2	359	S	C	2,0	1	2	1,0
KO-5	11	4	82	62			T	P	2,0			4,0
KO-6	11	4	85	201			T	P	2,0			4,0
KO-7	11	4	85	299			T	P	2,0			4,0
KO-8	11	4	85	280			T	P	2,0			4,0
KO-9	11	3	90	89	85	236		P	2,0	1	2	1,0

(Continued)

Table A1. (Continued)

Id	Properties		Fracture plane		Slip line/tension fracture		Slip	Conf.	Weight	Activ.	Striae	Subset
	Format	Type	Dip	Dip-Direction	Plunge/Dip	Azimuth/Dip-Dir.	Sense	Level	Factor	Type	Intens.	Index
KO-10	11	4	85	275			S	C	2,0	1		1,0
KO-11	11	3	85	62	90	225		P	2,0	1	2	1,0
KO-12	11	4	86	99			S	C	2,0	1		1,0
KO-13	11	3	86	99	80	235		P	2,0	1	2	1,0
KO-14	11	3	90	62	85	16		P	2,0	1	2	1,0
KO-15	11	4	80	102			T	P	2,0			4,0
KO-16	11	4	90	13			T	P	2,0			4,0
KO-17	11	4	70	250			T	P	2,0			4,0
KO-18	11	4	80	62			T	P	2,0			4,0
KO-19	11	4	90	115			T	P	2,0			4,0
KO-20	11	4	80	189			T	P	2,0			4,0
KO-21	11	4	85	274			T	P	2,0			4,0
KO-22	11	4	82	68			T	P	2,0			4,0
KO-23	11	4	82	68			T	P	2,0			4,0
KO-24	11	4	90	68			T	P	2,0			4,0
KO-25	11	4	90	99			T	P	2,0			4,0
KO-26	11	4	75	325			D	P	2,0	1		2,0
KO-27	11	3	80	253	85	23		P	2,0	1	2	1,0
KO-28	11	4	80	242			S	C	2,0	1		1,0
KO-29	11	4	88	245			S	C	2,0	1		1,0
KO-30	11	3	85	250	85	13		P	2,0	1	2	1,0
KO-31	11	3	85	250	85	209		P	2,0	1	2	1,0

(Continued)

Table A1. (Continued)

Id	Properties		Fracture plane		Slip line/tension fracture		Slip	Conf.	Weight	Activ.	Striae	Subset
	Format	Type	Dip	Dip-Direction	Plunge/Dip	Azimuth/Dip-Dir.	Sense	Level	Factor	Type	Intens.	Index
KO-32	11	4	85	175			D	P	2,0	1		4,0
KO-33	11	4	85	300			D	P	2,0	1		2,0
KO-34	11	4	85	300			D	P	2,0	1		2,0
KO-35	11	4	85	222			S	P	2,0	1		4,0
KO-36	11	4	90	125			D	C	2,0	1		2,0
KO-37	11	3	80	280	85	55		P	2,0	1	2	1,0
KO-38	11	4	65	89			S	C	2,0	1		1,0
KO-39	11	4	85	210			D	C	2,0	1		4,0
KO-40	11	4	85	45			T	C	2,0			3,0
KO-41	11	4	85	40			T	C	2,0			3,0
KO-42	11	4	85	252			D	C	2,0	1		1,0
KO-43	11	4	85	310			D	C	2,0	1		2,0
KO-44	11	1	85	130	1	40	D	C	2,0	1	2	2,0
KO-45	11	4	75	20			T	C	2,0			4,0
KO-46	11	1	85	245	2	155	S	C	2,0	1	2	1,0
KO-47	11	1	80	310	5	220	D	C	2,0	1	2	2,0
KO-48	11	1	80	325	2	235	D	C	2,0	1	2	2,0
KO-49	11	1	80	100	2	10	S	C	2,0	1	2	1,0
KO-50	11	3	80	100	80	55		P	2,0	1	2	1,0
KO-51	11	1	70	96	2	6	S	C	2,0	1	2	1,0
KO-52	11	4	90	235			D	C	2,0			4,0
KO-53	11	4	85	310			T	C	2,0			4,0

(Continued)

Table A1. (Continued)

	Properties		Fracture plane		Slip line/tension fracture		Slip	Conf.	Weight	Activ.	Stride	Subset
Id	Format	Type	Dip	Dip-Direction	Plunge/Dip	Azimuth/Dip-Dir.	Sense	Level	Factor	Type	Intens.	Index
KO-54	11	4	85	50			D	C	2,0	1		4,0
KO-55	11	4	75	75			T	P	2,0			4,0
KO-56	11	4	80	55			T	P	2,0			4,0
KO-57	11	4	90	135			T	P	2,0			4,0
KO-58	11	4	80	30			T	P	2,0			4,0
KO-59	11	4	85	142			T	P	2,0			4,0
KO-60	11	4	70	302			T	P	2,0			4,0
KO-61	11	4	75	64			T	P	2,0			4,0
KO-62	11	4	85	54			T	P	2,0			4,0
KO-63	11	4	85	20			T	P	2,0			4,0
KO-64	11	4	85	39			T	P	2,0			4,0
KO-65	11	4	90	50			T	P	2,0			4,0
KO-66	11	4	85	130			T	P	2,0			4,0
KO-67	11	4	85	130			T	P	2,0			4,0
KO-68	11	4	85	68			T	C	2,0			3,0
KO-69	11	4	70	145			T	C	2,0			3,0
KO-70	11	4	85	89			T	C	2,0			3,0
KO-71	11	4	65	49			T	C	2,0			4,0
KO-72	11	4	70	142			D	C	2,0	1		2,0
KO-73	11	4	70	62			S	C	2,0	1		1,0
KO-74	11	4	82	42			T	C	2,0			3,0
KO-75	11	4	90	142			T	C	2,0			4,0

(Continued)

Table A1. (Continued)

Id	Properties		Fracture plane		Slip line/tension fracture		Slip	Conf.	Weight	Activ.	Striae	Subset
	Format	Type	Dip	Dip-Direction	Plunge/Dip	Azimuth/Dip-Dir.	Sense	Level	Factor	Type	Intens.	Index
KO-76	11	4	85	135			D	C	2,0	1		2,0
KO-77	11	4	90	313			T	C	2,0			3,0
KO-78	11	4	85	135			T	P	2,0			4,0
KO-79	11	4	85	315			T	P	2,0			4,0
KO-80	11	4	90	75			T	P	2,0			4,0
KO-81	11	4	90	110			T	P	2,0			4,0
KO-82	11	4	85	74			T	P	2,0			4,0
KO-83	11	4	80	33			T	P	2,0			4,0
KO-84	11	4	85	104			S	C	2,0	1		1,0
KO-85	11	3	85	104	90	224		P	2,0	1	2	1,0
KO-86	11	4	90	34			D	P	2,0	1		4,0
KO-87	11	4	75	85			T	P	2,0			4,0
KO-88	11	4	89	241			T	P	2,0			4,0
KO-89	11	1	70	90	2	0	S	C	2,0	2	2	1,0
KO-90	11	4	85	286			S	C	2,0	2		1,0
KO-91	11	4	85	22			D	C	2,0	2		4,0
KO-92	11	4	80	45			D	C	2,0	2		4,0
KO-93	11	4	85	240			S	C	2,0	2		4,0
KO-94	11	3	70	140	60	203		P	2,0	1	2	2,0
KO-95	11	3	90	75	60	30		P	2,0	1	2	1,0
KO-96	11	4	90	322			D	C	2,0	1		2,0
KO-97	11	1	85	135	2	45	D	C	2,0	1	2	2,0

(Continued)

Table A1. (Continued)

Id	Properties		Fracture plane		Slip line/tension fracture		Slip	Conf.	Weight	Activ.	Striae	Subset
	Format	Type	Dip	Dip-Direction	Plunge/Dip	Azimuth/Dip-Dir.	Sense	Level	Factor	Type	Intens.	Index
KO-98	11	4	85	145			D	C	2,0	1		2,0
KO-99	11	4	85	138			D	C	2,0	1		2,0
KO-100	11	1	85	139	3	50	D	C	2,0	1	2	2,0
KO-101	11	4	90	130			D	C	2,0	1		2,0
KO-102	11	4	90	130			D	C	2,0	1		2,0
KO-103	11	4	70	75			D	C	2,0	1		1,0
KO-104	11	3	85	109	90	235		P	2,0	1	2	1,0
KO-105	11	3	80	260	80	205		P	2,0	1	2	1,0
KO-106	11	3	87	87	80	225		P	2,0	1	2	1,0
KO-107	11	3	87	87	80	210		P	2,0	1	2	1,0
KO-108	11	3	85	100	90	239		P	2,0	1	2	1,0
KO-109	11	1	90	100	10	10	S	C	2,0	1	2	1,0
KO-110	11	1	90	100	21	10	S	C	2,0	1	2	1,0
KO-111	11	4	80	60			T	C	2,0			3,0
KO-112	11	1	85	240	20	152	S	C	2,0	2	2	1,0
KO-113	11	1	90	255	15	165	S	C	2,0	2	2	1,0
KO-114	11	4	90	239			T	C	2,0			3,0
KO-115	11	4	90	100			T	C	2,0			3,0
KO-116	11	3	90	105	90	226		P	2,0	1	2	1,0
KO-117	11	3	80	258	90	200		P	2,0	1	2	1,0
KO-118	11	4	80	319			D	C	2,0	1		2,0
KO-119	11	3	90	87	90	229		P	2,0	1	2	1,0

(Continued)

Table A1. (Continued)

Id	Properties		Fracture plane		Slip line/tension fracture		Slip	Conf.	Weight	Activ.	Striae	Subset
	Format	Type	Dip	Dip-Direction	Plunge/Dip	Azimuth/Dip-Dir.	Sense	Level	Factor	Type	Intens.	Index
KO-120	11	1	90	254	6	164	S	C	2,0	1	2	1,0
KO-121	11	1	90	135	5	45	D	C	2,0	1	2	2,0
KO-122	11	1	90	145	5	55	D	C	2,0	1	2	2,0
KO-123	11	3	90	254	90	210		P	2,0	1	2	1,0
KO-124	11	3	90	264	90	209		P	2,0	1	2	1,0
KO-125	11	1	80	259	5	169	S	C	2,0	1	2	1,0
KO-126	11	1	80	269	3	180	S	C	2,0	1	2	1,0
KO-127	11	4	80	268			T	C	2,0			3,0
KO-128	11	4	90	258			T	C	2,0			3,0
KO-129	11	1	70	115	2	25	D	C	2,0	2	2	2,0
KO-130	11	3	85	322	90	185		P	2,0	1	2	2,0
KO-131	11	1	85	120	5		D	C	2,0	1	2	2,0
KO-132	11	1	70	135	2		D	C	2,0	1	2	2,0
KO-133	11	1	90	142	3	32	D	C	2,0	1	2	2,0
KO-134	11	1	68	166	2	45	D	C	2,0	1	2	2,0
KO-135	11	1	85	275	10	77	D	C	2,0	1	2	2,0

Characteristics of soil exchangeable potassium according to soil color and landscape in Ferralsols environment

Brahima Koné[1]*, Traoré Lassane[2], Sehi Zokagon Sylvain[1] and Kouassi Kouassi Jacques[1]

*Corresponding author: Brahima Koné, Earth Science Unit, Soil Science Department, Felix Houphouet-Boigny University, 22 BP 582 Abidjan, Côte d'Ivoire
E-mail: kbrahima@hotmail.com

Reviewing editor: Craig O'Neill, Macquarie University, Australia

Abstract: The use of soil color as Munsell data was explored for *in situ* indication of soil potassium (K) availability toward a friendly method of agricultural land survey. Soil contents of K, calcium, and magnesium were determined for 998 upland soil samples from Côte d'Ivoire (7–10°N). Soil depths (0–20, 20–60, 60–80, and 80–150 cm), redness ratio (RR), and redness factor (RF) were considered. Significant association was observed between K-levels (high, medium, and low) in topsoil and its color hue, and the highest cumulative frequency of 2.5YR in high and medium levels was characterizing the hill slope position (summit and upper slope). Deep horizon, foot slope, and yellowish color (7.5YR and 10YR) were more relevant to low K-level. Significant linear regressions of soil content of K were observed according to both redness indices indifferently to the topographic positions and soil depths in some extend. Of these finding in the line of folk knowledge, RR and RF are recommended for *in situ* measurement of soil K, and 2.5YR as color hue may be use as indicator of K-enriched soil at hill slope position.

Subjects: Earth Sciences; Environment & Agriculture; Environmental Studies & Management

Keywords: indicator of soil fertility; potassium; catena; soil color; folk knowledge

ABOUT THE AUTHORS

The data reported in the current paper were drawn from the thesis of Brahima Koné as first author. The work was carried out in Côte d'Ivoire in order to develop a tool for soil survey in Ferralsol environment. The co-authors of current paper have contributed to data analysis and interpretation as well as for the writing of the manuscript.

PUBLIC INTEREST STATEMENT

Soil use and management are important factors affecting agricultural production, while conventional standard methods of soil assessment are not well understood, especially by non-educated farmers of tropical Africa. In contrast, there were social and economical perceptions of farming including the land use as long as populations are practicing agriculture. So call folk knowledge, local populations around the word have friendly methods of soil classification in relation with its fertility and productivity. The soil color accounts for criteria in folk knowledge. Regarding to the importance of soil content of potassium in Ferralsol, the current study emphasized the relation between soil content of this nutrient and soil color in interaction with the topographic position of landscapes. The hill slope position characterized by reddish soils was found to be the most enriched land in potassium.

1. Introduction

Sustaining agriculture requires sound soil evaluation methods in concordance with morphopedology standards (Bertrand et al., 1985; Loukili et al., 2000). However, minimum data-set requirements are differing according to authors: soil attributes or both soil and plant parameters may be concerned (Larson and Pierce, 1991; Pearson et al., 1995; Pieri, 1992). This variance in methodology as a weakness was tackled by Riquier et al. (1970) when initiating the use of soil productivity index which was improved with data relevant to crop potential yielding (root development) as function of soil environment (Burger, 1996; Gale et al., 1991; Kiniry et al., 1983; Milner et al., 1996; Pierce et al., 1983). In the meantime, ecological specificity was considered by including soil contents of phosphorus (P) and organic matter (Neill, 1979) and typical model was suggested for tropical environment (Sys and Frankart, 1971). Of existing models, soil content of potassium (K) was missing as parameter though; this nutrient is among the most limiting of crop growth in tropical agro-ecologies (Koné, Fatogoma, Chérif, 2013).

The most recent approach of soil management is the fertility capability classification (FCC) system (Sanchez et al., 2003) based on five data-sets of soil as modifiers including soil content of K. However, this system is not popularly adopted yet. The wide number of required parameters, specially the chemical analytical data and the skill required for FCC may be of concerned. Therefore, a friendly method of soil chemical parameters estimation in field can contribute to wide adoption of such fertility classification of soil. For this purpose, the color of soil as a component of folk classification (Krasilnikov & Tabor, 2003), may have consistent contribution as friendly prediction method of Ferralsols K availability.

In fact, there is evidence of Ferralsols inherent fertility classification as poor for yellow and richer for reddish colors at a given topographic position, especially for soil contents of P, K, and magnesium (Mg) in addition to soil particle sizes (Koné, Yao-Kouamé, et al., 2009). Moreover, the opportunity of soil K supplying capacity was successfully explored using soil color by Koné, Bongoua-Devisme, Kouadio, Kouadio, and Traoré (2014).

Indeed, the colors of Ferralsols are relevant to difference in their mineralogy, organic matter content, and texture (Koné, Diatta, et al., 2009; Stoner et al., 1980) as major descriptors of their history.

The use of soil redness (Torrent et al., 1980, 1983; Santana, 1984) for estimating soil K availability on the basis of linear regression may be a fast and cheapest method, hence limiting constraint in agricultural soil capacity evaluation where soil content of K may be critical.

Soil survey was conducted in Côte d'Ivoire above the latitude 7°N applying randomized and unequal stratified soil sampling method. Soil color was determined by Munsell chart and soil exchangeable K was analyzed for exploring the accuracy of soil redness rate (RR) and redness factor (RF) in a linear regression of soil K. The aim was (i) to identify a soil color hue for a given level (high, medium, and low) of soil K in spatial distribution according to soil depth and topographic section, (ii) to defined the relation between soil exchangeable K and soil color, and (iii) to identify fit model of soil K among that using RR and RF. Definitively, a model of soil exchangeable K content should be recommended for landscape section and soil depth as a tool for most friendly method of soil survey.

2. Material and methods

2.1. Studied zone description

The study was carried out across 19 sites characterized by Ferralsols encountered between the latitudes 7–10°N in Côte d'Ivoire. Four major agro-ecologies were described in there by Koné (2007) as Sudan savannah with grassland, Guinea savannah with woodland, derived savannah (a transition between savannah and forest agro-ecologies), and mountainous zone located in the west of the country. Annual average rainfall amount ranged from 1,200 to 2,000 mm.

Dismantled or unaffected summit (SUM) ferruginous cuirass of plateau landscapes further characterized by slightly concave or convex sides were frequently encountered in the studied area beside
of hills with bedrock outcroppings. A few inselbergs were also observed and, the landside and length
of landscapes were variable accordingly. Upland soils were essentially Ferralsols plinthic belonging
to hyperdystric or dystric groups.

2.2. Soil sampling
Two hundred and eighty-nine soil profiles were surveyed along the toposequences of various landscapes at 19 sites. The soil profiles were unequally distributed (Webster and Oliver, 1990) on three
sub-groups of Ferralsols encountered in the studied area (Koné, Diatta, et al., 2009). The identified
horizons in the soil profiles were coded according to depth classes—H1 (0–20 cm), H2 (20–60 cm),
H3 (60–80 cm), and H4 (80–150 cm) dividing the soil profile into organic horizon (Diatta, 1996), minimum, medium, and maximum crop rooting depths (Böhn, 1976; Chopart, 1985), respectively. A total
of 995 samples (2 kg for each) were taken from soil horizons up to a maximum depth of 1.5 m when
possible (Table 1).

2.3. Soil sample characteristics
Soil sample size of a given color hue was variable according to soil depth across the studied zone.

The corresponding numbers of soil samples were 274, 325, 279, and 117 for the SUM, upper slope
(US), middle slope (MS), and foot slope (FS), respectively.

2.4. Laboratory analysis and classification of soil K contents
Soil samples were dried under forced air at room temperature. Then, they were crushed before siev-
ing through a 2.0-mm stainless steel sieve. Exchangeable K of soil was extracted by shaking 1 g in
10 ml of 1 M NH4OAc during 5 min before the use of atomic emission in a Perkin Elmer Analyst 100
spectrometer (Page, 1982). Three classes of soil contents of exchangeable K were defined as done
by Berryman et al. (1984) for tropical soils:

L = Low soil content of K ranging below 0.15 cmol kg^{-1}.

M = Moderate soil content of K ranging between 0.15 and 0.30 cmol kg^{-1}.

H = High soil content of K ranging over 0.30 cmol kg^{-1}.

Soil contents of calcium (Ca) and Mg were also analyzed using the same extraction method described above.

2.5. Soil color identification
The year 2000 revised washable edition of Munsell soil color charts (Gretagmacьeth, 2000) composed of 322 different standard color ships was used in field during the survey for soil color identification. Wet soil samples were compared with the standard color ships, respectively, and the three
components of the color were recorded as "Hue (He); Chroma (C)/Value (V)." The RF defined as RF by
Santana (1984) was calculated for each of the soil samples likewise the redness ratio (RR) according
to Torrent et al. (1980, 1983):

Table 1. Soil sample size of a given soil color hue as identified in soil depths

	Number of soil sample				
	H1	**H2**	**H3**	**H4**	**Total**
2.5YR	77	94	68	54	293
5YR	121	134	86	46	387
7.5YR	81	74	35	16	206
10YR	40	36	18	15	109
Total	319	338	207	131	995

$$RF = (10 - He) + C/V \tag{1}$$

$$RR = (10 - He) \times C/V \tag{2}$$

2.6. Statistical analysis

By descriptive analysis, average frequency of soil content of exchangeable K was determined in topsoil (H1 and H2) for each topographic section (S, US, MS, and FS) and for each soil color hue as encountered (2.5, 5, 7.5, and 10YR) using SPSS 10 package. Cross-table analysis was done to determine the frequency of soil K levels (H = high, M = moderate, and L = low) according to soil depths (H1, H2, H3, and H4) and the topographic sections for the identified soil color hue. Pearson correlation analysis was also performed between RR and soil contents of K, Ca, and Mg and likely for RF, respectively. Furthermore, soil content of exchangeable K was predicted by RR and RF separately using linear regression analysis step by step with constant term or not, and the most significant ($p < 0.05$) model was reported. The thickness of elementary soil horizon was considered as weighted variable. SAS (version 8) was used for these statistical analyses and the critical level of probability was fixed as 0.05 (α).

3. Results

3.1. Characterization of soil K levels

The frequencies of soil color hues (2.5, 5, 7.5, and 10YR) as determined for different soil depths (H1, H2, H3, and H4) according to K-levels are presented in Figure 1. The highest frequencies are related to the low K-level (L) throughout the soil profile with outstanding values for 5 and 7.5YR in the topsoil layers (H1 and H2), while similarly observed for 2.5 and 7.5YR in the subsoil (H3 and H4). Nevertheless, the highest cumulative frequencies of 2.5YR in medium (M) and high (H) levels of soil K is observed likewise for 5YR as soil color hue.

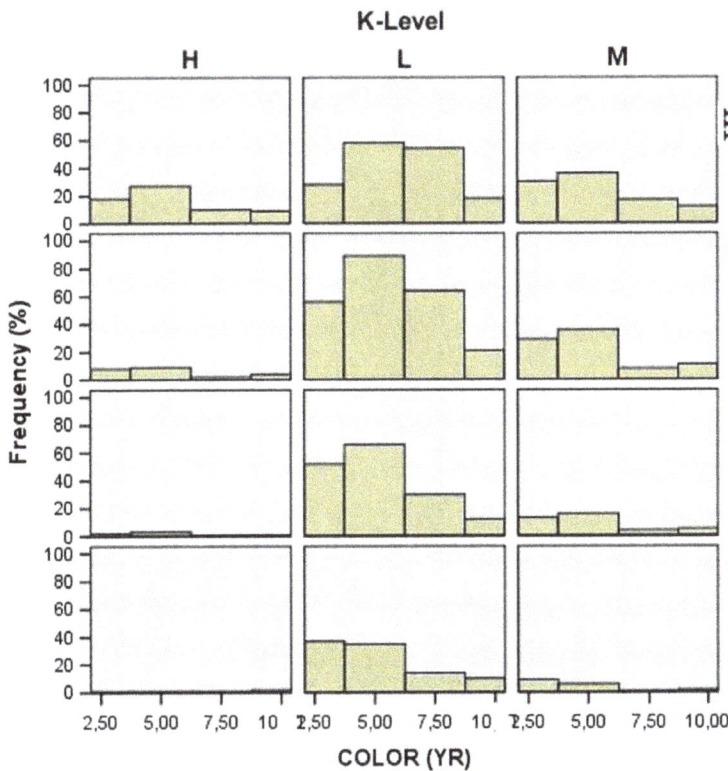

Figure 1. Frequency of soil potassium levels (H, M, and L) according encountered Munsell color hues (2.5, 5, 7.5, and 10YR) in different soil depths (H1, H2, H3, and H4).

In fact, there are 40, 37.3, and 22.6% of chance to observed 2.5YR as soil color hue in topsoil 0–20 cm for medium, low, and high soil K-levels, respectively (Figure 2a). Hence, the cumulative frequency of 2.5YR referring to medium and high K-levels is about 62.67% over the occurrence as low K-level. Similar results also account for the soil samples of 5YR (52.06% vs. 47.29%) and 10YR (53.15% vs. 46.15%) in color hues contrasting with the results observed for 7.5YR. Further contrast is observed in 20–60 cm soil depth showing the highest frequencies of low K-level (L) compared to the cumulative frequency (high (H) and medium (M) K-levels) indifferently to soil color (Figure 2b).

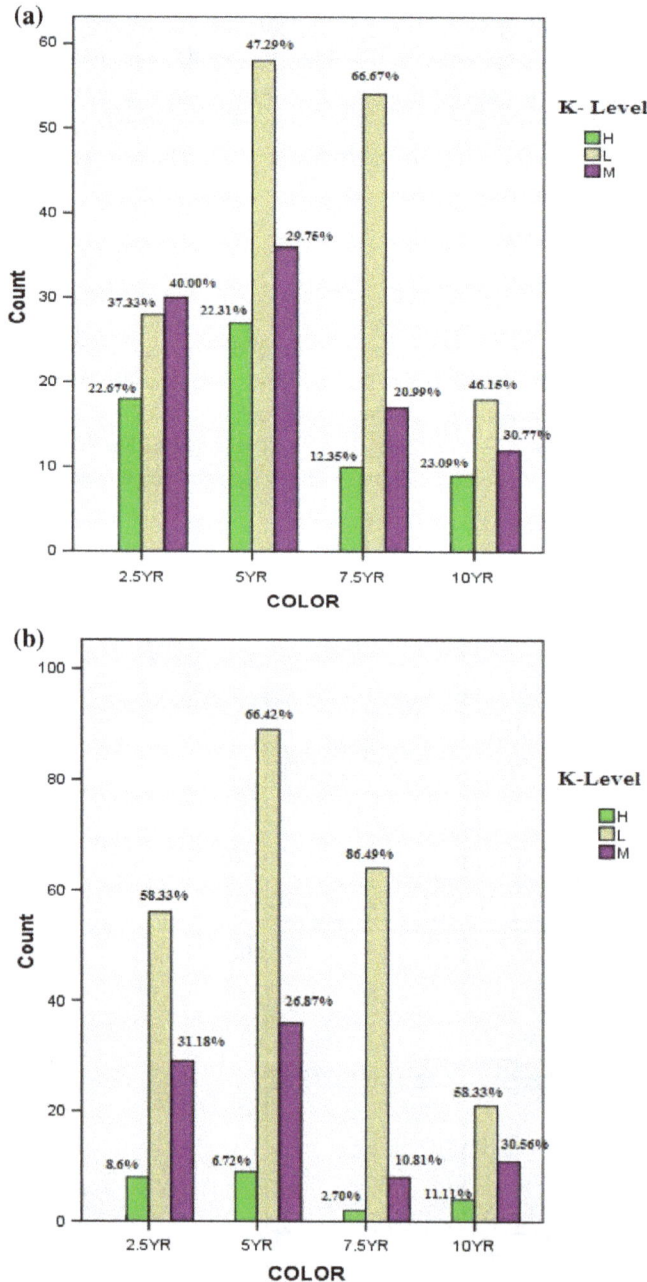

Figure 2. Frequency of potassium level in H1 (a) and H2 (b) according to encountered Munsell color hue (2.5, 5, 7.5, and 10YR).

More details of these results are presented in Figure 3 considering the topographic positions. The soils colored in 2.5YR are outstanding with the highest frequencies of K-levels in both soil depths (0–20 and 20–60 cm) at the SUM position, while similar observations account for 5, 7.5, and 10YR at the US, MS, and FS positions, respectively (Figure 3a): highest cumulative frequency of 2.5 YR in H and M levels of soil K is characterized by the soil samples of SUM and US, while almost equal chances are observed between this cumulative frequency and that related to L when referring to a soil sampled at US with 5YR in color. In turn, the highest frequencies characterizing the soils of 7.5 and 10YR in L are observed when sampled in 0–20 cm at MS and FS positions, respectively. Highest frequencies of L are also observed in 20–60 cm depth indifferently to soil color and topographic section even when compared with the cumulative frequency relative to H and M (Figure 3b). Hence, reddish (2.5YR) topsoil (0–20 cm) appeared to be most enriched (H and M) in exchangeable K, especially for the SUM and UP positions.

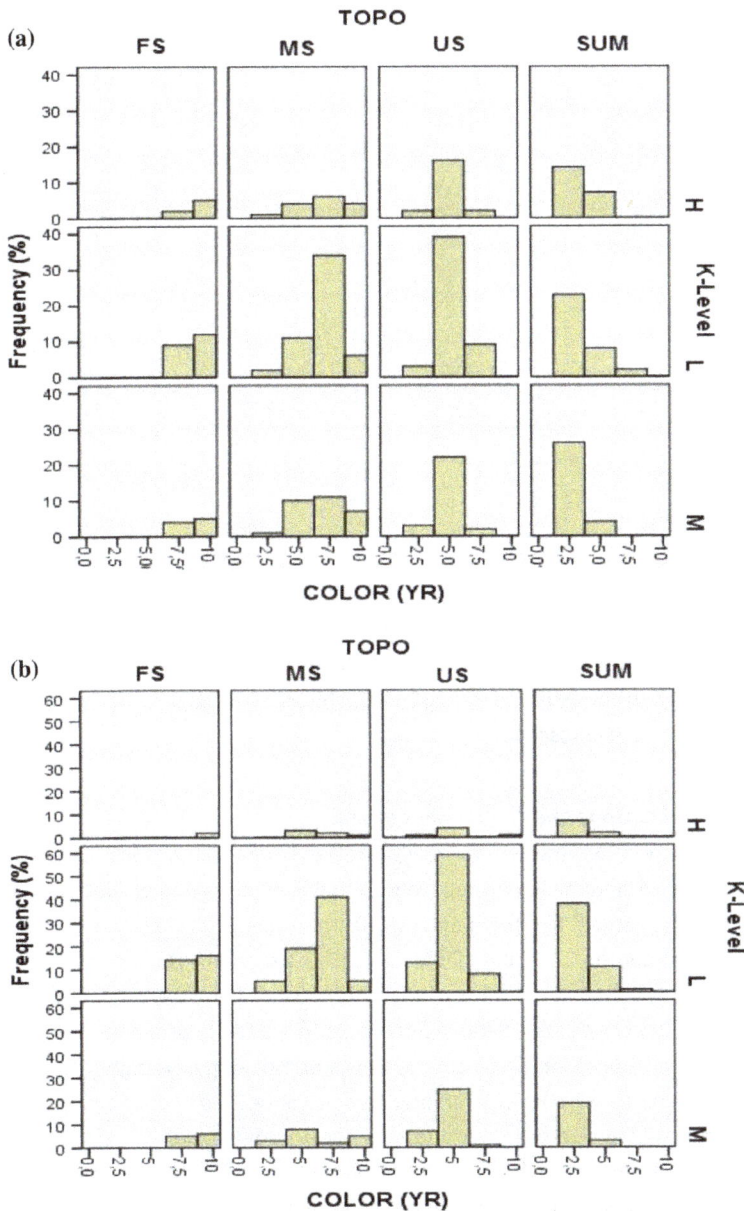

Figure 3. Frequency of potassium level in H1 (a) and H2 (b) depth according to topographic section (TOPO) and encountered Munsell color hue (2.5, 5, 7.5, and 10YR).

Table 2. Frequency of soil color hue as associated to soil K-level (H, M, and L) in 0–20 cm soil depth at different topographic positions

		Frequency (%)		
		H	M	L
Summit	2.5YR	22.22	41.27	36.51
	5YR	36.84	21.05	42.11
	7.5YR	0.00	0.00	100
	10YR	0.00	0.00	0.00
χ^2 probability				0.521
Upper slope	2.5YR	25.00	37.50	37.50
	5YR	20.78	28.57	50.65
	7.5YR	15.38	15.38	69.24
	10YR	0.00	0.00	0.00
χ^2 probability				0.054
Middle slope	2.5YR	25.00	25.00	50.00
	5YR	16.00	40.00	44.00
	7.5YR	11.00	21.57	66.43
	10YR	23.53	41.17	35.30
χ^2 probability				0.298
Foot slope	2.5YR	0.00	0.00	0.00
	5YR	0.00	0.00	0.00
	7.5YR	13.33	26.67	60.00
	10YR	22.73	22.73	54.56
χ^2 probability				0.412

These observations are strengthened by crossing case occurrences according to the topographic sections and soil color hues as determined in the topsoil 0–20 cm soil depth (Table 2): although not always significant, the highest frequencies of reddish soils are accounting for H and M levels of K at the SUM, US, and MS. Greater cumulative frequency of reddish soil relative to H and M levels of soil K is characterizing the S and US when compared to that of L. Arguably, soil K-level of L is more associated to 7.5 and 10YR at FS position of landscape.

3.2. Relationship between soil cations and the redness of soil

Tables 3 and 4 are showing the relations (correlation coefficient) between soil contents of cations and its redness referring to the RR and the RF, respectively. Except the negatives correlation coefficients observed in 80–150 cm at the SUM (−0.48) and 20–60 cm at the FS (−0.27), no significant relation is noticed between soil content of K and RR (Table 3). This statement is contrasting with the positive correlations observed for soil contents of Mg in 20–60 cm ($R = 0.22$; $p = 0.020$) and 60–80 cm ($R = 0.22$; $p = 0.09$) depths at MS position. Significant ($p = 0.040$) correlation between soil content of Ca and RR only accounts for 20–60 cm soil depth at foot slope. No significant correlation accounts for the topsoil (0–20 cm) indifferently to soil contents of cations and topographic sections. Overall, almost the significant correlations are observed in 20–60 cm depth at down slope (MS and FS) of landscape involving all the studied cations. In some extend, similar results are observed in Table 4 relative to the correlation between RF and soil contents of cations (K, Ca, and Mg).

Table 3. Correlation between soil contents of cations (K, Ca, and Mg) and its (RR) in soil depths according to topographic sections

		0–20 cm		20–60 cm		60–80 cm		80–150 cm	
		R	p-value	R	p-value	R	p-value	R	p-value
SUM	K	−0.14	0.199	0.02	0.828	−0.10	0.460	−0.48	0.001
	Ca	−0.12	0.272	0.04	0.689	−0.10	0.440	−0.003	0.818
	Mg	0.006	0.981	−0.14	0.218	−0.08	0.535	−0.22	0.127
US	K	−0.003	0.975	−0.08	0.344	−0.09	0.426	0.22	0.195
	Ca	0.020	0.837	0.03	0.747	0.009	0.961	0.07	0.666
	Mg	0.010	0.310	−0.003	0.967	−0.000	0.996	0.000	0.995
MS	K	0.01	0.877	−0.03	0.742	−0.025	0.851	−0.24	0.189
	Ca	0.008	0.931	0.14	0.828	−0.04	0.722	0.02	0.909
	Mg	−0.02	0.845	0.22	0.020	0.22	0.09	−0.04	0.789
FS	K	−0.14	0.398	−0.27	0.060	−0.17	0.404	−0.30	0.399
	Ca	−0.03	0.644	0.30	0.040	0.09	0.648	−0.07	0.822
	Mg	0.03	0.831	0.24	0.126	−0.02	0.922	−0.09	0.791

Pearson correlation for RR

Table 4. Correlation between soil contents of cations (K, Ca, and Mg) and its redness factor (RF) in soil depths according to topographic sections

		0–20 cm		20–60 cm		60–80 cm		80–150 cm	
		R	p-value	R	p-value	R	p-value	R	p-value
SUM	K	0.06	0.560	0.07	0.497	−0.11	0.112	−0.54	0.0002
	Ca	−0.04	0.653	0.13	0.227	0.02	0.861	0.03	0.811
	Mg	0.06	0.535	0.05	0.627	0.035	0.790	−0.09	0.503
US	K	0.06	0.50	−0.06	0.459	−0.09	0.455	0.30	0.070
	Ca	−0.01	0.862	0.04	0.612	0.005	0.961	0.15	0.342
	Mg	0.09	0.329	0.05	0.585	−0.000	0.998	0.08	0.617
MS	K	0.01	0.870	−0.02	0.828	0.08	0.548	−0.29	0.105
	Ca	−0.02	0.831	0.22	0.030	0.05	0.692	0.007	0.967
	Mg	−0.09	0.341	0.29	0.004	0.27	0.044	−0.06	0.729
FS	K	−0.19	0.260	−0.17	0.255	−0.12	0.542	−0.13	0.699
	Ca	−0.10	0.844	0.31	0.044	0.13	0.529	0.04	0.898
	Mg	0.036	0.830	0.32	0.038	−0.04	0.835	−0.08	0.808

Pearson correlation for RF

However, significant ($p < 0.0001$) linear regressions of soil content of K are observed according to RR and RF, respectively, except the soil sampled at 80–150 cm depth at middle (RF and RR) and foot (RR) slope positions (Table 5). The coefficients of these regressions are low (1/100 times) and characterized by an increasing trend with soil depth indifferently to the descriptive variables.

Table 5. Linear regression of soil exchangeable K according to RF and RR, respectively, according to topographic sections and soil depth

		Linear regression of soil K												
		0–20 cm			20–60 cm			60–80 cm			80–150 cm		
		Coef.	p-value	Err.	Coef.	p-value	Err.	Coef.	p-value	Err.	Coef.	p-value	Err.
SUM	RF	0.03	<0.0001	0.003	0.02	<0.0001	0.02	0.01	<0.0001	0.001	0.02	<0.0001	0.003
	RR	0.02	<0.0001	0.002	0.01	<0.0001	0.001	0.01	<0.0001	0.001	0.01	<0.0001	0.002
US	RF	0.05	<0.0001	0.004	0.02	<0.0001	0.024	0.01	<0.0001	0.001	0.01	<0.0001	0.001
	RR	0.04	<0.0001	0.004	0.02	<0.0001	0.001	0.01	<0.0001	0.001	0.01	<0.0001	0.001
MS	RF	0.05	<0.001	0.008	0.03	<0.0001	0.004	0.02	<0.0001	0.003	0.03	0.500	0.05
	RR	0.05	<0.0001	0.009	0.02	<0.0001	0.003	0.02	<0.0001	0.003	0.02	0.711	0.046
FS	RF	0.09	0.08	0.05	0.05	<0.0001	0.005	0.04	0.001	0.011	0.04	0.030	0.01
	RR	0.08	0.38	0.09	0.04	0.009	0.01	0.03	0.06	0.01	0.02	0.366	0.02

4. Discussion

4.1. Soil K appraisal by landscape and soil color

There was more evidence for identification of the soils characterized by low K-level using soil color across the surveyed area: deepest horizon, the FS position of landscape, and the soil color of 10YR (very pale brown-yellow–dark yellowish brown) were fund to be more relevant to this finding. The release of K by organic matters as surface K-source combined with a relative poor (vs. Na) mobility of this nutrient (Hem, 1992) and the downward gradient of bed rock weathering as inner source of K (Wedpohl, 1978) may account for this.

In turn, high and medium levels of soil K were often observed at the SUM and UP positions and characterized by 2.5YR (pink–red) as soil color hues, especially in the topsoil. However, there is chance to observed a variance of this result according to the landscape variability in the studied area somewhere including inselbergs and plateau with outcrop bedrocks (Eschenbrenner and Badarello, 1978; Poss, 1982): of course, encountered young soil (e.g. Regosols) with shallow depth (<40 cm) may have high level of K though, the color hue may range between 7.5 and 10YR. In fact, these yellowish colors are characterized by goethite (α-FeOOH) preceding the reddish soil matrix of lepidocrocite (γ-FeOOH) and hematite (α-Fe$_2$O$_3$) in the course of soil development (Buxbaum and Printzen, 1998; Schwertmann, 1985). As young soil, enriched-K parental material (e.g. feldspar and mica) may have influence in soil composition (Nahon, 1991). When excluding similar cases, reddish soil color hue of 2.5YR at the hill slope (SUM and US) is somewhat a consistent environmental indicator of enriched soil in exchangeable K. However, the occurrence of 2.5YR as soil color was found to be limited in topsoil and FS because of vertical and lateral gradients of tropical soil color (Koné, Yao-Kouamé, et al., 2009). Definitively, enriched soils in K are more limited in the UP position than the SUM on the basis of the highest cumulative frequency of 2.5YR in H and M levels of soil K and the lateral gradient of soil color.

Furthermore, the red pigmentation (e.g. Hematite) of 2.5YR is deriving from ferrihydrate (Schwertmann, 1985) requiring edaphic conditions which may include high temperature, low water activity, low organic matter content, nearly neutral pH, and high contents of Ca and Mg contrasting with the optimum conditions of goethite formation for yellowish pigmentation (Torrent & Barron, 2002). Almost these parameters involved in soil pigmentation are also accounting for major chemical modifiers of fertility capability soil classifications (FCC) defined by Sanchez et al. (2003) emphasizing an opportunity to make easier the adoption this classification system.

Indeed, degraded Ferralsols are richer in coarse particle with dominance of yellow pigmentation coupled with low content of soil exchangeable K even for enriched bedrock K-primary mineral (Koné, Amadji, Touré, Togola, et al., 2013; Koné, Diatta, et al., 2009; Koné, Touré, Amadji, Yao-Kouamé, et al., 2013).

In light of current analysis, soil color use in field is a potential friendly method for soil fertility classification gathering indigenous and scientific methods as additional reliable tool for participatory land use planning in tropical environment.

4.2. Predictability of exchangeable soil K

The linear regressions observed for soil content of exchangeable K according to soil RR and RF, respectively, were exceptionally significant in the same manner characterizing the relation between soil-exchangeable K and water-soluble K in kaolinitic and mixed mineralogy soils Sharpley (1989). In fact, kaolinit is dominant in Ferralsol, but change in mineralogy can occurred throughout soil profile and along a toposequence (Diatta, 1996) though, still remaining 1:1 clay mineral (kaolinit and illit) in topsoil, while smectite can be observed in deep horizons and down slope. Consequently, the lack of fit observed for the linear models of K prediction in 80–150 cm soil depth of middle and foot slopes (Table 5) could account for such neoformations due to soil moisture and geochemistry (Azizi et al., 2011). However, soil content of K can be predicted by soil redness within 0–80 cm depth indifferently to the topographic positions. This result can be considered as a significant advance in methodology of soil survey, especially for annual crops. Actually, up to date, rapid assessment of soil content of K was only referring to laboratory data including complex method as near-infrared measurement (He and Song, 2006).

Furthermore, the current investigation may have implication in the prediction of soil Ca using the model of Pal (1998) for kaolinitic soil toward a readily estimation of soil cation exchangeable capacity (Bigorre, 1999; Larson & Pierce, 1994). Hence, the use of soil redness may increase the adoption of existing method of soil CEC estimation though; the prediction models (RR and RF) of Mg and Na are still required. Nevertheless, the current study may have significant contribution in pedometry, especially when using pedotransfert functions for estimation of soil cation saturation ratio, hence for easily evaluation of soil fertility in field.

The sensitivity of soil color to wetness (Poss et al., 1991) and the change of soil physic and chemical characteristics (Vizier, 1971) as observed for soil content of K (Koné, Diatta, et al., 2009), are further supporting current finding and relevant assertions.

However, there was scant evidence of correlation between soil RR and RF with soil content of exchangeable K, respectively, throughout soil profile as much as observed for Ca and Mg (Tables 3 and 4) asserting that Ca and Mg may be better predicted by soil redness indices than soil K. In fact, divalent cations are more relevant to soil redness (Koné, Diatta, et al., 2009; Segalen, 1969), while colloids dispersion induced by monovalent cations (Grolimund et al., 1998; Kaplan et al., 1996) may alter the red pigmentation. Well, beside of hematite, the red pigmentation of soil is also inducing by amorphous component (Mauricio & Ildeu, 2005; Segalen, 1969) accounting for soil colloidal phase.

In light of these analyses, there is also chance to predict soil contents of Ca and Mg using soil redness indices in away to estimate soil CEC *in situ* during agricultural soil survey. Of such investigation in future, agricultural land assessment may be strengthening, especially for Ferralsols.

5. Conclusion

Reddish topsoil of 2.5YR in color observed at hill slope positions (SUM and US) is identified as most enriched soil in K when referring to a cumulative frequency in high and medium K-levels. In turn, low soil K-level was accounting for deep soil horizon, FS position, and yellowish (7.5 and 10YR) soil while characterizing 50% of 5YR as soil color. Furthermore, it was revealed a predictability of soil K content within 0–80 cm depth using soil redness indices also advocated for *in situ* prediction of soil contents of Ca and Mg respectively toward the estimation of soil cation exchangeable capacity. Hence, further challenge was outlined concerning the prediction of soil contents of Ca and Mg using Munsell chart in Ferralsol environment. Overall, soil color was deemed as promising tool for improvement of soil fertility management.

Acknowledgment
We are thankfull for DCGTx (BNETD) and for Sitapha Diatta for their respective contribution to this study.

Funding
This work was supported by DCGtx [grant number 1990].

Author details
Brahima Koné[1]
E-mail: kbrahima@hotmail.com
Traoré Lassane[2]
E-mail: tlassane@hotmail.com
Sehi Zokagon Sylvain[1]
E-mail: sehisylvain_nung@yahoo.fr
Kouassi Kouassi Jacques[1]
E-mail: kouassi.kouassijacques@yahoo.fr
[1] Earth Science Unit, Soil Science Department, Felix Houphouet-Boigny University, 22 BP 582, Abidjan, Côte d'Ivoire.
[2] Department of Economic Sciences, Biology Sciences, Peleforo Gon Coulibaly University, BP 1328 Korhogo, Côte d'Ivoire.

References
Azizi, P., Mahmood, S., Torabi, H., Masihabadi, M. H., & Homaee, M. (2011). Morphological, physic-chemical and clay mineralogy investigation on gypsiferous soils in southern of Tehran, Iran. *Middle-East Journal of Scientific Research, 7*, 153–161.

Berryman, C., Brower, R., Charteres, C., Davis, H., Davison, R., Eavis, B., ... Yates, R. A. (1984). *Booker tropical soil manual: A handbook for soil survey and agricultural land evaluation in the tropics and subtropics.* London: Longman.

Bertrand, R., Kilian, J., Raunet, M., Guillobez, S., & Bourgeon, G. (1985). Characterizing landscape systems, a prerequisite for environment protection. Methodological approach. *Recherche Agronomie Gembloux, 20*, 545–559.

Bigorre, F. (1999). Contribution of clays and organic matters to soil water holding. Mean and fondamental implication to exchangeable cation capacity. *CR Academic Science, 330*, 245–250.

Böhn, W. (1976). In situ estimation of root length at natural soil profiles. *The Journal of Agricultural Science, 87*, 365–368.

Burger, J. A. (1996). Limitations of bioassays for monitoring forest soil productivity: Rationale and example. *Soil Science Society of America Journal, 60*, 1674–1678. http://dx.doi.org/10.2136/sssaj1996.036159950060000 60010x

Buxbaum, G., & Printzen, H. (1998). Colored poigments: Iron oxide pigments. In G. Buxbaum (Ed.), *Industrial inorganic pigments* (pp. 85–107). Weiheim: VCH. http://dx.doi.org/10.1002/9783527612116

Chopart, J. L. (1985). Root development of some annual crops in west Africa and resistance to drought in intertropical zone. In *For integrated drought management*. Paris: CILF edition.

Diatta, S. (1996). *Grew soils of foot slope on granit-gneiss rock in the centre region of Côte d'Ivoire: Toposequential and spatial structures, hydrologic regime. Consequence for rice cropping* (Doctorate thesis). Henri Point Carré University, Nancy I.

Eschenbrenner, V., & Badarello, L. (1978). *Pedological study of the region of Odiénné (Côte d'Ivoire). Morpho-pedological mape* (p. 123). Paris: Manual N°74. ORSTOM .

Gale, M. R., Grigal, D. F., & Harding, R. B. (1991). Soil productivity index: Predictions of site quality for white spruce plantations. *Soil Science Society of America Journal, 55*, 1701–1709. http://dx.doi.org/10.2136/sssaj1991.03615995005500 060033x

Gretagmacьeth. (2000). *Munsell soil color charts: Year 2000 revised washable edition.* New York, NY: Author.

Grolimund, D., Elimelech, M., Borkovee, M., Barmettle, K., Kretzschmar, R., & Sticher, H. (1998). Transport of in situ mobilized colloidal particles in packed soil columns. *Environmental Science & Technology, 32*, 3562–3569.

He, Y., & Song, H. (2006). Prediction of soil content using near-infrared spectroscopy. Hangzhou: SPIE Newsroom (The International Society of Optical Engineering). doi: http://dx.doi.org/10.1117/2.1200604.0164

Hem, J. D. (1992). *Study and interpretation of the chemical characteristics of natural water* (3rd ed., p. 263). Alexandria, VA: United States Geological Survey Water Supply Paper 2254.

Kaplan, D. I., Sumner, M. E., Bertsch, P. M., & Adriano, D. C. (1996). Chemical conditions conducive to the release of mobile colloids from ultisol profiles. *Soil Science Society of America Journal, 60*, 269–274. http://dx.doi.org/10.2136/sssaj1996.03615995006000010041x

Kiniry, L. M., Scrivener, C. L. & Keener, M. E. (1983). *A soil productivity index based on water depletion and root growth.* Res. Bull. 105 Colombia University of Missouri. (p. 89).

Koné, B. (2007). *Color as indicator of the soils' fertility: Data use for assessing inherent fertility of ferralsols over the latitude 7°N of Côte d'Ivoire* (Doctorate thesis). Cocody University. 146p+annexe.

Koné B., Amadji, G. L., Toure A., Togola A., Mariko M., & Huat, J. (2013). A Case of *Cyperus* spp. and *Imperata cylindrica* occurrences on acrisol of the dahomey gap in South Benin as affected by soil characteristics: A strategy for soil and weed Management. *Applied and Environmental Soil Science, 2103*, Article ID 601058, 7 p.

Koné, B., Bongoua-Devisme, A. J., Kouadio, K. H., Kouadio, K. F., & Traoré, M. J. (2014). Potassium supplying capacity as indicated by soil colour in Ferralsol environment. *Basic Research Journal of Soil and Environmental Science, 2*, 46–55.

Koné, B., Diatta, S., Sylvester, O., Yoro, G., Camara, M., Dohm, D. D., & Assa, A. (2009). Assessment of ferralsol potential fertility by color: Color use in morpho pedology. *Canadian Journal of Soil Science, 89*, 331–342. http://dx.doi.org/10.4141/CJSS07119

Koné, B., Fatogoma, S., & Chérif, M. (2013). Diagnostic of mineral deficiencies and interactions in upland rice yield declining on foot slope soil in a humid forest zone. *International Journal of Agronomy and Agricultural Research, 3*, 11–20.

Koné, B., Touré, A., Amadji, G. L., Yao-Kouamé, A., Angui, P. T., & Huat, J. (2013). Soil characteristics and *Cyperus* spp. occurrence along a toposequence. *African Journal of Ecology, 51*, 402–408. http://dx.doi.org/10.1111/aje.2013.51.issue-3

Koné, B., Yao-Kouamé, A., Ettien, J. B., Oikeh, S., Yoro, G., & Diatta, S. (2009). Modelling the relationship between soil color and particle size for soil survey in Ferralsol environments. *Soil and Environment, 28*, 93–105.

Krasilnikov, P. V., & Tabor, J. A. (2003). Perspectives on utilitarian ethnopedology. *Geoderma, 111*, 197–215. http://dx.doi.org/10.1016/S0016-7061(02)00264-1

Larson, W. E., & Pierce, F. J. (1991). Conservation and enhancement of soil quality. In *Evaluation for sustainable land management in the developing world*. International Board for Soil Research and Management (IBSRAM) (Proceedings 12) (Vol. 2, pp. 175–203). Bangkok.

Larson, W. E., & Pierce, F. J. (1994). The dynamics of soil quality as a measure of sustainable management. In Doran, J. W., et al. (Eds.), *Defining soil quality for a sustainable environment* (pp 37–52). Madison, WI: SSSA. Pub. 35.

Loukili, M., Bock, L., Engles, P., & Mathieu, L. (2000). Geomorphological approach and Geographic Information System (GIS) for land management in Moroco. *Etude et Gestion des Sols, 7*, 37–52.

Mauricio, P., & Ildeu, A. (2005). Color attributes and mineralogical characteristics, evaluated by radiometry of highly weathered tropical soils. *Soil Science Society of America Journal, 69*, 1162–1172.

Milner, K. S., Running, S. W., & Coble, D. W. (1996). Biopsical soil—site model for estimating potential productivity of forested landscape. *Canadian Journal of Soil Sciences, 55*, 228–234.

Nahon, D. B. (1991). *Introduction to the petrology of soils and chemical weathering*. New York, NY: John Wiley and Sons.

Neill, L. L. (1979). *An evaluation of soil productivity based on root growth and water depletion* (M.Sc. Thesis). University of Missouri, Columbia, MO.

Page, A. L. (1982). *Methods of soil analysis. Part 2. Chemical and microbiological properties* (2nd ed.). Madison, WI: SSA, ASA.

Pal, S. K. (1998). Prediction of plant available potassium in kaolinitic soils of India. *Agropedology, 8*, 94–100.

Pearson, C. J., Norman, D. W., & Dixon, J. (1995). *Sustainable dryland cropping in relation to soil productivity - FAO soils bulletin 72*. Rome: FAO.

Pieri, C. J. M. G. (1992). *Fertility of soils. A future for farming in the West African savannah*. Berlin: Springer-Verlag.

Pierce, F. J., Larson, W. E., Dowdy, R. H., & Graham, W. A. P. (1983). Productivity of soils: Assessing long term changes due to erosion. *Journal of Soil and Water Conservation, 38*, 39–44.

Poss, R. (1982). *Etude morphopédologique de la région de Katiola (Côte d'Ivoire)* (p. 142). Paris: ORSTOM. Note explicative N°94.

Poss, R., Fardeau, J. C., Saragonit, H., & Quantin, P. (1991). Potassium release and fixation in Ferralsols (Oxisols) from Southern Togo. *Journal of Soil Science, 42*, 649–660. http://dx.doi.org/10.1111/ejs.1991.42.issue-4

Riquier, J., Cornet, J. P., & Braniao, D. L. (1970). A new system of soil appraisal in terms of actual and potential productivity (1st Approx). *World Soil Res* (p. 44). Rome: FAO.

Torrent, J., Schwertmann, U., Fechter, H., & Alferez, F. (1983). Quantitative relationships between soil color and hematite content. *Soil Science, 136*, 354–358. http://dx.doi.org/10.1097/00010694-198312000-00004

Torrent, J., Schwertmann, U., & Schulze, D. G. (1980). Iron oxide mineralogy of some soils of two river terrace sequences in Spain. *Geoderma, 23*, 191–208. http://dx.doi.org/10.1016/0016-7061(80)90002-6

Sanchez, P. A., Palm, C. A., & Buol, S. W. (2003). Fertility capability soil classification: a tool to help assess soil quality in the tropics. *Geoderma, 114*, 157–185. http://dx.doi.org/10.1016/S0016-7061(03)00040-5

Santana, D. P. (1984). *Soil formation in a toposequence of oxisols from Patos de Minas region* (PhD. Thesis), Minas Gerais State, Brazil, Purdi Univ, West Lafayette, IN.

Schwertmann, U. (1985). The effect of pedogenic environments on iron oxide minerals. *Advances in soil science, 1*, 171–200.

Segalen, P. (1969). Contribution to knowledge of sesquioxyd soil colors in intertropical zone: Yellow soils and red soils. *Cah ORSTOM, Ser Pedol, 7*, 225–236.

Sharpley, A. N. (1989). Relationship between soil potassium forms and mineralogy. *Soil Science Society of America Journal, 52*, 1023–1028.

Stoner, E. R., Baumgardner, M. F., Weismiller, R. A., Beilh, L. L., & Robbinson, B. F. (1980). *Atlas of soil reflectance properties*. Purdue Uni., West Lafayette, IN. Res. Bull. 962 Agric. Exp. Stn.

Sys, C., & Frankart, R. (1971). Soils' assessment in humid tropical zones. *African Soils, 15*, 177–199.

Torrent, J., & Barrón, V. (2002). *Iron oxides in relation to the colour of Mediterranean soils. Applied Study of Cultural Heritage and Clays* (pp. 377–386). Madrid: Consejo Superior de Investigaciones Científicas.

Vizier, J. F. (1971). Study of soil oxydo reduction statut and consequences on iron dynamic in hydromorphic soils. *Cah. ORSTOM. Ser. Pedol., 4*, 373–398.

Webster, R., & Oliver, M. O. (1990). *Statistical methods in soil and land resources survey*. New York, NY: Oxford University Press.

Wedpohl, K. H. (1978). *Handbook of Geochemistry*. Berlin: Springer Verlag.

PERMISSIONS

LIST OF CONTRIBUTORS

Snehal Rajeev Pathak
Department of Civil Engineering, College of Engineering Pune, Pune, Maharashtra, India

Asita Nilesh Dalvi
Department of Civil Engineering, SITS, Pune, Maharashtra, India

Akram Afifi and Ahmed El-Rabbany
Department of Civil Engineering, Ryerson University, Toronto, Ontario, Canada

Robert J. Holm
Frogtech Geoscience, 2 King Street, Deakin West ACT 2600, Australia
Geosciences, College of Science & Engineering, James Cook University, Townsville, Queensland 4811, Australia

Benny Poke
Geological Survey Division, Mineral Resources Authority, Port Moresby 121, Papua New Guinea

Kotaro Iizuka
Center for Spatial Information Science (CSIS), University of Tokyo, Chiba, Japan
Center for Southeast Asian Studies (CSEAS), Kyoto University, Kyoto, Japan

Masayuki Itoh
Center for Southeast Asian Studies (CSEAS), Kyoto University, Kyoto, Japan
School of Human Science and Environment, University of Hyogo, Hyogo, Japan

Kazuo Watanabe and Satomi Shiodera
Center for Southeast Asian Studies (CSEAS), Kyoto University, Kyoto, Japan
National Institutes for the Humanities, Research Institute for Humanity and Nature (RIHN), Kyoto, Japan

Takashi Matsubara
Technical Research Institute, Obayashi Corporation, Tokyo, Kiyose, 204-8558, Japan

Mark Dohar
PT Austindo Aufwind New Energy, Jakarta, Indonesia

Hongwei Kuang
Institute of Geology, Chinese Academy of Geological Sciences, Beijing 100037, China

Guangchun Jin
Oil & Gas Survey of China Geological Survey, Beijing 100029, China

Zhenzhong Gao
School of Earth Science of Yangtze University, Wuhan 434023, Hubei, China

Kandala Rajsekhar
Department of Civil Engineering, Indian Institute of Technology, Roorkee, Roorkee 247667, India

Pramod Kumar Sharma
Department of Civil Engineering, Indian Institute of Technology, Roorkee, Roorkee 247667, India
Discipline of Civil and Environmental Engineering, School of Engineering, Edith Cowan University, Perth 6027, Australia

Sanjay Kumar Shukla
Discipline of Civil and Environmental Engineering, School of Engineering, Edith Cowan University, Perth 6027, Australia

Farzad Habibbeygi and Hamid Nikraz
Faculty of Science and Engineering, Department of Civil Engineering, Curtin University, Perth, Australia

María Teresa Ramírez-Herrera and Krzysztof Gaidzik
Laboratorio Universitario de Geofísica Ambiental & Instituto de Geografía, Universidad Nacional Autónoma de México, Ciudad Universitaria, Coyoacán, 04510 Ciudad de México, México

Huatao Yuan, Xiaoxiao Li, Yue Chu, Changmei Lu and Bin Lian
Jiangsu Key Laboratory for Microbes and Functional Genomics, Jiangsu Engineering and Technology Research Center for Microbiology, College of Life Sciences, Nanjing Normal University, Nanjing 210023, China

Leilei Xiao
Jiangsu Key Laboratory for Microbes and Functional Genomics, Jiangsu Engineering and Technology Research Center for Microbiology, College of Life Sciences, Nanjing Normal University, Nanjing 210023, China
Key Laboratory of Coastal Biology and Utilization, Yantai Institute of Coastal Zone Research, Chinese Academy of Sciences, Yantai 264003, China
Key Laboratory of Karst Environment and Geological Hazard Prevention, Ministry of Education, Guizhou University, Guiyang 550003, China

Qibiao Sun
Jiangsu Key Laboratory for Microbes and Functional Genomics, Jiangsu Engineering and Technology Research Center for Microbiology, College of Life Sciences, Nanjing Normal University, Nanjing 210023, China
Key Laboratory of Karst Environment and Geological Hazard Prevention, Ministry of Education, Guizhou University, Guiyang 550003, China

Yulong Ruan
Key Laboratory of Karst Environment and Geological Hazard Prevention, Ministry of Education, Guizhou University, Guiyang 550003, China

Farzad Habibbeygi and Hamid Nikraz
Department of Civil Engineering, Science and Engineering, Curtin University, Perth, Australia

Bill K Koul
4DGeotechnics Pty Ltd, West Leederville, Perth, WA, Australia

Timothée Miyouna, Hardy Medry Dieu-Veill Nkodia and Olivier Florent Essouli
Marien NGOUABI University, Brazzaville, Republic of Congo

Florent Boudzoumou
Marien NGOUABI University, Brazzaville, Republic of Congo
Geosciences Department, National Research Institute in Exact and Natural Sciences (IRSEN) of Brazzaville, Brazzaville, Republic of Congo

Moussa Dabo
Cheikh Anta DIOP University, Dakar, Republic of Senegal

Damien Delvaux
Africa Museum, Leuvensesteenweg 13, Tervuren, B-3080, Belgique

Brahima Koné, Sehi Zokagon Sylvain and Kouassi Kouassi Jacques
Earth Science Unit, Soil Science Department, Felix Houphouet-Boigny University, 22 BP 582, Abidjan, Côte d'Ivoire

Traoré Lassane
Department of Economic Sciences, Biology Sciences, Peleforo Gon Coulibaly University, BP 1328 Korhogo, Côte d'Ivoire

Index

A

Advanced Land Observation Satellite-2, 61, 63

Air-liquid Interface, 91-94, 97, 101

B

Biogeography, 141

C

Calcite, 81, 133-134, 136, 138-140, 142, 165-166, 172, 174

Carbon Fixation, 132, 136

Clay Mineralogy, 143-144, 146, 151-153, 156-158, 160

Cloudy Bay Volcanics, 35-39, 41, 43-47, 49-57, 59

D

Damage Zones, 161-162, 166-167, 176-178

Deep-seated Slide, 125

Digital Surface Model, 61, 66

Dolomite, 132-141

Dredged Slurry Sedimentation, 143

E

Empirical Liquefaction Model, 1-3, 5, 7, 9, 11, 13, 15, 17, 19, 21

F

Fertility Capability Classification, 193

Finite-volume Method, 91-95, 101

Fluvial-dominated Fan Delta Depositional System, 75-76, 84

G

Galileo System, 23, 31

Galileo Time Offset, 23

Galileo Time System, 23

Geochemistry, 35, 57-60, 134, 141, 201, 203

Geochronology, 35-36, 43, 45, 57-58

Geodesy, 22, 34

Geology, 35-36, 54, 57-60, 75-77, 89-90, 103, 111, 114, 129-130, 141-143, 161, 177-178

Geomorphology, 113, 129-131

Geophysics, 58-60, 130, 178

Geotechnical Engineers, 143-144

Global Navigation Satellite System (GNSS) Constellations, 22

Global Positioning System, 23

Gray Mid-lower Resistivity Matrix, 79

Grey-level Co-occurrence Matrix, 64

Ground Motion Parameters, 1-2, 10-11

H

Hill Slope Position, 192

Hurricane Manuel, 112-113, 121, 125-128, 130

Hydrology, 102

K

K-feldspar, 132, 134, 136, 138-140

Kappa Index of Agreement, 66

L

Landslides, 104, 111-113, 118-119, 121-122, 125-131

Liquid-solid Interface, 91, 93, 97-98, 101

Local Landscape Information, 62, 71-72

Lower Urho Formation, 75-79, 81-90

M

Mineral Weathering, 132, 139

Mineralogy, 36, 39, 57-60, 105, 143-144, 146, 151-153, 156-158, 160-161, 193, 201-203

Multi-criteria Decision-making Tools, 2

N

Net Carbon Sequestration, 132, 134

O

Oedometer (Consolidation) Tests, 144

Owen Stanley Metamorphic Complex, 37-38, 55-58

P

Paleostress, 161-162, 164, 167, 176-178

Palygorskite, 165-166

Petrology, 35-37, 39, 41, 43, 45, 47, 49-51, 53, 55-60, 161, 203

Phased Array L-band Synthetic Aperture Radar-2, 61, 63

Photogrammetry, 61-63, 65, 67, 69, 71-73

Potassium, 58-59, 132, 134-135, 137-141, 192-193, 195-197, 199, 201-203

Precise Point Positioning, 22-23, 25, 29, 31, 33-34

R
Redbeds Sequence, 179
Rotational Stress Optimization, 164

S
Satellite Sensors, 62
Sedimentary Facies, 75-79, 81, 83, 85, 87, 89-90
Sedimentology, 36, 75, 90, 161
Seismic Soil Liquefaction, 1, 12
Shallow Soil Slips, 125-126
Shear Rate, 103-107, 109-111
Soil Color Identification, 194
Soil Inorganic Carbon, 133
Soil Organic Carbon, 133
Soil Science, 142, 192, 202-203
Strike-slip Deformation, 161, 163, 165, 167, 169, 171, 173, 175, 177, 179, 181, 183, 185, 187, 189, 191
Strike-slip Faults Systems, 161, 170
Subaqueous Interdistributary Channel, 80-82
Subsurface Aquifer System, 91

T
Tropospheric Zenith Path Delay, 29-30

U
Unmanned Aerial Vehicle, 61-63, 65, 67, 69, 71, 73
Unsaturated Porous Media, 91-93, 95, 97, 101-102
Upper Permian, 75-76, 78, 84, 89-90

V
Virus Transport, 91-95, 97, 101-102
Virus Transport Equations, 91
Void Ratio, 105, 109, 111, 143-145, 149, 151, 156, 158, 160
Volcanic Suite, 56
Volcano, 58-59

W
World Stress Map Project, 164

Z
Zircon, 35-36, 38, 40-46, 50, 52, 54-59, 163, 177
Zircon U-pb Dating, 35, 44, 46, 54

www.ingramcontent.com/pod-product-compliance
Lightning Source LLC
Chambersburg PA
CBHW082028190326
41458CB00010B/3306